不只做一个技术者，
更要做一个思考者！

谭勇德/Tom

咕泡学院 Java架构师成长丛书
gupaoedu.com

Spring 5 核心原理
与30个类手写实战

谭勇德（Tom）◎著

电子工业出版社
Publishing House of Electronics Industry
北京·BEIJING

内 容 简 介

本书基于编程开发实践，不仅深度解析 Spring 5 的原理与新特性，更从环境准备、顶层结构设计、数据访问等方面一步步地推导出 Spring 的设计原理。在每个知识点上，均以大量的经典代码案例辅助讲解，使理论紧密联系实际。最后手写 30 个类，以体会 Spring 作者的创作过程，让每一位读者学以致用。

对于立志成为 Java 架构师的技术人员，以及对以 Spring 为核心的 Java Web 开发感兴趣的计算机专业高校生、在职 Java 开发人员来说，本书是一本具备超强实战意义的技术升级指南。读者通过本书可以看源码不再"晕车"，轻松找到入口；系统学习设计思想，提高解决问题的效率；培养架构思维能力，以及自驱学习能力。

未经许可，不得以任何方式复制或抄袭本书之部分或全部内容。
版权所有，侵权必究。

图书在版编目（CIP）数据

Spring 5 核心原理与 30 个类手写实战 / 谭勇德著. —北京：电子工业出版社，2019.7
（咕泡学院 Java 架构师成长丛书）
ISBN 978-7-121-36741-0

Ⅰ.①S… Ⅱ.①谭… Ⅲ.①JAVA 语言－程序设计 Ⅳ.①TP312.8

中国版本图书馆 CIP 数据核字（2019）第 111737 号

责任编辑：董　英
印　　刷：北京天宇星印刷厂
装　　订：北京天宇星印刷厂
出版发行：电子工业出版社
　　　　　北京市海淀区万寿路 173 信箱　　　邮编：100036
开　　本：787×980　1/16　　印张：35　　字数：778 千字
版　　次：2019 年 7 月第 1 版
印　　次：2023 年 5 月第 9 次印刷
印　　数：17001~17500 册　　定价：118.00 元

凡所购买电子工业出版社图书有缺损问题，请向购买书店调换。若书店售缺，请与本社发行部联系，联系及邮购电话：（010）88254888，88258888。
质量投诉请发邮件至 zlts@phei.com.cn，盗版侵权举报请发邮件至 dbqq@phei.com.cn。
本书咨询联系方式：010-51260888-819，faq@phei.com.cn。

序　　言

在 1996 年，Java 还只是一门新兴的、初出茅庐的编程语言。2004 年 3 月 24 日，Spring 正式对外发布 1.0 版本。我在 2009 年开始接触 Spring 2.0，从此爱不释手。Spring 的出现，恰如其名，开启了全世界 Java 程序员的春天。如今，Spring 已然成为一个生态，使用 Spring 简直就是一种享受。

本书结合我多年的 Spring 使用经验，整理了珍藏多年的 Spring 学习笔记，采用类图和代码片段的形式，加以中文注释，通俗、生动、全面、深入地剖析了 Spring 源码的重要细节。要想练就"降龙十八掌"，先得修炼内功。本书从设计模式开始讲解，以帮助大家更好地理解 Spring，让大家知其然，且知其所以然。

如果你现在的工作经验与应有的能力无法匹配，

如果你在自学路上找不到方向，

如果你现在是初级程序员，想往高级程序员晋级，

如果你希望成为架构师，

……

本书都能帮到你。

编写本书旨在使大家进入不一样的思维境界，手写源码不是为了重复造轮子，也不是为了装高大上，其实只是我推荐给大家的一种学习方式。

书中个人观点若有不妥之处，恳望纠正！

关于本书

适用对象	• 具有Spring开发经验，想深入了解Spring实现原理的人 • 看源码不知如何下手的人 • 希望理解设计模式在源码中的应用的人 • 立志成为Java架构师的人
源码版本	Spring 5.0.2.RELEASE
IDE版本	IntelliJ IDEA 2017.1.4
JDK版本	JDK 1.8及以上
Gradle	Gradle 4.0及以上
Maven版本	3.5.0及以上

读者服务

微信扫码回复：36741

- 获取本书配套素材
- 获取更多技术专家分享视频与学习资源
- 加入读者交流群，与更多读者互动

关于我

为什么说我来自文艺界？

我自幼爱好书法和美术，长了一双能书会画的手，而且手指又长又白，因此以前的艺名叫"玉手藝人"。中学期间，曾获市级书法竞赛一等奖，校园美术竞赛一等奖，校园征文比赛二等奖。担任过学生会宣传部长，负责校园黑板报、校园刊物的编辑、排版、设计。

2008年参加工作后，做过家具建模、平面设计等工作，亲自设计了咕泡学院的Logo。做讲师之后，给自己起了一个跟姓氏谐音的英文名字"Tom"，江湖人称"编程界写字写得最好的、书法界编程最牛的文艺汤"。

我的技术生涯

我的IT技术生涯应该算是从2009年开始的，在此之前做过UI设计，做过前端网页，到2009年才真正开始参与Java后台开发。在这里要感谢所有帮助过我入门编程的同事和老师。从2010年至2014年担任过项目组长、项目经理、架构师、技术总监，对很多的开源框架建立了自己的独特见解。我会习惯性地用形象思维来理解抽象世界。譬如：看到二进制0和1，我会想到《周易》中的两仪——阴和阳；看到颜色值用RGB表示，我会想到美术理论中的太阳光折射三原色；下班回家看到炒菜流程，我会想到模板方法模式；坐公交车看到学生卡、老人卡、爱心卡，我会想到策略模式；等等。大家看到的这本书，很多地方都融入了这种形象思维。

为什么写书？

其实一开始我没想过要写书，写书的初衷主要是满足学员的诉求。大家认为我个人的学习方法、思维模式、教学方式通俗易懂，很容易让人接受，但是通过视频形式传播受众有限，学员建议我把这些宝贵的经验以纸质书的形式奉献给大家，这样定会给社会带来更大的价值。

借此机会，特别感谢责任编辑董英及电子社的团队成员为本书审稿纠错；感谢我老婆在无数个加班的夜晚给我默默的支持；感谢咕泡学院的学员给本书内容提出宝贵的修改意见。

谭勇德（Tom）

2019年5月 于 长沙

目　录

第 1 篇　Spring 内功心法

第 1 章　软件架构设计原则 2
- 1.1　开闭原则 2
- 1.2　依赖倒置原则 4
- 1.3　单一职责原则 7
- 1.4　接口隔离原则 10
- 1.5　迪米特原则 12
- 1.6　里氏替换原则 14
- 1.7　合成复用原则 19
- 1.8　设计原则总结 20

第 2 章　Spring 中常用的设计模式 21
- 2.1　为什么要从设计模式开始 21
  - 2.1.1　写出优雅的代码 22
  - 2.1.2　更好地重构项目 24
  - 2.1.3　经典框架都在用设计模式解决问题 36
- 2.2　工厂模式详解 36
  - 2.2.1　工厂模式的由来 36
  - 2.2.2　简单工厂模式 37
  - 2.2.3　工厂方法模式 41
  - 2.2.4　抽象工厂模式 43
  - 2.2.5　利用工厂模式重构的实践案例 47
- 2.3　单例模式详解 53
  - 2.3.1　单例模式的应用场景 53
  - 2.3.2　饿汉式单例模式 53

目录

- 2.3.3 懒汉式单例模式 54
- 2.3.4 反射破坏单例 60
- 2.3.5 序列化破坏单例 61
- 2.3.6 注册式单例模式 68
- 2.3.7 线程单例实现 ThreadLocal 74
- 2.3.8 单例模式小结 75

2.4 原型模式详解 75
- 2.4.1 原型模式的应用场景 75
- 2.4.2 浅克隆 77
- 2.4.3 深克隆 79
- 2.4.4 克隆破坏单例模式 81
- 2.4.5 clone()方法的源码 82

2.5 代理模式详解 82
- 2.5.1 代理模式的应用场景 82
- 2.5.2 静态代理 83
- 2.5.3 动态代理 88
- 2.5.4 代理模式与 Spring 107
- 2.5.5 静态代理和动态代理的本质区别 108
- 2.5.6 代理模式的优缺点 109

2.6 委派模式详解 109
- 2.6.1 委派模式的定义及应用场景 109
- 2.6.2 委派模式在源码中的体现 111

2.7 策略模式详解 114
- 2.7.1 策略模式的应用场景 114
- 2.7.2 用策略模式实现选择支付方式的业务场景 114
- 2.7.3 策略模式在 JDK 源码中的体现 122
- 2.7.4 策略模式的优缺点 125
- 2.7.5 委派模式与策略模式综合应用 125

2.8 模板模式详解 129
- 2.8.1 模板模式的应用场景 129
- 2.8.2 利用模板模式重构 JDBC 操作业务场景 132
- 2.8.3 模板模式在源码中的体现 136
- 2.8.4 模板模式的优缺点 138

2.9 适配器模式详解 139
- 2.9.1 适配器模式的应用场景 139
- 2.9.2 重构第三方登录自由适配的业务场景 141
- 2.9.3 适配器模式在源码中的体现 149

  2.9.4 适配器模式的优缺点 153
 2.10 装饰者模式详解 153
  2.10.1 装饰者模式的应用场景 153
  2.10.2 装饰者模式和适配器模式对比 163
  2.10.3 装饰者模式在源码中的应用 163
  2.10.4 装饰者模式的优缺点 165
 2.11 观察者模式详解 165
  2.11.1 观察者模式的应用场景 165
  2.11.2 观察者模式在源码中的应用 175
  2.11.3 基于 Guava API 轻松落地观察者模式 176
  2.11.4 观察者模式的优缺点 177
 2.12 各设计模式的总结与对比 177
  2.12.1 GoF 23 种设计模式简介 177
  2.12.2 设计模式之间的关联关系 178
  2.12.3 Spring 中常用的设计模式 182
 2.13 Spring 中的编程思想总结 183

第 2 篇 Spring 环境预热

第 3 章 Spring 的前世今生 186
 3.1 一切从 Bean 开始 187
 3.2 Spring 的设计初衷 188
 3.3 BOP 编程伊始 188
 3.4 理解 BeanFactory 189
 3.5 AOP 编程理念 189

第 4 章 Spring 5 系统架构 191
 4.1 核心容器 192
 4.2 AOP 和设备支持 192
 4.3 数据访问与集成 193
 4.4 Web 组件 194
 4.5 通信报文 194
 4.6 集成测试 194
 4.7 集成兼容 194
 4.8 各模块之间的依赖关系 194

第 5 章 Spring 版本命名规则 196

5.1 常见软件的版本命名 196
5.2 语义化版本命名通行规则 197
5.3 商业软件中常见的修饰词 197
5.4 软件版本号使用限定 198
5.5 Spring 版本命名规则 199

第 6 章 Spring 源码下载及构建技巧 200

6.1 Spring 5 源码下载 200
6.2 基于 Gradle 的源码构建技巧 201
6.3 Gradle 构建过程中的坑 207

第 3 篇 Spring 核心原理

第 7 章 用 300 行代码手写提炼 Spring 核心原理 210

7.1 自定义配置 210
7.1.1 配置 application.properties 文件 210
7.1.2 配置 web.xml 文件 210
7.1.3 自定义注解 211
7.1.4 配置注解 212
7.2 容器初始化 213
7.2.1 实现 1.0 版本 213
7.2.2 实现 2.0 版本 216
7.2.3 实现 3.0 版本 223
7.3 运行效果演示 227

第 8 章 一步一步手绘 Spring IoC 运行时序图 228

8.1 Spring 核心之 IoC 容器初体验 228
8.1.1 再谈 IoC 与 DI 228
8.1.2 Spring 核心容器类图 229
8.1.3 Web IoC 容器初体验 232
8.2 基于 XML 的 IoC 容器的初始化 237
8.2.1 寻找入口 238
8.2.2 获得配置路径 238
8.2.3 开始启动 240

8.2.4 创建容器 ... 242
8.2.5 载入配置路径 ... 243
8.2.6 分配路径处理策略 ... 244
8.2.7 解析配置文件路径 ... 247
8.2.8 开始读取配置内容 ... 249
8.2.9 准备文档对象 ... 250
8.2.10 分配解析策略 ... 251
8.2.11 将配置载入内存 ... 252
8.2.12 载入<bean>元素 ... 257
8.2.13 载入<property>元素 ... 261
8.2.14 载入<property>子元素 ... 264
8.2.15 载入<list>子元素 ... 266
8.2.16 分配注册策略 ... 267
8.2.17 向容器注册 ... 267

8.3 基于注解的 IoC 初始化 ... 270
8.3.1 注解的前世今生 ... 270
8.3.2 定位 Bean 扫描路径 ... 271
8.3.3 读取注解的元数据 ... 273
8.3.4 扫描指定包并解析为 BeanDefinition ... 277
8.3.5 注册注解 BeanDefinition ... 283

8.4 IoC 容器初始化小结 ... 285

第 9 章 一步一步手绘 Spring DI 运行时序图ㅤ287

9.1 Spring 自动装配之依赖注入 ... 287
9.1.1 依赖注入发生的时间 ... 287
9.1.2 寻找获取 Bean 的入口 ... 288
9.1.3 开始实例化 ... 293
9.1.4 选择 Bean 实例化策略 ... 297
9.1.5 执行 Bean 实例化 ... 299
9.1.6 准备依赖注入 ... 301
9.1.7 解析属性依赖注入规则 ... 306
9.1.8 注入赋值 ... 310

9.2 Spring IoC 容器中那些鲜为人知的细节 ... 314
9.2.1 关于延时加载 ... 314
9.2.2 关于 FactoryBean 和 BeanFactory ... 317
9.2.3 再述 autowiring ... 322

第 10 章　一步一步手绘 Spring AOP 运行时序图 326

10.1　Spring AOP 初体验 326
　　10.1.1　再述 Spring AOP 应用场景 326
　　10.1.2　AOP 中必须明白的几个概念 327
　　10.1.3　使用 Spring AOP 的两种方式 329
　　10.1.4　切入点表达式的配置规则 333
10.2　Spring AOP 源码分析 334
　　10.2.1　寻找入口 334
　　10.2.2　选择代理策略 338
　　10.2.3　调用代理方法 341
　　10.2.4　触发通知 347

第 11 章　一步一步手绘 Spring MVC 运行时序图 352

11.1　初探 Spring MVC 请求处理流程 352
11.2　Spring MVC 九大组件 353
　　11.2.1　HandlerMapping 353
　　11.2.2　HandlerAdapter 353
　　11.2.3　HandlerExceptionResolver 354
　　11.2.4　ViewResolver 354
　　11.2.5　RequestToViewNameTranslator 354
　　11.2.6　LocaleResolver 354
　　11.2.7　ThemeResolver 355
　　11.2.8　MultipartResolver 355
　　11.2.9　FlashMapManager 355
11.3　Spring MVC 源码分析 355
　　11.3.1　初始化阶段 356
　　11.3.2　运行调用阶段 359
11.4　Spring MVC 优化建议 367

第 4 篇　Spring 手写实战

第 12 章　环境准备 370

12.1　IDEA 集成 Lombok 插件 370
　　12.1.1　安装插件 370
　　12.1.2　配置注解处理器 373

	12.1.3 使用插件 374
12.2	从 Servlet 到 ApplicationContext 375
12.3	准备基础配置 376
	12.3.1 application.properties 配置 377
	12.3.2 pom.xml 配置 377
	12.3.3 web.xml 配置 378
	12.3.4 GPDispatcherServlet 378

第 13 章 IoC 顶层结构设计 380

13.1	Annotation（自定义配置）模块 380
	13.1.1 @GPService 380
	13.1.2 @GPAutowired 381
	13.1.3 @GPController 381
	13.1.4 @GPRequestMapping 382
	13.1.5 @GPRequestParam 382
13.2	core（顶层接口）模块 382
	13.2.1 GPFactoryBean 382
	13.2.2 GPBeanFactory 383
13.3	beans（配置封装）模块 383
	13.3.1 GPBeanDefinition 383
	13.3.2 GPBeanWrapper 384
13.4	context（IoC 容器）模块 385
	13.4.1 GPAbstractApplicationContext 385
	13.4.2 GPDefaultListableBeanFactory 385
	13.4.3 GPApplicationContext 385
	13.4.4 GPBeanDefinitionReader 388
	13.4.5 GPApplicationContextAware 391

第 14 章 完成 DI 模块的功能 392

14.1	从 getBean()方法开始 393
14.2	GPBeanPostProcessor 395

第 15 章 完成 MVC 模块的功能 396

15.1	MVC 顶层设计 396
	15.1.1 GPDispatcherServlet 396

- 15.1.2 GPHandlerMapping ... 402
- 15.1.3 GPHandlerAdapter ... 403
- 15.1.4 GPModelAndView ... 406
- 15.1.5 GPViewResolver ... 406
- 15.1.6 GPView ... 407
- 15.2 业务代码实现 ... 409
  - 15.2.1 IQueryService ... 409
  - 15.2.2 QueryService ... 410
  - 15.2.3 IModifyService ... 410
  - 15.2.4 ModifyService ... 411
  - 15.2.5 MyAction ... 412
  - 15.2.6 PageAction ... 413
- 15.3 定制模板页面 ... 414
  - 15.3.1 first.html ... 414
  - 15.3.2 404.html ... 414
  - 15.3.3 500.html ... 415
- 15.4 运行效果演示 ... 415

第 16 章 完成 AOP 代码织入 ... 417

- 16.1 基础配置 ... 417
- 16.2 完成 AOP 顶层设计 ... 418
  - 16.2.1 GPJoinPoint ... 418
  - 16.2.2 GPMethodInterceptor ... 419
  - 16.2.3 GPAopConfig ... 419
  - 16.2.4 GPAdvisedSupport ... 420
  - 16.2.5 GPAopProxy ... 422
  - 16.2.6 GPCglibAopProxy ... 423
  - 16.2.7 GPJdkDynamicAopProxy ... 423
  - 16.2.8 GPMethodInvocation ... 425
- 16.3 设计 AOP 基础实现 ... 427
  - 16.3.1 GPAdvice ... 427
  - 16.3.2 GPAbstractAspectJAdvice ... 427
  - 16.3.3 GPMethodBeforeAdvice ... 428
  - 16.3.4 GPAfterReturningAdvice ... 429
  - 16.3.5 GPAfterThrowingAdvice ... 430
  - 16.3.6 接入 getBean()方法 ... 430

16.4　织入业务代码 432
　　16.4.1　LogAspect 432
　　16.4.2　IModifyService 433
　　16.4.3　ModifyService 434
16.5　运行效果演示 435

第 5 篇　Spring 数据访问

第 17 章　数据库事务原理详解 438
17.1　从 Spring 事务配置说起 438
17.2　事务的基本概念 439
17.3　事务的基本原理 439
17.4　Spring 事务的传播属性 440
17.5　数据库事务隔离级别 441
17.6　Spring 中的事务隔离级别 441
17.7　事务的嵌套 442
17.8　Spring 事务 API 架构图 444
17.9　浅谈分布式事务 444

第 18 章　Spring JDBC 源码初探 446
18.1　异常处理 447
18.2　config 模块 448
18.3　core 模块 450
18.4　DataSource 456
18.5　object 模块 457
18.6　JdbcTemplate 458
18.7　NamedParameterJdbcTemplate 458

第 19 章　基于 Spring JDBC 手写 ORM 框架 459
19.1　实现思路概述 459
　　19.1.1　从 ResultSet 说起 459
　　19.1.2　为什么需要 ORM 框架 464
19.2　搭建基础架构 467
　　19.2.1　Page 467
　　19.2.2　ResultMsg 470

19.2.3	BaseDao	471
19.2.4	QueryRule	473
19.2.5	Order	479

19.3 基于 Spring JDBC 实现关键功能 480
    19.3.1 ClassMappings 480
    19.3.2 EntityOperation 483
    19.3.3 QueryRuleSqlBuilder 488
    19.3.4 BaseDaoSupport 498

19.4 动态数据源切换的底层原理 507
    19.4.1 DynamicDataSource 508
    19.4.2 DynamicDataSourceEntry 509

19.5 运行效果演示 510
    19.5.1 创建 Member 实体类 510
    19.5.2 创建 Order 实体类 511
    19.5.3 创建 MemberDao 512
    19.5.4 创建 OrderDao 512
    19.5.5 修改 db.properties 文件 514
    19.5.6 修改 application-db.xml 文件 515
    19.5.7 编写测试用例 516

第 6 篇 Spring 经验分享

第 20 章 Spring 5 新特性总结 520

20.1 升级到 Java SE 8 和 Java EE 7 520
20.2 反应式编程模型 521
20.3 使用注解进行编程 521
20.4 函数式编程 522
20.5 使用 REST 端点执行反应式编程 523
20.6 支持 HTTP/2 523
20.7 Kotlin 和 Spring WebFlux 523
20.8 使用 Lambda 表达式注册 Bean 524
20.9 Spring Web MVC 支持最新的 API 524
20.10 使用 JUnit 5 执行条件和并发测试 525
20.11 包清理和弃用 526
20.12 Spring 核心和容器的一般更新 526
20.13 我如何看 Spring 5 527

第 21 章　关于 Spring 的经典高频面试题 528

21.1　什么是 Spring 框架，Spring 框架有哪些主要模块 528
21.2　使用 Spring 框架能带来哪些好处 528
21.3　什么是控制反转（IoC），什么是依赖注入 529
21.4　在 Java 中依赖注入有哪些方式 529
21.5　BeanFactory 和 ApplicationContext 有什么区别 530
21.6　Spring 提供几种配置方式来设置元数据 530
21.7　如何使用 XML 配置方式配置 Spring 531
21.8　Spring 提供哪些配置形式 532
21.9　怎样用注解的方式配置 Spring 533
21.10　请解释 Spring Bean 的生命周期 534
21.11　Spring Bean 作用域的区别是什么 535
21.12　什么是 Spring Inner Bean 535
21.13　Spring 中的单例 Bean 是线程安全的吗 536
21.14　请举例说明如何在 Spring 中注入一个 Java 集合 536
21.15　如何向 Spring Bean 中注入 java.util.Properties 537
21.16　请解释 Spring Bean 的自动装配 538
21.17　自动装配有哪些局限性 538
21.18　请解释各种自动装配模式的区别 539
21.19　请举例解释@Required 注解 539
21.20　请举例说明@Qualifier 注解 540
21.21　构造方法注入和设值注入有什么区别 540
21.22　Spring 中有哪些不同类型的事件 541
21.23　FileSystemResource 和 ClassPathResource 有什么区别 542
21.24　Spring 中用到了哪些设计模式 542
21.25　在 Spring 中如何更有效地使用 JDBC 543
21.26　请解释 Spring 中的 IoC 容器 543
21.27　在 Spring 中可以注入 null 或空字符串吗 543

第 1 篇
Spring 内功心法

第 1 章　软件架构设计原则
第 2 章　Spring 中常用的设计模式

第 1 章 软件架构设计原则

1.1 开闭原则

开闭原则（Open-Closed Principle，OCP）是指一个软件实体（如类、模块和函数）应该对扩展开放，对修改关闭。所谓的开闭，也正是对扩展和修改两个行为的一个原则。它强调的是用抽象构建框架，用实现扩展细节，可以提高软件系统的可复用性及可维护性。开闭原则是面向对象设计中最基础的设计原则，它指导我们如何建立稳定、灵活的系统。例如版本更新，我们尽可能不修改源代码，但是可以增加新功能。

在现实生活中开闭原则也有体现。比如，很多互联网公司都实行弹性作息时间，只规定每天工作 8 小时。意思就是说，对于每天工作 8 小时这个规定是关闭的，但是你什么时候来、什么时候走是开放的。早来早走，晚来晚走。

开闭原则的核心思想就是面向抽象编程，接下来我们来看一段代码。

以咕泡学院的课程体系为例，首先创建一个课程接口 ICourse：

```
public interface ICourse {
    Integer getId();
```

```
    String getName();
    Double getPrice();
}
```

整个课程生态有 Java 架构、大数据、人工智能、前端、软件测试等，我们来创建一个 Java 架构课程的类 JavaCourse：

```java
public class JavaCourse implements ICourse{
    private Integer Id;
    private String name;
    private Double price;
    public JavaCourse(Integer id, String name, Double price) {
        this.Id = id;
        this.name = name;
        this.price = price;
    }
    public Integer getId() {
        return this.Id;
    }
    public String getName() {
        return this.name;
    }
    public Double getPrice() {
        return this.price;
    }
}
```

现在我们要给 Java 架构课程做活动，价格优惠。如果修改 JavaCourse 中的 getPrice() 方法，则存在一定的风险，可能影响其他地方的调用结果。我们如何在不修改原有代码的前提前下，实现价格优惠这个功能呢？现在，我们再写一个处理优惠逻辑的类 JavaDiscountCourse（思考一下为什么要叫 JavaDiscountCourse，而不叫 DiscountCourse）：

```java
public class JavaDiscountCourse extends JavaCourse {
    public JavaDiscountCourse(Integer id, String name, Double price) {
        super(id, name, price);
    }
    public Double getOriginPrice(){
        return super.getPrice();
    }
    public Double getPrice(){
        return super.getPrice() * 0.61;
    }
}
```

回顾一下，简单看一下类结构图，如下图所示。

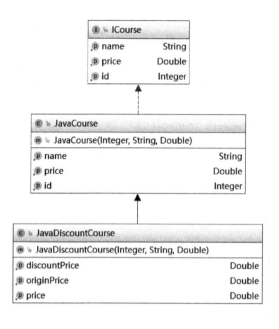

1.2 依赖倒置原则

依赖倒置原则（Dependence Inversion Principle，DIP）是指设计代码结构时，高层模块不应该依赖低层模块，二者都应该依赖其抽象。抽象不应该依赖细节，细节应该依赖抽象。通过依赖倒置，可以减少类与类之间的耦合性，提高系统的稳定性，提高代码的可读性和可维护性，并且能够降低修改程序所造成的风险。接下来看一个案例，还是以 Course（课程）为例，先来创建一个类 Tom：

```
public class Tom {
    public void studyJavaCourse(){
        System.out.println("Tom 在学习 Java 的课程");
    }

    public void studyPythonCourse(){
        System.out.println("Tom 在学习 Python 的课程");
    }
}
```

来调用一下：

```java
public static void main(String[] args) {
    Tom tom = new Tom();
    tom.studyJavaCourse();
    tom.studyPythonCourse();
}
```

Tom 热爱学习，目前正在学习 Java 课程和 Python 课程。大家都知道，学习也是会上瘾的。随着学习兴趣的"暴涨"，现在 Tom 还想学习 AI（人工智能）的课程。这时候，因为业务扩展，要从低层到高层（调用层）依次修改代码。在 Tom 类中增加 studyAICourse()方法，在高层也要追加调用。如此一来，系统发布以后，实际上是非常不稳定的，在修改代码的同时也会带来意想不到的风险。接下来我们优化代码，创建一个课程的抽象 ICourse 接口：

```java
public interface ICourse {
    void study();
}
```

然后编写 JavaCourse 类：

```java
public class JavaCourse implements ICourse {
    @Override
    public void study() {
        System.out.println("Tom 在学习 Java 课程");
    }
}
```

再实现 PythonCourse 类：

```java
public class PythonCourse implements ICourse {
    @Override
    public void study() {
        System.out.println("Tom 在学习 Python 课程");
    }
}
```

修改 Tom 类：

```java
public class Tom {
    public void study(ICourse course){
        course.study();
    }
}
```

来看调用代码：

```java
public static void main(String[] args) {
    Tom tom = new Tom();
```

```
    tom.study(new JavaCourse());
    tom.study(new PythonCourse());
}
```

这时候再看来代码，Tom 的兴趣无论怎么暴涨，对于新的课程，只需要新建一个类，通过传参的方式告诉 Tom，而不需要修改底层代码。实际上这是一种大家非常熟悉的方式，叫依赖注入。注入的方式还有构造器方式和 Setter 方式。我们来看构造器注入方式：

```
public class Tom {

    private ICourse course;

    public Tom(ICourse course){
        this.course = course;
    }

    public void study(){
        course.study();
    }
}
```

看调用代码：

```
public static void main(String[] args) {
    Tom tom = new Tom(new JavaCourse());
    tom.study();
}
```

根据构造器方式注入，在调用时，每次都要创建实例。如果 Tom 是全局单例，则我们就只能选择用 Setter 方式来注入，继续修改 Tom 类的代码：

```
public class Tom {
    private ICourse course;
    public void setCourse(ICourse course) {
        this.course = course;
    }
    public void study(){
        course.study();
    }
}
```

看调用代码：

```
public static void main(String[] args) {
    Tom tom = new Tom();
    tom.setCourse(new JavaCourse());
    tom.study();
}
```

```
tom.setCourse(new PythonCourse());
tom.study();
}
```

现在我们再来看最终的类图，如下图所示。

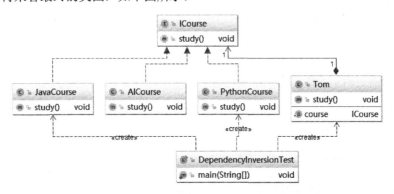

大家要切记：以抽象为基准比以细节为基准搭建起来的架构要稳定得多，因此在拿到需求之后，要面向接口编程，先顶层再细节地设计代码结构。

1.3 单一职责原则

单一职责（Simple Responsibility Pinciple，SRP）是指不要存在多于一个导致类变更的原因。假设我们有一个类负责两个职责，一旦发生需求变更，修改其中一个职责的逻辑代码，有可能导致另一个职责的功能发生故障。这样一来，这个类就存在两个导致类变更的原因。如何解决这个问题呢？将两个职责用两个类来实现，进行解耦。后期需求变更维护互不影响。这样的设计，可以降低类的复杂度，提高类的可读性，提高系统的可维护性，降低变更引起的风险。总体来说，就是一个类、接口或方法只负责一项职责。

接下来，我们来看代码实例，还是用课程举例，我们的课程有直播课和录播课。直播课不能快进和快退，录播课程可以任意地反复观看，功能职责不一样。还是先创建一个 Course 类：

```
public class Course {
    public void study(String courseName){
        if("直播课".equals(courseName)){
            System.out.println(courseName + "不能快进");
        }else{
            System.out.println(courseName + "可以反复回看");
        }
    }
}
```

看调用代码：

```java
public static void main(String[] args) {
    Course course = new Course();
    course.study("直播课");
    course.study("录播课");
}
```

从上面的代码来看，Course类承担了两种处理逻辑。假如现在要对课程进行加密，直播课程和录播课程的加密逻辑不一样，必须修改代码。而修改代码的逻辑势必会相互影响，容易带来不可控的风险。我们对职责进行解耦，来看代码，分别创建两个类：LiveCourse和ReplayCourse。

LiveCourse类的代码如下：

```java
public class LiveCourse {
    public void study(String courseName){
        System.out.println(courseName + "不能快进看");
    }
}
```

ReplayCourse类的代码如下：

```java
public class ReplayCourse {
    public void study(String courseName){
        System.out.println(courseName + "可以反复回");
    }
}
```

调用代码如下：

```java
public static void main(String[] args) {
    LiveCourse liveCourse = new LiveCourse();
    liveCourse.study("直播课");

    ReplayCourse replayCourse = new ReplayCourse();
    replayCourse.study("录播课");
}
```

业务继续发展，课程要做权限。没有付费的学员可以获取课程基本信息，已经付费的学员可以获得视频流，即学习权限。那么在控制课程层面上至少有两个职责。我们可以把展示职责和管理职责分离开来，都实现同一个抽象依赖。设计一个顶层接口，创建ICourse接口：

```java
public interface ICourse {
    //获得基本信息
```

```java
    String getCourseName();

    //获得视频流
    byte[] getCourseVideo();

    //学习课程
    void studyCourse();
    //退款
    void refundCourse();
}
```

我们可以把这个接口拆成两个接口：ICourseInfo 和 ICourseManager。

ICourseInfo 接口的代码如下：

```java
public interface ICourseInfo {
    String getCourseName();
    byte[] getCourseVideo();
}
```

ICourseManager 接口的代码如下：

```java
public interface ICourseManager {
    void studyCourse();
    void refundCourse();
}
```

来看一下类图，如下图所示。

ICourse			ICourseInfo			ICourseManager	
studyCourse()	void		courseName	String		studyCourse()	void
refundCourse()	void		courseVideo	byte[]		refundCourse()	void
courseName	String						
courseVideo	byte[]						

CourseImpl	
studyCourse()	void
refundCourse()	void
courseName	String
courseVideo	byte[]

下面我们来看一下方法层面的单一职责设计。有时候我们会偷懒，把一个方法写成下面这样：

```java
private void modifyUserInfo(String userName,String address){
    userName = "Tom";
    address = "Changsha";
}
```

还可能写成这样：

```java
private void modifyUserInfo(String userName,String... fileds){
    userName = "Tom";
//      address = "Changsha";
}
private void modifyUserInfo(String userName,String address,boolean bool){
    if(bool){

    }else{

    }
    userName = "Tom";
    address = "Changsha";
}
```

显然，上面的modifyUserInfo()方法承担了多个职责，既可以修改userName，也可以修改address，甚至更多，明显不符合单一职责。我们做如下修改，把这个方法拆成两个方法：

```java
private void modifyUserName(String userName){
    userName = "Tom";
}
private void modifyAddress(String address){
    address = "Changsha";
}
```

修改之后，开发起来简单，维护起来也容易。我们在实际开发中会有项目依赖、组合、聚合这些关系，还有项目的规模、周期、技术人员的水平、对进度的把控，很多类都不符合单一职责。但是，我们在编写代码的过程，尽可能地让接口和方法保持单一职责，对项目后期的维护是有很大帮助的。

1.4 接口隔离原则

接口隔离原则（Interface Segregation Principle, ISP）是指用多个专门的接口，而不使用单一的总接口，客户端不应该依赖它不需要的接口。这个原则指导我们在设计接口时应当注意以下几点：

（1）一个类对另一个类的依赖应该建立在最小的接口之上。

（2）建立单一接口，不要建立庞大臃肿的接口。

（3）尽量细化接口，接口中的方法尽量少（不是越少越好，一定要适度）。

接口隔离原则符合我们常说的高内聚、低耦合的设计思想，可以使类具有很好的可读性、可扩展性和可维护性。我们在设计接口的时候，要多花时间去思考，要考虑业务模型，包括对以后有可能发生变更的地方还要做一些预判。所以，对于抽象、对于业务模型的理解是非常重要的。下面我们来看一段代码，对一个动物行为进行抽象描述。

IAnimal 接口的代码如下：

```java
public interface IAnimal {
    void eat();
    void fly();
    void swim();
}
```

Bird 类的代码如下：

```java
public class Bird implements IAnimal {
    @Override
    public void eat() {}
    @Override
    public void fly() {}
    @Override
    public void swim() {}
}
```

Dog 类的代码如下：

```java
public class Dog implements IAnimal {
    @Override
    public void eat() {}
    @Override
    public void fly() {}
    @Override
    public void swim() {}
}
```

可以看出，Bird 的 swim()方法可能只能空着，但 Dog 的 fly()方法显然是不可能的。这时候，我们针对不同动物行为来设计不同的接口，分别设计 IEatAnimal、IFlyAnimal 和 ISwimAnimal 接口，来看代码。

IEatAnimal 接口的代码如下：

```java
public interface IEatAnimal {
    void eat();
}
```

IFlyAnimal 接口的代码如下：

```java
public interface IFlyAnimal {
    void fly();
}
```

ISwimAnimal 接口的代码如下：

```java
public interface ISwimAnimal {
    void swim();
}
```

Dog 只实现 IEatAnimal 和 ISwimAnimal 接口，代码如下：

```java
public class Dog implements ISwimAnimal,IEatAnimal {
    @Override
    public void eat() {}
    @Override
    public void swim() {}
}
```

来看一下两种类图的对比，如下图所示，还是非常清晰明了的。

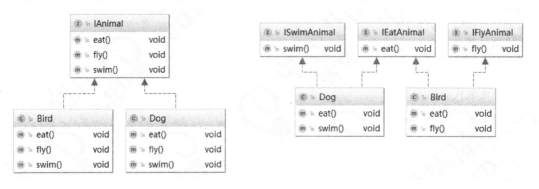

1.5 迪米特原则

迪米特原则（Law of Demeter LoD）是指一个对象应该对其他对象保持最少的了解，又叫最少知道原则（Least Knowledge Principle，LKP），尽量降低类与类之间的耦合度。迪米特原则主要强调：只和朋友交流，不和陌生人说话。出现在成员变量、方法的输入、输出参数中的类都可以称为成员朋友类，而出现在方法体内部的类不属于朋友类。

现在来设计一个权限系统，Boss 需要查看目前发布到线上的课程数量。这时候，Boss 要找到 TeamLeader 去进行统计，TeamLeader 再把统计结果告诉 Boss。接下来我们还是来看代码。

Course 类的代码如下：

```java
public class Course {
}
```

TeamLeader 类的代码如下：

```java
public class TeamLeader {
    public void checkNumberOfCourses(List<Course> courseList){
        System.out.println("目前已发布的课程数量是："+courseList.size());
    }
}
```

Boss 类的代码如下：

```java
public class Boss {
    public void commandCheckNumber(TeamLeader teamLeader){
        //模拟 Boss 一页一页往下翻页，TeamLeader 实时统计
        List<Course> courseList = new ArrayList<Course>();
        for (int i= 0; i < 20 ;i ++){
            courseList.add(new Course());
        }
        teamLeader.checkNumberOfCourses(courseList);
    }
}
```

测试代码如下：

```java
public static void main(String[] args) {
    Boss boss = new Boss();
    TeamLeader teamLeader = new TeamLeader();
    boss.commandCheckNumber(teamLeader);
}
```

写到这里，其实功能都已经实现，代码看上去也没什么问题。根据迪米特原则，Boss 只想要结果，不需要跟 Course 直接交流。而 TeamLeader 统计需要引用 Course 对象。Boss 和 Course 并不是朋友，从下面的类图就可以看出来。

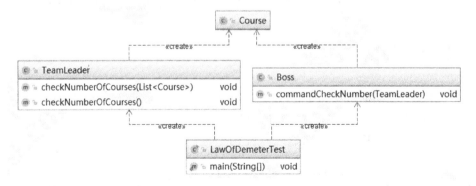

下面对代码进行改造。

TeamLeader 类的代码如下：

```
public class TeamLeader {
    public void checkNumberOfCourses(){
        List<Course> courseList = new ArrayList<Course>();
        for(int i = 0 ;i < 20;i++){
            courseList.add(new Course());
        }
        System.out.println("目前已发布的课程数量是："+courseList.size());
    }
}
```

Boss 类的代码如下：

```
public class Boss {
    public void commandCheckNumber(TeamLeader teamLeader){
        teamLeader.checkNumberOfCourses();
    }
}
```

再来看下面的类图，Course 和 Boss 已经没有关联了。

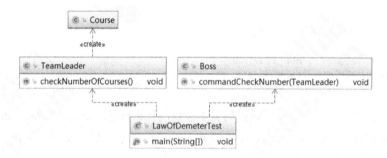

学习软件设计原则，千万不能形成强迫症。碰到业务复杂的场景，我们需要随机应变。

1.6 里氏替换原则

里氏替换原则（Liskov Substitution Principle，LSP）是指如果对每一个类型为 T1 的对象 O1，都有类型为 T2 的对象 O2，使得以 T1 定义的所有程序 P 在所有的对象 O1 都替换成 O2 时，程序 P 的行为没有发生变化，那么类型 T2 是类型 T1 的子类型。

这个定义看上去还是比较抽象的，我们重新理解一下。可以理解为一个软件实体如果适用于一个父类，那么一定适用于其子类，所有引用父类的地方必须能透明地使用其子类的对象，子类对象能够替换父类对象，而程序逻辑不变。根据这个理解，引申含义为：子类可以扩展父类的功能，但不能改变父类原有的功能。

（1）子类可以实现父类的抽象方法，但不能覆盖父类的非抽象方法。

（2）子类可以增加自己特有的方法。

（3）当子类的方法重载父类的方法时，方法的前置条件（即方法的输入/入参）要比父类方法的输入参数更宽松。

（4）当子类的方法实现父类的方法时（重写/重载或实现抽象方法），方法的后置条件（即方法的输出/返回值）要比父类更严格或与父类一样。

在讲开闭原则的时候我埋下了一个伏笔，在获取折扣时重写覆盖了父类的 getPrice()方法，增加了一个获取源码的方法 getOriginPrice()，显然就违背了里氏替换原则。我们修改一下代码，不应该覆盖 getPrice()方法，增加 getDiscountPrice()方法：

```java
public class JavaDiscountCourse extends JavaCourse {
    public JavaDiscountCourse(Integer id, String name, Double price) {
        super(id, name, price);
    }
    public Double getDiscountPrice(){
        return super.getPrice() * 0.61;
    }
}
```

使用里氏替换原则有以下优点：

（1）约束继承泛滥，是开闭原则的一种体现。

（2）加强程序的健壮性，同时变更时也可以做到非常好的兼容性，提高程序的可维护性和扩展性，降低需求变更时引入的风险。

现在来描述一个经典的业务场景，用正方形、矩形和四边形的关系说明里氏替换原则，我们都知道正方形是一个特殊的长方形，所以就可以创建一个父类 Rectangle：

```java
public class Rectangle {
    private long height;
    private long width;
    @Override
    public long getWidth() {
        return width;
    }
    @Override
    public long getHeight() {
        return height;
```

```
    }
    public void setHeight(long height) {
        this.height= height;
    }
    public void setWidth(long width) {
        this.width = width;
    }
}
```

创建正方形类 Square 继承 Rectangle 类：

```
public class Square extends Rectangle {
    private long length;
    public long getLength() {
        return length;
    }
    public void setLength(long length) {
        this.length = length;
    }
    @Override
    public long getWidth() {
        return getLength();
    }
    @Override
    public long getHeight() {
        return getLength();
    }
    @Override
    public void setHeight(long height) {
        setLength(height);
    }
    @Override
    public void setWidth(long width) {
        setLength(width);
    }
}
```

在测试类中创建 resize()方法，长方形的宽应该大于等于高，我们让高一直自增，直到高等于宽，变成正方形：

```
public static void resize(Rectangle rectangle){
    while (rectangle.getWidth() >= rectangle.getHeight()){
        rectangle.setHeight(rectangle.getHeight() + 1);
        System.out.println("width:"+rectangle.getWidth() + ",height:"+rectangle.getHeight());
    }
    System.out.println("resize 方法结束" +
        "\nwidth:"+rectangle.getWidth() + ",height:"+rectangle.getHeight());
}
```

测试代码如下:

```java
public static void main(String[] args) {
    Rectangle rectangle = new Rectangle();
    rectangle.setWidth(20);
    rectangle.setHeight(10);
    resize(rectangle);
}
```

运行结果如下图所示。

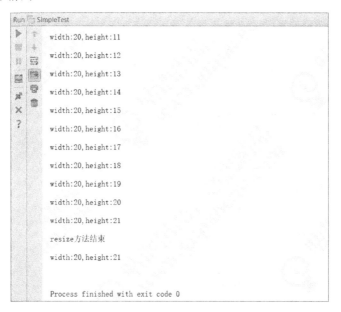

我们发现高比宽还大了,这在长方形中是一种非常正常的情况。现在我们把 Rectangle 类替换成它的子类 Square,修改测试代码:

```java
public static void main(String[] args) {
    Square square = new Square();
    square.setLength(10);
    resize(square);
}
```

上述代码运行时出现了死循环,违背了里氏替换原则,将父类替换为子类后,程序运行结果没有达到预期。因此,我们的代码设计是存在一定风险的。里氏替换原则只存在于父类与子类之间,约束继承泛滥。我们再来创建一个基于长方形与正方形共同的抽象四边形接口 Quadrangle:

```java
public interface Quadrangle {
    long getWidth();
```

```
    long getHeight();
}
```

修改长方形类 Rectangle：

```
public class Rectangle implements Quadrangle {
    private long height;
    private long width;
    @Override
    public long getWidth() {
        return width;
    }
    public long getHeight() {
        return height;
    }
    public void setHeight(long height) {
        this.height = height;
    }
    public void setWidth(long width) {
        this.width = width;
    }
}
```

修改正方形类 Square：

```
public class Square implements Quadrangle {
    private long length;
    public long getLength() {
        return length;
    }
    public void setLength(long length) {
        this.length = length;
    }
    @Override
    public long getWidth() {
        return length;
    }
    @Override
    public long getHeight() {
        return length;
    }
}
```

此时，如果我们把 resize() 方法的参数换成四边形接口 Quadrangle，方法内部就会报错。因为正方形类 Square 已经没有了 setWidth() 和 setHeight() 方法。因此，为了约束继承泛滥，resize() 方法的参数只能用 Rectangle 类。当然，我们在后面的设计模式的内容中还会继续深入讲解。

1.7 合成复用原则

合成复用原则（Composite/Aggregate Reuse Principle，CARP）是指尽量使用对象组合（has-a）/聚合（contanis-a）而不是继承关系达到软件复用的目的。可以使系统更加灵活，降低类与类之间的耦合度，一个类的变化对其他类造成的影响相对较少。

继承叫作白箱复用，相当于把所有的实现细节暴露给子类。组合/聚合称为黑箱复用，我们是无法获取到类以外的对象的实现细节的。虽然我们要根据具体的业务场景来做代码设计，但也需要遵循 OOP 模型。以数据库操作为例，先来创建 DBConnection 类：

```java
public class DBConnection {
    public String getConnection(){
        return "MySQL 数据库连接";
    }
}
```

创建 ProductDao 类：

```java
public class ProductDao{
    private DBConnection dbConnection;
    public void setDbConnection(DBConnection dbConnection) {
        this.dbConnection = dbConnection;
    }
    public void addProduct(){
        String conn = dbConnection.getConnection();
        System.out.println("使用"+conn+"增加产品");
    }
}
```

这就是一种非常典型的合成复用原则的应用场景。但是，就目前的设计来说，DBConnection 还不是一种抽象，不便于系统扩展。目前的系统支持 MySQL 数据库连接，假设业务发生变化，数据库操作层要支持 Oracle 数据库。当然，我们可以在 DBConnection 中增加对 Oracle 数据库的支持，但是这违背了开闭原则。其实，我们可以不修改 Dao 的代码，而将 DBConnection 修改为"abstract"的，来看代码：

```java
public abstract class DBConnection {
    public abstract String getConnection();
}
```

然后将 MySQL 的逻辑抽离：

```java
public class MySQLConnection extends DBConnection {
    @Override
    public String getConnection() {
```

```
        return "MySQL 数据库连接";
    }
}
```

再创建 Oracle 支持：

```
public class OracleConnection extends DBConnection {
    @Override
    public String getConnection() {
        return "Oracle 数据库连接";
    }
}
```

具体选择交给应用层，来看一下类图，如下图所示。

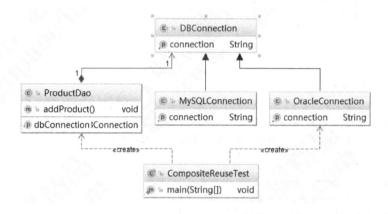

1.8 设计原则总结

学习设计原则是学习设计模式的基础。在实际开发过程中，并不要求所有代码都遵循设计原则，我们要考虑人力、时间、成本、质量，不能刻意追求完美，但要在适当的场景遵循设计原则，这体现的是一种平衡取舍，可以帮助我们设计出更加优雅的代码结构。

第 2 章
Spring 中常用的设计模式

2.1 为什么要从设计模式开始

先来看一个生活案例，当我们开心时，总会寻求"表达"的方式。在学设计模式之前，你可能会如下图所示这样感叹。

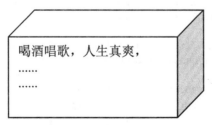

学完设计模式之后，你可能会如下图所示这样感叹。

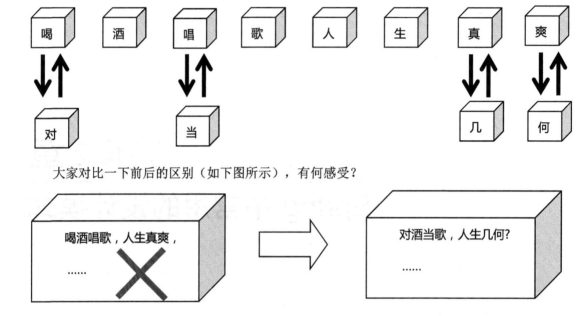

大家对比一下前后的区别（如下图所示），有何感受？

回到代码中，我们来思考一下，设计模式能帮我们解决哪些问题？

2.1.1 写出优雅的代码

来看一段很多年前写的代码：

```
public void setCurForm(Gw_exammingForm curForm,String parameters)throws BaseException {
    JSONObject jsonObj = new JSONObject(parameters);
    //试卷主键
    if(jsonObj.getString("examinationPaper_id")!= null && 
(!jsonObj.getString("examinationPaper_id").equals("")))
        curForm.setExaminationPaper_id(jsonObj.getLong("examinationPaper_id"));
    //剩余时间
    if(jsonObj.getString("leavTime") != null && (!jsonObj.getString("leavTime").equals("")))
        curForm.setLeavTime(jsonObj.getInt("leavTime"));
    //单位主键
    if(jsonObj.getString("organization_id")!= null && 
(!jsonObj.getString("organization_id").equals("")))
        curForm.setOrganization_id(jsonObj.getLong("organization_id"));
    //考试主键
    if(jsonObj.getString("id")!= null && (!jsonObj.getString("id").equals("")))
        curForm.setId(jsonObj.getLong("id"));
    //考场主键
    if(jsonObj.getString("examroom_id")!= null && (!jsonObj.getString("examroom_id").equals("")))
```

```java
    curForm.setExamroom_id(jsonObj.getLong("examroom_id"));
//用户主键
if(jsonObj.getString("user_id")!= null && (!jsonObj.getString("user_id").equals("")))
    curForm.setUser_id(jsonObj.getLong("user_id"));
//专业代码
if(jsonObj.getString("specialtyCode")!= null && (!jsonObj.getString("specialtyCode").equals("")))
    curForm.setSpecialtyCode(jsonObj.getLong("specialtyCode"));
//报考岗位
if(jsonObj.getString("postionCode")!= null && (!jsonObj.getString("postionCode").equals("")))
    curForm.setPostionCode(jsonObj.getLong("postionCode"));
//报考等级
if(jsonObj.getString("gradeCode")!= null && (!jsonObj.getString("gradeCode").equals("")))
    curForm.setGradeCode(jsonObj.getLong("gradeCode"));
//考试开始时间
curForm.setExamStartTime(jsonObj.getString("examStartTime"));
//考试结束时间
curForm.setExamEndTime(jsonObj.getString("examEndTime"));
//考试时长
if(jsonObj.getString("examTime")!= null && (!jsonObj.getString("examTime").equals("")))
    curForm.setExamTime(jsonObj.getInt("examTime"));
//总分
if(jsonObj.getString("fullScore")!= null && (!jsonObj.getString("fullScore").equals("")))
    curForm.setFullScore(jsonObj.getLong("fullScore"));
//及格分
if(jsonObj.getString("passScore")!= null && (!jsonObj.getString("passScore").equals("")))
    curForm.setPassScore(jsonObj.getLong("passScore"));
//学员姓名
curForm.setUserName(jsonObj.getString("user_name"));
//考试得分
if(jsonObj.getString("score")!= null && (!jsonObj.getString("score").equals("")))
    curForm.setScore(jsonObj.getLong("score"));
//是否及格
curForm.setResult(jsonObj.getString("result"));
curForm.setIsPassed(jsonObj.getString("result"));
//单选题答对数量
if(jsonObj.getString("single_ok_count")!= null && (!jsonObj.getString("single_ok_count").equals("")))
    curForm.setSingle_ok_count(jsonObj.getInt("single_ok_count"));
//多选题答对数量
if(jsonObj.getString("multi_ok_count")!= null && (!jsonObj.getString("multi_ok_count").equals("")))
    curForm.setMulti_ok_count(jsonObj.getInt("multi_ok_count"));
//判断题答对数量
if(jsonObj.getString("judgement_ok_count")!= null && (!jsonObj.getString("judgement_ok_count").equals("")))
    curForm.setJudgement_ok_count(jsonObj.getInt("judgement_ok_count"));
}
```

优化之后的代码如下：

```java
public class ExamPaper extends Gw_exammingForm{

    private String examinationPaperId;//试卷主键
    private String leavTime;//剩余时间
    private String organizationId;//单位主键
    private String id;//考试主键
    private String examRoomId;//考场主键
    private String userId;//用户主键
    private String specialtyCode;//专业代码
    private String postionCode;//报考岗位
    private String gradeCode;//报考等级
    private String examStartTime;//考试开始时间
    private String examEndTime;//考试结束时间
    private String examTime;//考试时长
    private String fullScore;//总分
    private String passScore;//及格分
    private String userName;//学员姓名
    private String score;//考试得分
    private String resut;//是否及格
    private String singleOkCount;//单选题答对数量
    private String multiOkCount;//多选题答对数量
    private String judgementOkCount;//判断题答对数量

}

public void setCurForm(Gw_exammingForm curForm,String parameters)throws BaseException {
    try {
        JSONObject jsonObj = new JSONObject(parameters);
        ExamPaper examPaper = JSONObject.parseObject(parameters,ExamPaper.class);

        curForm = examPaper;

    }catch (Exception e){
        e.printStackTrace();
    }

}
```

2.1.2 更好地重构项目

平时我们写的代码虽然满足了需求，但有可能不利于项目的开发与维护，以下面的 JDBC 代码为例：

```java
public void save(Student stu){
    String sql="INSERT INTO t_student(name,age) VALUES(?,?)";
    Connection conn=null;
    Statement st=null;
    try{
        // 1. 加载注册驱动
        Class.forName("com.mysql.jdbc.Driver");
        // 2. 获取数据库连接
        conn=DriverManager.getConnection("jdbc:mysql:///jdbcdemo","root","root");
        // 3. 创建语句对象
        PreparedStatement ps=conn.prepareStatement(sql);
        ps.setObject(1,stu.getName());
        ps.setObject(2,stu.getAge());
        // 4. 执行 SQL 语句
        ps.executeUpdate();
        // 5. 释放资源
    }catch(Exception e){
        e.printStackTrace();
    }finally{
        try{
            if(st!=null)
                st.close();
        }catch(SQLException e){
            e.printStackTrace();
        }finally{
            try{
                if(conn!=null)
                    conn.close();
            }catch(SQLException e){
                e.printStackTrace();
            }
        }
    }
}

// 删除学生信息
public void delete(Long id){
    String sql="DELETE FROM t_student WHERE id=?";
    Connection conn=null;
    Statement st=null;
    try{
        // 1. 加载注册驱动
        Class.forName("com.mysql.jdbc.Driver");
        // 2. 获取数据库连接
        conn=DriverManager.getConnection("jdbc:mysql:///jdbcdemo","root","root");
        // 3. 创建语句对象
        PreparedStatement ps=conn.prepareStatement(sql);
```

```java
            ps.setObject(1,id);
            // 4. 执行SQL语句
            ps.executeUpdate();
            // 5. 释放资源
        }catch(Exception e){
            e.printStackTrace();
        }finally{
            try{
                if(st!=null)
                    st.close();
            }catch(SQLException e){
                e.printStackTrace();
            }finally{
                try{
                    if(conn!=null)
                        conn.close();
                }catch(SQLException e){
                    e.printStackTrace();
                }
            }
        }
}

// 修改学生信息
public void update(Student stu){
    String sql="UPDATE t_student SET name=?,age=? WHERE id=?";
    Connection conn=null;
    Statement st=null;
    try{
        // 1. 加载注册驱动
        Class.forName("com.mysql.jdbc.Driver");
        // 2. 获取数据库连接
        conn=DriverManager.getConnection("jdbc:mysql:///jdbcdemo","root","root");
        // 3. 创建语句对象
        PreparedStatement ps=conn.prepareStatement(sql);
        ps.setObject(1,stu.getName());
        ps.setObject(2,stu.getAge());
        ps.setObject(3,stu.getId());
        // 4. 执行SQL语句
        ps.executeUpdate();
        // 5. 释放资源
    }catch(Exception e){
        e.printStackTrace();
    }finally{
        try{
            if(st!=null)
                st.close();
```

```java
        }catch(SQLException e){
            e.printStackTrace();
        }finally{
            try{
                if(conn!=null)
                    conn.close();
            }catch(SQLException e){
                e.printStackTrace();
            }
        }
    }
}
```

上述代码的功能没问题，但是重复代码太多，我们可以进行抽取，把重复的代码放到一个工具类 JdbcUtil 里。

```java
//工具类
public class JdbcUtil {
    private JdbcUtil() { }
    static {
        // 1. 加载注册驱动
        try {
            Class.forName("com.mysql.jdbc.Driver");
        } catch (Exception e) {
            e.printStackTrace();
        }
    }

    public static Connection getConnection() {
        try {
            // 2. 获取数据库连接
            return DriverManager.getConnection("jdbc:mysql:///jdbcdemo", "root", "root");
        } catch (Exception e) {
            e.printStackTrace();
        }
        return null;
    }

    //释放资源
    public static void close(ResultSet rs, Statement st, Connection conn) {
        try {
            if (rs != null)
                rs.close();
        } catch (SQLException e) {
            e.printStackTrace();
        } finally {
            try {
```

```
            if (st != null)
                st.close();
        } catch (SQLException e) {
            e.printStackTrace();
        } finally {
            try {
                if (conn != null)
                    conn.close();
            } catch (SQLException e) {
                e.printStackTrace();
            }
        }
    }
}
```

在实现类中直接调用工具类 JdbcUtil 中的方法即可:

```
// 增加学生信息
public void save(Student stu) {
    String sql = "INSERT INTO t_student(name,age) VALUES(?,?)";
    Connection conn = null;
    PreparedStatement ps=null;
    try {
        conn = JDBCUtil.getConnection();
        // 3. 创建语句对象
        ps = conn.prepareStatement(sql);
        ps.setObject(1, stu.getName());
        ps.setObject(2, stu.getAge());
        // 4. 执行 SQL 语句
        ps.executeUpdate();
        // 5. 释放资源
    } catch (Exception e) {
        e.printStackTrace();
    } finally {
        JDBCUtil.close(null, ps, conn);
    }

}

// 删除学生信息
public void delete(Long id) {
    String sql = "DELETE FROM t_student WHERE id=?";
    Connection conn = null;
    PreparedStatement ps = null;
    try {
        conn=JDBCUtil.getConnection();
```

```java
            // 3. 创建语句对象
            ps = conn.prepareStatement(sql);
            ps.setObject(1, id);
            // 4. 执行 SQL 语句
            ps.executeUpdate();
            // 5. 释放资源
        } catch (Exception e) {
            e.printStackTrace();
        } finally {
            JDBCUtil.close(null, ps, conn);
        }
    }

    // 修改学生信息
    public void update(Student stu) {
        String sql = "UPDATE t_student SET name=?,age=? WHERE id=?";
        Connection conn = null;
        PreparedStatement ps = null;
        try {
            conn=JDBCUtil.getConnection();
            // 3. 创建语句对象
            ps = conn.prepareStatement(sql);
            ps.setObject(1, stu.getName());
            ps.setObject(2, stu.getAge());
            ps.setObject(3, stu.getId());
            // 4. 执行 SQL 语句
            ps.executeUpdate();
            // 5. 释放资源
        } catch (Exception e) {
            e.printStackTrace();
        } finally {
            JDBCUtil.close(null, ps, conn);
        }
    }

    public Student get(Long id) {
        String sql = "SELECT * FROM t_student WHERE id=?";
        Connection conn = null;
        Statement st = null;
        ResultSet rs = null;
        PreparedStatement ps=null;
        try {
            conn = JDBCUtil.getConnection();
            // 3. 创建语句对象
            ps = conn.prepareStatement(sql);
```

```java
            ps.setObject(1, id);
            // 4. 执行 SQL 语句
            rs = ps.executeQuery();
            if (rs.next()) {
                String name = rs.getString("name");
                int age = rs.getInt("age");
                Student stu = new Student(id, name, age);
                return stu;
            }
            // 5. 释放资源
        } catch (Exception e) {
            e.printStackTrace();
        } finally {
            JDBCUtil.close(rs, ps, conn);
        }
        return null;
    }

    public List<Student> list() {
        List<Student> list = new ArrayList<>();
        String sql = "SELECT * FROM t_student ";
        Connection conn = null;
        Statement st = null;
        ResultSet rs = null;
        PreparedStatement ps=null;
        try {
            conn=JDBCUtil.getConnection();
            // 3. 创建语句对象
            ps = conn.prepareStatement(sql);
            // 4. 执行 SQL 语句
            rs = ps.executeQuery();
            while (rs.next()) {
                long id = rs.getLong("id");
                String name = rs.getString("name");
                int age = rs.getInt("age");
                Student stu = new Student(id, name, age);
                list.add(stu);
            }
            // 5. 释放资源
        } catch (Exception e) {
            e.printStackTrace();
        } finally {
            JDBCUtil.close(rs, ps, conn);
        }
        return list;
    }
}
```

虽然完成了重复代码的抽取,但数据库中的账号和密码等直接显示在代码中,不利于后期账户和密码的维护,我们可以建立一个 db.propertise 文件来存储这些信息:

```
driverClassName =com.mysql.jdbc.Driver
url =jdbc:mysql:///jdbcdemo
username =root
password =root
```

只需在工具类 JdbcUtil 中获取里面的信息即可:

```java
static {
    // 1. 加载注册驱动
    try {
        ClassLoader loader = Thread.currentThread().getContextClassLoader();
        InputStream inputStream = loader.getResourceAsStream("db.properties");
        p = new Properties();
        p.load(inputStream);
        Class.forName(p.getProperty("driverClassName"));
    } catch (Exception e) {
        e.printStackTrace();
    }
}

public static Connection getConnection() {
    try {
        // 2. 获取数据库连接
        return DriverManager.getConnection(p.getProperty("url"), p.getProperty("username"),
            p.getProperty("password"));
    } catch (Exception e) {
        e.printStackTrace();
    }
    return null;
}
```

抽取到这里貌似已经完成,但在实现类中依然存在部分重复代码,在 DML 操作中,除了 SQL 和设置值不同,其他都相同,将相同的部分抽取出去,不同的部分通过参数传递进来,无法直接放在工具类中,这时我们可以创建一个模板类 JdbcTemplate,创建一个 DML 和 DQL 的模板来对代码进行重构:

```java
// 查询统一模板
public static List<Student> query(String sql,Object...params){
    List<Student> list=new ArrayList<>();
    Connection conn = null;
    PreparedStatement ps=null;
    ResultSet rs = null;
    try {
```

```java
        conn=JDBCUtil.getConnection();
        ps=conn.prepareStatement(sql);
        // 设置值
        for (int i = 0; i < params.length; i++) {
            ps.setObject(i+1, params[i]);
        }
        rs = ps.executeQuery();
        while (rs.next()) {
            long id = rs.getLong("id");
            String name = rs.getString("name");
            int age = rs.getInt("age");
            Student stu = new Student(id, name, age);
            list.add(stu);
        }
        // 释放资源
    } catch (Exception e) {
        e.printStackTrace();
    } finally {
        JDBCUtil.close(rs, ps, conn);
    }
    return list;
}
```

实现类直接调用方法即可：

```java
// 增加学生信息
public void save(Student stu) {
    String sql = "INSERT INTO t_student(name,age) VALUES(?,?)";
    Object[] params=new Object[]{stu.getName(),stu.getAge()};
    JdbcTemplate.update(sql, params);
}

// 删除学生信息
public void delete(Long id) {
    String sql = "DELETE FROM t_student WHERE id = ?";
    JdbcTemplate.update(sql, id);
}

// 修改学生信息
public void update(Student stu) {
    String sql = "UPDATE t_student SET name = ?,age = ? WHERE id = ?";
    Object[] params=new Object[]{stu.getName(),stu.getAge(),stu.getId()};
    JdbcTemplate.update(sql, params);
}

public Student get(Long id) {
    String sql = "SELECT * FROM t_student WHERE id=?";
```

```
    List<Student> list = JDBCTemplate.query(sql, id);
    return list.size()>0? list.get(0):null;
}
public List<Student> list() {
    String sql = "SELECT * FROM t_student ";
    return JDBCTemplate.query(sql);
}
```

　　这样重复的代码基本就处理好了，但有个很严重的问题就是，这个程序的 DQL 操作中只能处理 Student 类和 t_student 表的相关数据，无法处理其他数据，如 Teacher 类和 t_teacher 表。不同表（不同的对象）就应该有不同的列，不同列处理结果集的代码就应该不一样，处理结果集的操作只有 DAO 自己最清楚。也就是说，处理结果的方法压根就不应该放在模板方法中，应该由每个 DAO 自己来处理。因此我们可以创建一个 IRowMapper 接口来处理结果集：

```
public interface IRowMapper {
    //处理结果集
    List rowMapper(ResultSet rs) throws Exception;
}
```

　　在 DQL 模板类中调用 IRowMapper 接口中的 handle 方法，提醒实现类自己去实现 mapping 方法：

```
public static List<Student> query(String sql,IRowMapper rsh, Object...params){
    List<Student> list = new ArrayList<>();
    Connection conn = null;
    PreparedStatement ps=null;
    ResultSet rs = null;
    try {
        conn = JdbcUtil.getConnection();
        ps = conn.prepareStatement(sql);
        // 设置值
        for (int i = 0; i < params.length; i++) {
            ps.setObject(i+1, params[i]);
        }
        rs = ps.executeQuery();
        return rsh.mapping(rs);
        // 释放资源
    } catch (Exception e) {
        e.printStackTrace();
    } finally {
        JdbcUtil.close(rs, ps, conn);
    }
    return list ;
}
```

实现类自己去实现 IRowMapper 接口的 mapping 方法，想要处理什么类型数据在里面定义即可：

```java
public Student get(Long id) {
    String sql = "SELECT * FROM t_student WHERE id = ?";
    List<Student> list = JdbcTemplate.query(sql,new StudentRowMapper(), id);
    return list.size()>0? list.get(0):null;
}
public List<Student> list() {
    String sql = "SELECT * FROM t_student ";
    return JdbcTemplate.query(sql,new StudentRowMapper());
}
class StudentRowMapper implements IRowMapper{
    public List mapping(ResultSet rs) throws Exception {
        List<Student> list=new ArrayList<>();
        while(rs.next()){
            long id = rs.getLong("id");
            String name = rs.getString("name");
            int age = rs.getInt("age");
            Student stu=new Student(id, name, age);
            list.add(stu);
        }
        return list;
    }
}
```

好了，基本快大功告成了，但是 DQL 查询不只查询学生信息（List 类型），还查询学生数量，这时就要通过泛型来完成：

```java
public interface IRowMapper<T> {
    // 处理结果集
    T mapping(ResultSet rs) throws Exception;
}


public static <T> T query(String sql,IRowMapper<T> rsh, Object...params){
    Connection conn = null;
    PreparedStatement ps=null;
    ResultSet rs = null;
    try {
        conn = JdbcUtil.getConnection();
        ps = conn.prepareStatement(sql);
        // 设置值
        for (int i = 0; i < params.length; i++) {
            ps.setObject(i+1, params[i]);
        }
        rs = ps.executeQuery();
```

```
            return rsh.mapping(rs);
            // 释放资源
        } catch (Exception e) {
            e.printStackTrace();
        } finally {
            JdbcUtil.close(rs, ps, conn);
        }
        return null;
    }
```

StudentRowMapper 类的代码如下：

```
class StudentRowMapper implements IRowMapper<List<Student>>{
    public List<Student> mapping(ResultSet rs) throws Exception {
        List<Student> list=new ArrayList<>();
        while(rs.next()){
            long id = rs.getLong("id");
            String name = rs.getString("name");
            int age = rs.getInt("age");
            Student stu=new Student(id, name, age);
            list.add(stu);
        }
        return list;
    }
}
```

这样不仅可以查询学生信息，还可以查询学生数量：

```
public Long getCount(){
    String sql = "SELECT COUNT(*) total FROM t_student";
    Long totalCount = (Long) JdbcTemplate.query(sql,
            new IRowMapper<Long>() {
                public Long mapping(ResultSet rs) throws Exception {
                    Long totalCount = null;
                    if(rs.next()){
                        totalCount = rs.getLong("total");
                    }
                    return totalCount;
                }
            });
    return totalCount;
}
```

好了，重构设计已经完成，好的代码会让我们以后维护更方便，因此学会对代码的重构是非常重要的。

2.1.3 经典框架都在用设计模式解决问题

Spring 就是一个把设计模式用得淋漓尽致的经典框架，其实从类的命名就能看出来，下面一一列举，如下表所示。

设计模式名称	举例
工厂模式	BeanFactory
装饰者模式	BeanWrapper
代理模式	AopProxy
委派模式	DispatcherServlet
策略模式	HandlerMapping
适配器模式	HandlerAdapter
模板模式	JdbcTemplate
观察者模式	ContextLoaderListener

需要特别声明的是，设计模式从来都不是单个模式独立使用的。在实际应用中，通常是将多个设计模式混合使用，你中有我，我中有你。本书会围绕 Spring 的 IoC、AOP、MVC、JDBC，根据其设计类型来设计讲解顺序，如下表所示。

类型	名称	英文
创建型模式	工厂模式	Factory Pattern
	单例模式	Singleton Pattern
	原型模式	Prototype Pattern
结构型模式	适配器模式	Adapter Pattern
	装饰者模式	Decorator Pattern
	代理模式	Proxy Pattern
行为型模式	策略模式	Strategy Pattern
	模板模式	Template Pattern
	委派模式	Delegate Pattern
	观察者模式	Observer Pattern

2.2 工厂模式详解

2.2.1 工厂模式的由来

在现实生活中我们都知道，原始社会自给自足（没有工厂）、农耕社会有了小作坊（简单工厂，如民间酒坊）、工业革命后有了流水线（工厂方法，自产自销）、现代产业链中有代工厂（抽象工厂,如富士康)。

我们的项目代码同样也是由简到繁一步一步迭代而来的,但对于调用者来说却越来越简单化了。

2.2.2 简单工厂模式

简单工厂模式(Simple Factory Pattern)是指由一个工厂对象决定创建哪一种产品类的实例,但它不属于 GoF 的 23 种设计模式。简单工厂模式适用于工厂类负责创建的对象较少的场景,且客户端只需要传入工厂类的参数,对于如何创建对象不需要关心。

我们来看代码,还是以课程为例。咕泡学院目前开设有 Java 架构、大数据、人工智能等课程,已经形成了一个生态。我们可以定义一个课程标准 ICourse 接口:

```java
public interface ICourse {
    /** 录制视频 */
    public void record();
}
```

创建一个 Java 课程的实现 JavaCourse 类:

```java
public class JavaCourse implements ICourse {
    public void record() {
        System.out.println("录制 Java 课程");
    }
}
```

我们会这样写客户端调用代码:

```java
public static void main(String[] args) {
    ICourse course = new JavaCourse();
    course.record();
}
```

在上面的代码中,父类 ICourse 指向子类 JavaCourse 的引用,应用层代码需要依赖 JavaCourse,如果业务扩展,继续增加 PythonCourse 甚至更多课程,那么客户端的依赖会变得越来越臃肿。因此,我们要想办法把这种依赖减弱,把创建细节隐藏起来。虽然在目前的代码中,创建对象的过程并不复杂,但从代码设计的角度来讲不易于扩展。现在,我们用简单工厂模式对代码进行优化。先增加课程类 PythonCourse:

```java
public class PythonCourse implements ICourse {
    public void record() {
        System.out.println("录制 Python 课程");
    }
}
```

创建工厂类 CourseFactory：

```java
public class CourseFactory {
    public ICourse create(String name){
        if("java".equals(name)){
            return new JavaCourse();
        }else if("python".equals(name)){
            return new PythonCourse();
        }else {
            return null;
        }
    }
}
```

修改客户端调用代码如下：

```java
public class SimpleFactoryTest {
    public static void main(String[] args) {
        CourseFactory factory = new CourseFactory();
        factory.create("java");
    }
}
```

当然，为了调用方便，可将工厂中的 create() 方法改为静态方法，下面来看一下类图，如下图所示。

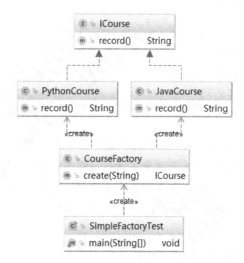

客户端调用变简单了，但如果我们的业务继续扩展，要增加前端课程，那么工厂中的 create() 方法就要每次都根据产品的增加修改代码逻辑，不符合开闭原则。因此，我们还可以对简单工厂模式继续优化，采用反射技术：

```java
public class CourseFactory {
    public ICourse create(String className){
        try {
            if (!(null == className || "".equals(className))) {
                return (ICourse) Class.forName(className).newInstance();
            }
        }catch (Exception e){
            e.printStackTrace();
        }
        return null;
    }
}
```

修改客户端调用代码：

```java
public static void main(String[] args) {
    CourseFactory factory = new CourseFactory();
    ICourse course =
        factory.create("com.gupaoedu.vip.pattern.factory.simplefactory.JavaCourse");
    course.record();
}
```

优化之后，产品不断丰富的过程中不需要修改 CourseFactory 中的代码。但还有个问题是，方法参数是字符串，可控性有待提升，而且还需要强制转型。再修改一下代码：

```java
public ICourse create(Class<? extends ICourse> clazz){
    try {
        if (null != clazz) {
            return clazz.newInstance();
        }
    }catch (Exception e){
        e.printStackTrace();
    }
    return null;
}
```

优化客户端代码：

```java
public static void main(String[] args) {
    CourseFactory factory = new CourseFactory();
    ICourse course = factory.create(JavaCourse.class);
    course.record();
}
```

再看一下类图，如下图所示。

简单工厂模式在 JDK 源码中也无处不在,现在我们来举个例子。例如 Calendar 类,看一下 Calendar.getInstance()方法,下面是 Calendar 的具体创建类:

```java
private static Calendar createCalendar(TimeZone zone,
                    Locale aLocale)
{
    CalendarProvider provider =
        LocaleProviderAdapter.getAdapter(CalendarProvider.class, aLocale)
                    .getCalendarProvider();
    if (provider != null) {
        try {
            return provider.getInstance(zone, aLocale);
        } catch (IllegalArgumentException iae) {
        }
    }

    Calendar cal = null;

    if (aLocale.hasExtensions()) {
        String caltype = aLocale.getUnicodeLocaleType("ca");
        if (caltype != null) {
            switch (caltype) {
            case "buddhist":
            cal = new BuddhistCalendar(zone, aLocale);
                break;
            case "japanese":
                cal = new JapaneseImperialCalendar(zone, aLocale);
                break;
            case "gregory":
                cal = new GregorianCalendar(zone, aLocale);
                break;
            }
        }
    }
    if (cal == null) {
        if (aLocale.getLanguage() == "th" && aLocale.getCountry() == "TH") {
```

```java
        cal = new BuddhistCalendar(zone, aLocale);
    } else if (aLocale.getVariant() == "JP" && aLocale.getLanguage() == "ja"
            && aLocale.getCountry() == "JP") {
        cal = new JapaneseImperialCalendar(zone, aLocale);
    } else {
        cal = new GregorianCalendar(zone, aLocale);
    }
}
    return cal;
}
```

还有一个大家经常使用的 logback，我们可以看到 LoggerFactory 中有多个重载的方法 getLogger()：

```java
public static Logger getLogger(String name) {
    ILoggerFactory iLoggerFactory = getILoggerFactory();
    return iLoggerFactory.getLogger(name);
}

public static Logger getLogger(Class clazz) {
    return getLogger(clazz.getName());
}
```

简单工厂模式也有它的缺点：工厂类的职责相对过重，不易于扩展过于复杂的产品结构。

2.2.3 工厂方法模式

工厂方法模式（Fatory Method Pattern）是指定义一个创建对象的接口，但让实现这个接口的类来决定实例化哪个类，工厂方法模式让类的实例化推迟到子类中进行。在工厂方法模式中用户只需要关心所需产品对应的工厂，无须关心创建细节，而且加入新的产品时符合开闭原则。

工厂方法模式主要解决产品扩展的问题。在简单工厂模式中，随着产品链的丰富，如果每个课程的创建逻辑有区别，则工厂的职责会变得越来越多，有点像万能工厂，不便于维护。根据单一职责原则我们将职能继续拆分，专人干专事。Java 课程由 Java 工厂创建，Python 课程由 Python 工厂创建，对工厂本身也做一个抽象。来看代码，先创建 ICourseFactory 接口：

```java
public interface ICourseFactory {
    ICourse create();
}
```

再分别创建子工厂，JavaCourseFactory 类的代码如下：

```java
import com.gupaoedu.vip.pattern.factory.ICourse;
```

```java
import com.gupaoedu.vip.pattern.factory.JavaCourse;

public class JavaCourseFactory implements ICourseFactory {
    public ICourse create() {
        return new JavaCourse();
    }
}
```

PythonCourseFactory 类的代码如下：

```java
import com.gupaoedu.vip.pattern.factory.ICourse;
import com.gupaoedu.vip.pattern.factory.PythonCourse;

public class PythonCourseFactory implements ICourseFactory {
    public ICourse create() {
        return new PythonCourse();
    }
}
```

测试代码如下：

```java
public static void main(String[] args) {
    ICourseFactory factory = new PythonCourseFactory();
    ICourse course = factory.create();
    course.record();

    factory = new JavaCourseFactory();
    course = factory.create();
    course.record();
}
```

现在再来看一下类图，如下图所示。

再来看看 logback 中工厂方法模式的应用，看看类图就可以了，如下图所示。

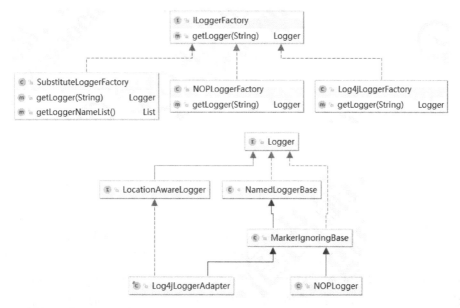

工厂方法模式适用于以下场景：

（1）创建对象需要大量重复的代码。

（2）客户端（应用层）不依赖于产品类实例如何被创建、如何被实现等细节。

（3）一个类通过其子类来指定创建哪个对象。

工厂方法模式也有缺点：

（1）类的个数容易过多，增加复杂度。

（2）增加了系统的抽象性和理解难度。

2.2.4　抽象工厂模式

抽象工厂模式（Abstract Factory Pattern）是指提供一个创建一系列相关或相互依赖对象的接口，无须指定它们的具体类。客户端（应用层）不依赖于产品类实例如何被创建、如何被实现等细节，强调的是一系列相关的产品对象（属于同一产品族）一起使用创建对象需要大量重复的代码。需要提供一个产品类的库，所有的产品以同样的接口出现，从而使客户端不依赖于具体实现。

讲解抽象工厂模式之前，我们要了解两个概念：产品等级结构和产品族。看下面这张图。

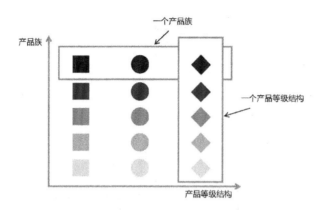

上图中有正方形、圆形和菱形三种图形，相同颜色深浅的代表同一个产品族，相同形状的代表同一个产品等级结构。同样可以做个类比，比如，美的电器生产多种家用电器，上图中颜色最深的正方形代表美的洗衣机、颜色最深的圆形代表美的空调、颜色最深的菱形代表美的热水器，颜色最深的一排都属于美的品牌，都属于美的电器这个产品族。再看最右侧的菱形，颜色最深的我们指定它代表美的热水器，第二排颜色稍微浅一点的菱形可代表海信的热水器。同理，同一产品结构下还有格力热水器、格力空调、格力洗衣机。

再看下面的这张图，最左侧的小房子我们就认为是具体的工厂，有美的工厂，有海信工厂，有格力工厂，每个品牌的工厂都生产洗衣机、热水器和空调。

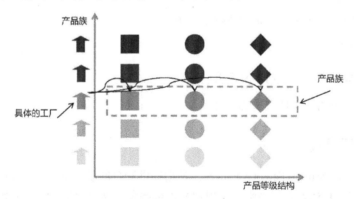

通过两张图的对比，相信大家对抽象工厂模式有了非常形象的理解。接下来我们来看一个具体的业务场景并用代码来实现。还是以课程为例，咕泡学院第三期课程有了新的标准，每个课程不仅要提供课程的录播视频，还要提供老师的课堂笔记。相当于现在的业务变更为同一个课程不单纯包含一个课程信息，同时要包含录播视频、课堂笔记，甚至还要提供源码才能构成一个完整的课程。在产品等级中增加两个产品：IVideo 录播视频和 INote 课堂笔记。

IVideo 接口如下：

```java
public interface IVideo {
    void record();
}
```

INote 接口如下：

```java
public interface INote {
    void edit();
}
```

然后创建一个抽象工厂类 CourseFactory：

```java
import com.gupaoedu.vip.pattern.factory.INote;
import com.gupaoedu.vip.pattern.factory.IVideo;
/**
 * 抽象工厂是用户的主入口
 * 是 Spring 中应用得最广泛的一种设计模式
 * 易于扩展
 */
public interface CourseFactory {
    INote createNote();
    IVideo createVideo();
}
```

接下来，创建 Java 产品族的 Java 视频类 JavaVideo：

```java
public class JavaVideo implements IVideo {
    public void record() {
        System.out.println("录制 Java 视频");
    }
}
```

扩展产品等级 Java 课堂笔记类 JavaNote：

```java
public class JavaNote implements INote {
    public void edit() {
        System.out.println("编写 Java 笔记");
    }
}
```

创建 Java 产品族的具体工厂 JavaCourseFactory：

```java
public class JavaCourseFactory implements CourseFactory {
    public INote createNote() {
        return new JavaNote();
    }
    public IVideo createVideo() {
        return new JavaVideo();
```

```
        }
    }
```

然后创建 Python 产品的 Python 视频类 PythonVideo：

```
public class PythonVideo implements IVideo {
    public void record() {
        System.out.println("录制 Python 视频");
    }
}
```

扩展产品等级 Python 课堂笔记类 PythonNote：

```
public class PythonNote implements INote {
    public void edit() {
        System.out.println("编写 Python 笔记");
    }
}
```

创建 Python 产品族的具体工厂 PythonCourseFactory：

```
public class PythonCourseFactory implements CourseFactory {
    public INote createNote() {
        return new PythonNote();
    }
    public IVideo createVideo() {
        return new PythonVideo();
    }
}
```

来看客户端调用代码：

```
public static void main(String[] args) {
    JavaCourseFactory factory = new JavaCourseFactory();
    factory.createNote().edit();
    factory.createVideo().record();
}
```

上面的代码完整地描述了两个产品族：Java 课程和 Python 课程，也描述了两个产品等级视频和笔记。抽象工厂模式非常完美清晰地描述了这样一层复杂的关系。但是，不知道大家有没有发现，如果我们再继续扩展产品等级，将源码 Source 也加入课程，那么我们的代码从抽象工厂到具体工厂要全部调整，很显然不符合开闭原则。由此可知抽象工厂模式也是有缺点的：

（1）规定了所有可能被创建的产品集合，产品族中扩展新的产品困难，需要修改抽象工厂的接口。

（2）增加了系统的抽象性和理解难度。

但在实际应用中,我们千万不能"犯强迫症"甚至"有洁癖"。在实际需求中,产品等级结构升级是非常正常的一件事情。只要不频繁升级,根据实际情况可以不遵循开闭原则。代码每半年升级一次或者每年升级一次又有何不可呢?

2.2.5 利用工厂模式重构的实践案例

下面演示 JDBC 操作案例,我们每次操作都需要重新创建数据库连接,每次创建其实都非常耗费性能,消耗业务调用时间。我们利用工厂模式将数据库连接先创建好,放到容器中缓存,在业务调用时就只需现取现用。接下来我们来看代码。

Pool 抽象类的代码如下:

```java
package org.jdbc.sqlhelper;

import java.io.IOException;
import java.io.InputStream;
import java.sql.*;
import java.util.Properties;

/**
 * 自定义连接池 getInstance(),返回 Pool 的唯一实例,第一次调用时将执行构造函数
 * 构造函数 Pool()调用驱动装载 loadDrivers()函数;createPool()函数创建连接池,loadDrivers()装载驱动
 * getConnection()返回一个连接实例,getConnection(long time)添加时间限制
 * freeConnection(Connection con)将 con 连接实例返回连接池,getNum()返回空闲连接数
 * getNumActive()返回当前使用的连接数
 */
public abstract class Pool {
    public String propertiesName = "connection-INF.properties";

    private static Pool instance = null; // 定义唯一实例

    protected int maxConnect = 100; // 最大连接数

    protected int normalConnect = 10; // 保持连接数

    protected String driverName = null; // 驱动字符串

    protected Driver driver = null; // 驱动变量

    // 私有构造函数,不允许外界访问
    protected Pool() {
        try
        {
```

```java
      init();
      loadDrivers(driverName);
   }catch(Exception e)
   {
      e.printStackTrace();
   }
}

// 初始化所有从配置文件中读取的成员变量
private void init() throws IOException {
   InputStream is = Pool.class.getResourceAsStream(propertiesName);
   Properties p = new Properties();
   p.load(is);
   this.driverName = p.getProperty("driverName");
   this.maxConnect = Integer.parseInt(p.getProperty("maxConnect"));
   this.normalConnect = Integer.parseInt(p.getProperty("normalConnect"));
}

// 装载和注册所有 JDBC 驱动程序
protected void loadDrivers(String dri) {
   String driverClassName = dri;
   try {
      driver = (Driver) Class.forName(driverClassName).newInstance();
      DriverManager.registerDriver(driver);
      System.out.println("成功注册 JDBC 驱动程序" + driverClassName);
   } catch (Exception e) {
      System.out.println("无法注册 JDBC 驱动程序:" + driverClassName + ",错误:" + e);
   }
}

// 创建连接池
public abstract void createPool();
/**
 * （单例模式）返回数据库连接池 Pool 的实例
 *
 * @param driverName 数据库驱动字符串
 * @return
 * @throws IOException
 * @throws ClassNotFoundException
 * @throws IllegalAccessException
 * @throws InstantiationException
 */
public static synchronized Pool getInstance() throws IOException,
      InstantiationException, IllegalAccessException,
      ClassNotFoundException {

   if (instance != null) {
```

```java
            instance = (Pool) Class.forName("org.jdbc.sqlhelper.Pool")
                    .newInstance();
        }
        return instance;
    }

    // 获得一个可用的连接,如果没有则创建一个连接,且小于最大连接限制
    public abstract Connection getConnection();

    // 获得一个连接,有时间限制
    public abstract Connection getConnection(long time);

    // 将连接对象返回给连接池
    public abstract void freeConnection(Connection con);

    // 返回当前空闲连接数
    public abstract int getNum();

    // 返回当前工作的连接数
    public abstract int getNumActive();

    protected synchronized void release() {
        // 撤销驱动
        try {
            DriverManager.deregisterDriver(driver);
            System.out.println("撤销 JDBC 驱动程序 " + driver.getClass().getName());
        } catch (SQLException e) {
            System.out
                    .println("无法撤销 JDBC 驱动程序的注册:" + driver.getClass().getName());
        }
    }
}
```

DBConnectionPool 数据库连接池:

```java
package org.jdbc.sqlhelper;

import java.io.IOException;
import java.io.InputStream;
import java.sql.*;
import java.util.*;
import java.util.Date;
// 数据库链接池管理类
public final class DBConnectionPool extends Pool {
    private int checkedOut; //正在使用的连接数
    private Vector<Connection> freeConnections = new Vector<Connection>(); //存放产生的连接对象容器
```

```java
private String passWord = null; // 密码
private String url = null; // 连接字符串
private String userName = null; // 用户名
private static int num = 0;// 空闲连接数
private static int numActive = 0;// 当前可用的连接数
private static DBConnectionPool pool = null;// 连接池实例变量

// 产生数据连接池
public static synchronized DBConnectionPool getInstance()
{
    if(pool == null)
    {
        pool = new DBConnectionPool();
    }
    return pool;
}

// 获得一个数据库连接池的实例
private DBConnectionPool() {
    try
    {
        init();
        for (int i = 0; i < normalConnect; i++) { //初始 normalConn 个连接
            Connection c = newConnection();
            if (c != null) {
                freeConnections.addElement(c); //往容器中添加一个连接对象
                num++; //记录总连接数
            }
        }
    }catch(Exception e)
    {
        e.printStackTrace();
    }
}

// 初始化
private void init() throws IOException
{
    InputStream is = DBConnectionPool.class.getResourceAsStream(propertiesName);
    Properties p = new Properties();
    p.load(is);
    this.userName = p.getProperty("userName");
    this.passWord = p.getProperty("passWord");
    this.driverName = p.getProperty("driverName");
    this.url = p.getProperty("url");
    this.maxConnect = Integer.parseInt(p.getProperty("maxConnect"));
```

```java
    this.normalConnect = Integer.parseInt(p.getProperty("normalConnect"));
}

// 如果不再使用某个连接对象，可调此方法将该对象释放到连接池
public synchronized void freeConnection(Connection con) {
    freeConnections.addElement(con);
    num++;
    checkedOut--;
    numActive--;
    notifyAll(); //解锁
}

// 创建一个新连接
private Connection newConnection() {
    Connection con = null;
    try {
        if (userName == null) { // 用户、密码都为空
            con = DriverManager.getConnection(url);
        } else {
            con = DriverManager.getConnection(url, userName, passWord);
        }
        System.out.println("连接池创建一个新的连接");
    } catch (SQLException e) {
        System.out.println("无法创建这个 URL 的连接" + url);
        return null;
    }
    return con;
}

// 返回当前空闲连接数
public int getNum() {
    return num;
}

// 返回当前连接数
public int getNumActive() {
    return numActive;
}

// （单例模式）获取一个可用连接
public synchronized Connection getConnection() {
    Connection con = null;
    if (freeConnections.size() > 0) { // 还有空闲的连接
        num--;
        con = (Connection) freeConnections.firstElement();
        freeConnections.removeElementAt(0);
        try {
```

```java
            if (con.isClosed()) {
                System.out.println("从连接池中删除一个无效连接");
                con = getConnection();
            }
        } catch (SQLException e) {
            System.out.println("从连接池中删除一个无效连接");
            con = getConnection();
        }
    } else if (maxConnect == 0 || checkedOut < maxConnect) { // 没有空闲连接且当前连接小于最
                                                            // 大允许值，最大值为0则不限制
        con = newConnection();
    }
    if (con != null) { // 当前连接数加1
        checkedOut++;
    }
    numActive++;
    return con;
}

// 获取一个连接，并加上等待时间限制，单位为毫秒
public synchronized Connection getConnection(long timeout) {
    long startTime = new Date().getTime();
    Connection con;
    while ((con = getConnection()) == null) {
        try {
            wait(timeout); // 线程等待
        } catch (InterruptedException e) {
        }
        if ((new Date().getTime() - startTime) >= timeout) {
            return null; // 如果超时，则返回
        }
    }
    return con;
}

// 关闭所有连接
public synchronized void release() {
    try {
        //将当前连接赋值到枚举中
        Enumeration allConnections = freeConnections.elements();
        //使用循环关闭所用连接
        while (allConnections.hasMoreElements()) {
            //如果此枚举对象至少还有一个可提供的元素，则返回此枚举对象的下一个元素
            Connection con = (Connection) allConnections.nextElement();
            try {
                con.close();
                num--;
```

```
            } catch (SQLException e) {
                System.out.println("无法关闭连接池中的连接");
            }
        }
        freeConnections.removeAllElements();
        numActive = 0;
    } finally {
        super.release();
    }
}

// 建立连接池
public void createPool() {
    pool = new DBConnectionPool();
    if (pool != null) {
        System.out.println("创建连接池成功");
    } else {
        System.out.println("创建连接池失败");
    }
}
```

2.3 单例模式详解

2.3.1 单例模式的应用场景

单例模式（Singleton Pattern）是指确保一个类在任何情况下都绝对只有一个实例，并提供一个全局访问点。单例模式是创建型模式。单例模式在现实生活中应用也非常广泛，例如，公司 CEO、部门经理等。J2EE 标准中的 ServletContext、ServletContextConfig 等、Spring 框架应用中的 ApplicationContext、数据库的连接池等也都是单例形式。

2.3.2 饿汉式单例模式

先来看单例模式的类结构图，如下图所示。

C ≜ Singleton	
ℱ ≜ singleton	Singleton
m ≜ Singleton()	
m ≜ getInstance()	Singleton

饿汉式单例模式在类加载的时候就立即初始化，并且创建单例对象。它绝对线程安全，在线

程还没出现以前就实例化了，不可能存在访问安全问题。

优点：没有加任何锁、执行效率比较高，用户体验比懒汉式单例模式更好。

缺点：类加载的时候就初始化，不管用与不用都占着空间，浪费了内存，有可能"占着茅坑不拉屎"。

Spring 中 IoC 容器 ApplicationContext 本身就是典型的饿汉式单例模式。接下来看一段代码：

```java
public class HungrySingleton {
    //先静态、后动态
    //先属性、后方法
    //先上后下
    private static final HungrySingleton hungrySingleton = new HungrySingleton();

    private HungrySingleton(){}

    public static HungrySingleton getInstance(){
        return hungrySingleton;
    }
}
```

还有另外一种写法，利用静态代码块的机制：

```java
//饿汉式静态块单例模式
public class HungryStaticSingleton {
    private static final HungryStaticSingleton hungrySingleton;
    static {
        hungrySingleton = new HungryStaticSingleton();
    }
    private HungryStaticSingleton(){}
    public static HungryStaticSingleton getInstance(){
        return hungrySingleton;
    }
}
```

这两种写法都非常简单，也非常好理解，饿汉式单例模式适用于单例对象较少的情况。下面我们来看性能更优的写法。

2.3.3 懒汉式单例模式

懒汉式单例模式的特点是：被外部类调用的时候内部类才会加载。下面看懒汉式单例模式的简单实现 LazySimpleSingleton：

```java
//懒汉式单例模式在外部需要使用的时候才进行实例化
public class LazySimpleSingleton {
```

```java
    private LazySimpleSingleton(){}
    //静态块，公共内存区域
    private static LazySimpleSingleton lazy = null;
    public static LazySimpleSingleton getInstance(){
        if(lazy == null){
            lazy = new LazySimpleSingleton();
        }
        return lazy;
    }
}
```

然后写一个线程类 ExectorThread：

```java
public class ExectorThread implements Runnable{
    @Override
    public void run() {
        LazySimpleSingleton singleton = LazySimpleSingleton.getInstance();
        System.out.println(Thread.currentThread().getName() + ":" + singleton);
    }
}
```

客户端测试代码如下：

```java
public class LazySimpleSingletonTest {
    public static void main(String[] args) {
        Thread t1 = new Thread(new ExectorThread());
        Thread t2 = new Thread(new ExectorThread());
        t1.start();
        t2.start();
        System.out.println("End");
    }
}
```

运行结果如下图所示。

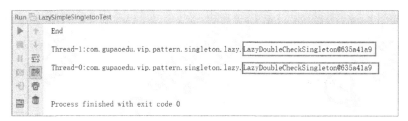

上面的代码有一定概率出现两种不同结果，这意味着上面的单例存在线程安全隐患。我们通过调试运行再具体看一下。这里教大家一种新技能，用线程模式调试，手动控制线程的执行顺序来跟踪内存的变化。先给 ExectorThread 类打上断点，如下图所示。

```
10    public class ExectorThread implements Runnable{
11
12        public void run() {
13            LazySimpleSingleton singleton = LazySimpleSingleton.getInstance();
14            System.out.println(Thread.currentThread().getName() + ":" + singleton);
15        }
16    }
```

使用鼠标右键单击断点，切换为 Thread 模式，如下图所示。

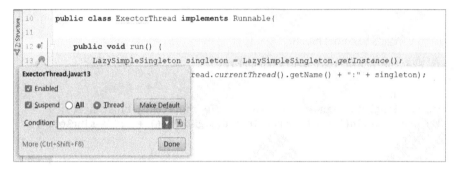

然后给 LazySimpleSingleton 类打上断点，同样标记为 Thread 模式，如下图所示。

```
 9    public class LazySimpleSingleton {
10        private LazySimpleSingleton(){}
11        //静态块，公共内存区域
12        private static LazySimpleSingleton lazy = null;
13        public static LazySimpleSingleton getInstance(){
14            if(null == lazy){
15                lazy = new LazySimpleSingleton();
16            }
17            return lazy;
18        }
19    }
```

切回客户端测试代码，同样也打上断点，同时改为 Thread 模式，如下图所示。

```
 6    public class LazySimpleSingletonTest {
 7        public static void main(String[] args) {
 8            Thread t1 = new Thread(new ExectorThread());
 9            Thread t2 = new Thread(new ExectorThread());
10            t1.start();
11            t2.start();
12            System.out.println("End");
13        }
14    }
```

开始"Debug"之后,会看到 Debug 控制台可以自由切换 Thread 的运行状态,如下图所示。

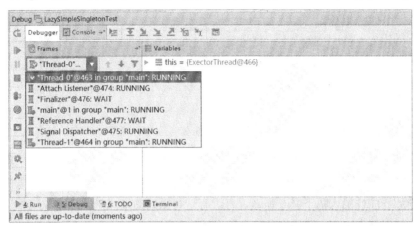

通过不断切换线程,并观测其内存状态,我们发现在线程环境下 LazySimpleSingleton 被实例化了两次。有时我们得到的运行结果可能是相同的两个对象,实际上是被后面执行的线程覆盖了,我们看到了一个假象,线程安全隐患依旧存在。那么,我们如何来优化代码,使得懒汉式单例模式在线程环境下安全呢?来看下面的代码,给 getInstance()加上 synchronized 关键字,使这个方法变成线程同步方法:

```
public class LazySimpleSingleton {
    private LazySimpleSingleton(){}
    //静态块,公共内存区域
    private static LazySimpleSingleton lazy = null;
    public synchronized static LazySimpleSingleton getInstance(){
        if(lazy == null){
            lazy = new LazySimpleSingleton();
        }
        return lazy;
    }
}
```

我们再来调试。当执行其中一个线程并调用 getInstance()方法时,另一个线程在调用 getInstance()方法,线程的状态由 RUNNING 变成了 MONITOR,出现阻塞。直到第一个线程执行完,第二个线程才恢复到 RUNNING 状态继续调用 getInstance()方法,如下图所示。

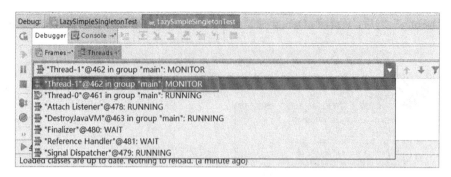

上图完美地展现了 synchronized 监视锁的运行状态，线程安全的问题解决了。但是，用 synchronized 加锁时，在线程数量比较多的情况下，如果 CPU 分配压力上升，则会导致大批线程阻塞，从而导致程序性能大幅下降。那么，有没有一种更好的方式，既能兼顾线程安全又能提升程序性能呢？答案是肯定的。我们来看双重检查锁的单例模式：

```java
public class LazyDoubleCheckSingleton {
    private volatile static LazyDoubleCheckSingleton lazy = null;

    private LazyDoubleCheckSingleton(){}
    public static LazyDoubleCheckSingleton getInstance(){
        if(lazy == null){
            synchronized (LazyDoubleCheckSingleton.class){
                if(lazy == null){
                    lazy = new LazyDoubleCheckSingleton();
                    //1.分配内存给这个对象
                    //2.初始化对象
                    //3.设置 lazy 指向刚分配的内存地址
                }
            }
        }
        return lazy;
    }
}
```

现在，我们来进行断点调试，如下图所示。

```java
public class LazyDoubleCheckSingleton {
    private volatile static LazyDoubleCheckSingleton lazy = null;

    private LazyDoubleCheckSingleton(){}
    public static LazyDoubleCheckSingleton getInstance(){
        if(lazy == null){
            synchronized (LazyDoubleCheckSingleton.class){
                if(lazy == null){
                    lazy = new LazyDoubleCheckSingleton();
                    //1.分配内存给这个对象
                    //2.初始化对象
                    //3.设置lazy指向刚分配的内存地址
                    //4.初次访问对象
                }
            }
        }
        return lazy;
    }
}
```

当第一个线程调用 getInstance()方法时，第二个线程也可以调用。当第一个线程执行到 synchronized 时会上锁，第二个线程就会变成 MONITOR 状态，出现阻塞。此时，阻塞并不是基于整个 LazySimpleSingleton 类的阻塞，而是在 getInstance()方法内部的阻塞，只要逻辑不太复杂，对于调用者而言感知不到。

但是，用到 synchronized 关键字总归要上锁，对程序性能还是存在一定影响的。难道就真的没有更好的方案吗？当然有。我们可以从类初始化的角度来考虑，看下面的代码，采用静态内部类的方式：

```java
//这种形式兼顾饿汉式单例模式的内存浪费问题和 synchronized 的性能问题
//完美地屏蔽了这两个缺点
public class LazyInnerClassSingleton {
    //使用 LazyInnerClassGeneral 的时候，默认会先初始化内部类
    //如果没使用，则内部类是不加载的
    private LazyInnerClassSingleton(){}

    //每一个关键字都不是多余的,static 是为了使单例的空间共享，保证这个方法不会被重写、重载
    public static final LazyInnerClassSingleton getInstance(){
        //在返回结果以前，一定会先加载内部类
        return LazyHolder.LAZY;
    }

    //默认不加载
    private static class LazyHolder{
```

```
        private static final LazyInnerClassSingleton LAZY = new LazyInnerClassSingleton();
    }
}
```

这种方式兼顾了饿汉式单例模式的内存浪费问题和 synchronized 的性能问题。内部类一定是要在方法调用之前初始化，巧妙地避免了线程安全问题。由于这种方式比较简单，我们就不带大家一步一步调试了。

2.3.4 反射破坏单例

大家有没有发现，上面介绍的单例模式的构造方法除了加上 private 关键字，没有做任何处理。如果我们使用反射来调用其构造方法，再调用 getInstance()方法，应该有两个不同的实例。现在来看一段测试代码，以 LazyInnerClassSingleton 为例：

```java
public class LazyInnerClassSingletonTest {
    public static void main(String[] args) {
        try{
            //在很无聊的情况下，进行破坏
            Class<?> clazz = LazyInnerClassSingleton.class;

            //通过反射获取私有的构造方法
            Constructor c = clazz.getDeclaredConstructor(null);
            //强制访问
            c.setAccessible(true);

            //暴力初始化
            Object o1 = c.newInstance();

            //调用了两次构造方法，相当于"new"了两次，犯了原则性错误
            Object o2 = c.newInstance();

            System.out.println(o1 == o2);
        }catch (Exception e){
            e.printStackTrace();
        }
    }
}
```

运行结果如下图所示。

```
Run  LazyInnerClassSingletonTest
     false

     Process finished with exit code 0
```

显然，创建了两个不同的实例。现在，我们在其构造方法中做一些限制，一旦出现多次重复创建，则直接抛出异常。来看优化后的代码：

```java
//自认为史上最牛的单例模式的实现方式
public class LazyInnerClassSingleton {
    //使用 LazyInnerClassGeneral 的时候，默认会先初始化内部类
    //如果没使用，则内部类是不加载的
    private LazyInnerClassSingleton(){
        if(LazyHolder.LAZY != null){
            throw new RuntimeException("不允许创建多个实例");
        }
    }

    //每一个关键字都不是多余的，static 是为了使单例的空间共享，保证这个方法不会被重写、重载
    public static final LazyInnerClassSingleton getInstance(){
        //在返回结果以前，一定会先加载内部类
        return LazyHolder.LAZY;
    }

    //默认不加载
    private static class LazyHolder{
        private static final LazyInnerClassSingleton LAZY = new LazyInnerClassSingleton();
    }
}
```

再运行测试代码，会得到如下图所示结果。

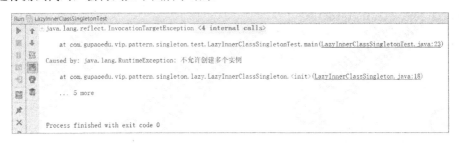

至此，自认为史上最牛的单例模式的实现方式便大功告成。

2.3.5 序列化破坏单例

一个单例对象创建好后，有时候需要将对象序列化然后写入磁盘，下次使用时再从磁盘中读取对象并进行反序列化，将其转化为内存对象。反序列化后的对象会重新分配内存，即重新创建。如果序列化的目标对象为单例对象，就违背了单例模式的初衷，相当于破坏了单例，来看一段代码：

```java
//反序列化导致破坏单例模式
public class SeriableSingleton implements Serializable {
    //序列化就是把内存中的状态通过转换成字节码的形式
    //从而转换一个I/O流，写入其他地方（可以是磁盘、网络I/O）
    //内存中的状态会永久保存下来

    //反序列化就是将已经持久化的字节码内容转换为I/O流
    //通过I/O流的读取，进而将读取的内容转换为Java对象
    //在转换过程中会重新创建对象new

    public  final static SeriableSingleton INSTANCE = new SeriableSingleton();
    private SeriableSingleton(){}

    public static SeriableSingleton getInstance(){
        return INSTANCE;
    }
}
```

编写测试代码：

```java
package com.gupaoedu.vip.pattern.singleton.test;

import com.gupaoedu.vip.pattern.singleton.seriable.SeriableSingleton;
import java.io.FileInputStream;
import java.io.FileOutputStream;
import java.io.ObjectInputStream;
import java.io.ObjectOutputStream;

public class SeriableSingletonTest {
    public static void main(String[] args) {
        SeriableSingleton s1 = null;
        SeriableSingleton s2 = SeriableSingleton.getInstance();

        FileOutputStream fos = null;
        try {
            fos = new FileOutputStream("SeriableSingleton.obj");
            ObjectOutputStream oos = new ObjectOutputStream(fos);
            oos.writeObject(s2);
            oos.flush();
            oos.close();

            FileInputStream fis = new FileInputStream("SeriableSingleton.obj");
            ObjectInputStream ois = new ObjectInputStream(fis);
            s1 = (SeriableSingleton)ois.readObject();
            ois.close();

            System.out.println(s1);
            System.out.println(s2);
            System.out.println(s1 == s2);
```

```
        } catch (Exception e) {
            e.printStackTrace();
        }
    }
}
```

运行结果如下图所示。

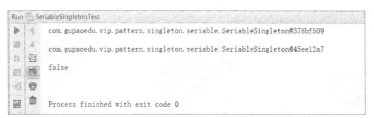

从运行结果可以看出，反序列化后的对象和手动创建的对象是不一致的，实例化了两次，违背了单例模式的设计初衷。那么，我们如何保证在序列化的情况下也能够实现单例模式呢？其实很简单，只需要增加 readResolve() 方法即可。来看优化后的代码：

```
package com.gupaoedu.vip.pattern.singleton.seriable;

import java.io.Serializable;

public class SeriableSingleton implements Serializable {

    public final static SeriableSingleton INSTANCE = new SeriableSingleton();
    private SeriableSingleton(){}

    public static SeriableSingleton getInstance(){
        return INSTANCE;
    }
    private Object readResolve(){
        return INSTANCE;
    }
}
```

再看运行结果，如下图所示。

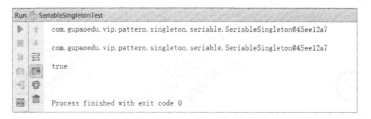

大家一定会想：这是什么原因呢？为什么要这样写？看上去很神奇的样子，也让人有些费解。不如我们一起来看看 JDK 的源码实现以了解清楚。我们进入 ObjectInputStream 类的 readObject() 方法，代码如下：

```java
public final Object readObject()
    throws IOException, ClassNotFoundException
{
    if (enableOverride) {
        return readObjectOverride();
    }

    int outerHandle = passHandle;
    try {
        Object obj = readObject0(false);
        handles.markDependency(outerHandle, passHandle);
        ClassNotFoundException ex = handles.lookupException(passHandle);
        if (ex != null) {
            throw ex;
        }
        if (depth == 0) {
            vlist.doCallbacks();
        }
        return obj;
    } finally {
        passHandle = outerHandle;
        if (closed && depth == 0) {
            clear();
        }
    }
}
```

我们发现，在 readObject() 方法中又调用了重写的 readObject0() 方法。进入 readObject0() 方法，代码如下：

```java
private Object readObject0(boolean unshared) throws IOException {
    ...

    case TC_OBJECT:
        return checkResolve(readOrdinaryObject(unshared));

    ...
}
```

我们看到 TC_OBJECT 中调用了 ObjectInputStream 的 readOrdinaryObject() 方法，看源码：

```java
private Object readOrdinaryObject(boolean unshared)
    throws IOException
```

```
{
    if (bin.readByte() != TC_OBJECT) {
        throw new InternalError();
    }

    ObjectStreamClass desc = readClassDesc(false);
    desc.checkDeserialize();

    Class<?> cl = desc.forClass();
    if (cl == String.class || cl == Class.class
            || cl == ObjectStreamClass.class) {
        throw new InvalidClassException("invalid class descriptor");
    }

    Object obj;
    try {
        obj = desc.isInstantiable() ? desc.newInstance() : null;
    } catch (Exception ex) {
        throw (IOException) new InvalidClassException(
            desc.forClass().getName(),
            "unable to create instance").initCause(ex);
    }

    ...

    return obj;
}
```

我们发现调用了 ObjectStreamClass 的 isInstantiable()方法，而 isInstantiable()方法的代码如下：

```
boolean isInstantiable() {
    requireInitialized();
    return (cons != null);
}
```

上述代码非常简单，就是判断一下构造方法是否为空，构造方法不为空就返回 true。这意味着只要有无参构造方法就会实例化。

这时候其实还没有找到加上 readResolve()方法就避免了单例模式被破坏的真正原因。再回到 ObjectInputStream 的 readOrdinaryObject()方法，继续往下看：

```
private Object readOrdinaryObject(boolean unshared)
    throws IOException
{
    if (bin.readByte() != TC_OBJECT) {
        throw new InternalError();
    }
```

```java
ObjectStreamClass desc = readClassDesc(false);
desc.checkDeserialize();

Class<?> cl = desc.forClass();
if (cl == String.class || cl == Class.class
        || cl == ObjectStreamClass.class) {
    throw new InvalidClassException("invalid class descriptor");
}

Object obj;
try {
    obj = desc.isInstantiable() ? desc.newInstance() : null;
} catch (Exception ex) {
    throw (IOException) new InvalidClassException(
        desc.forClass().getName(),
        "unable to create instance").initCause(ex);
}

...

if (obj != null &&
    handles.lookupException(passHandle) == null &&
    desc.hasReadResolveMethod())
{
    Object rep = desc.invokeReadResolve(obj);
    if (unshared && rep.getClass().isArray()) {
        rep = cloneArray(rep);
    }
    if (rep != obj) {
        if (rep != null) {
            if (rep.getClass().isArray()) {
                filterCheck(rep.getClass(), Array.getLength(rep));
            } else {
                filterCheck(rep.getClass(), -1);
            }
        }
        handles.setObject(passHandle, obj = rep);
    }
}

return obj;
}
```

判断无参构造方法是否存在之后，又调用了 hasReadResolveMethod()方法，来看代码：

```java
boolean hasReadResolveMethod() {
```

```
    requireInitialized();
    return (readResolveMethod != null);
}
```

上述代码逻辑非常简单，就是判断 readResolveMethod 是否为空，不为空就返回 true。那么 readResolveMethod 是在哪里赋值的呢？通过全局查找知道，在私有方法 ObjectStreamClass()中给 readResolveMethod 进行了赋值，来看代码：

```
readResolveMethod = getInheritableMethod(
    cl, "readResolve", null, Object.class);
```

上面的逻辑其实就是通过反射找到一个无参的 readResolve()方法，并且保存下来。现在回到 ObjectInputStream 的 readOrdinaryObject()方法继续往下看，如果 readResolve()方法存在则调用 invokeReadResolve()方法，来看代码：

```
Object invokeReadResolve(Object obj)
    throws IOException, UnsupportedOperationException
{
    requireInitialized();
    if (readResolveMethod != null) {
        try {
            return readResolveMethod.invoke(obj, (Object[]) null);
        } catch (InvocationTargetException ex) {
            Throwable th = ex.getTargetException();
            if (th instanceof ObjectStreamException) {
                throw (ObjectStreamException) th;
            } else {
                throwMiscException(th);
                throw new InternalError(th);
            }
        } catch (IllegalAccessException ex) {
            throw new InternalError(ex);
        }
    } else {
        throw new UnsupportedOperationException();
    }
}
```

我们可以看到，在 invokeReadResolve()方法中用反射调用了 readResolveMethod 方法。

通过 JDK 源码分析我们可以看出，虽然增加 readResolve()方法返回实例解决了单例模式被破坏的问题，但是实际上实例化了两次，只不过新创建的对象没有被返回而已。如果创建对象的动作发生频率加快，就意味着内存分配开销也会随之增大，难道真的就没办法从根本上解决问题吗？下面讲的注册式单例也许能帮助到你。

2.3.6 注册式单例模式

注册式单例模式又称为登记式单例模式,就是将每一个实例都登记到某一个地方,使用唯一的标识获取实例。注册式单例模式有两种:一种为枚举式单例模式,另一种为容器式单例模式。

1. 枚举式单例模式

先来看枚举式单例模式的写法,来看代码,创建 EnumSingleton 类:

```java
public enum EnumSingleton {
    INSTANCE;
    private Object data;
    public Object getData() {
        return data;
    }
    public void setData(Object data) {
        this.data = data;
    }
    public static EnumSingleton getInstance(){
        return INSTANCE;
    }
}
```

来看测试代码:

```java
public class EnumSingletonTest {
    public static void main(String[] args) {
        try {
            EnumSingleton instance1 = null;

            EnumSingleton instance2 = EnumSingleton.getInstance();
            instance2.setData(new Object());

            FileOutputStream fos = new FileOutputStream("EnumSingleton.obj");
            ObjectOutputStream oos = new ObjectOutputStream(fos);
            oos.writeObject(instance2);
            oos.flush();
            oos.close();

            FileInputStream fis = new FileInputStream("EnumSingleton.obj");
            ObjectInputStream ois = new ObjectInputStream(fis);
            instance1 = (EnumSingleton) ois.readObject();
            ois.close();

            System.out.println(instance1.getData());
            System.out.println(instance2.getData());
            System.out.println(instance1.getData() == instance2.getData());
```

```
        }catch (Exception e){
            e.printStackTrace();
        }
    }
}
```

运行结果如下图所示。

没有做任何处理，我们发现运行结果和预期的一样。那么枚举式单例模式如此神奇，它的神秘之处在哪里体现呢？下面通过分析源码来揭开它的神秘面纱。

下载一个非常好用的 Java 反编译工具 Jad（下载地址：https://varaneckas.com/jad/），解压后配置好环境变量（这里不做详细介绍），就可以使用命令行调用了。找到工程所在的 Class 目录，复制 EnumSingleton.class 所在的路径，如下图所示。

然后切换到命令行,切换到工程所在的 Class 目录,输入命令 jad 并在后面输入复制好的路径,在 Class 目录下会多出一个 EnumSingleton.jad 文件。打开 EnumSingleton.jad 文件我们惊奇地发现有如下代码:

```
static
{
    INSTANCE = new EnumSingleton("INSTANCE", 0);
    $VALUES = (new EnumSingleton[] {
        INSTANCE
    });
}
```

原来,枚举式单例模式在静态代码块中就给 INSTANCE 进行了赋值,是饿汉式单例模式的实现。至此,我们还可以试想,序列化能否破坏枚举式单例模式呢?不妨再来看一下 JDK 源码,还是回到 ObjectInputStream 的 readObject0()方法:

```
private Object readObject0(boolean unshared) throws IOException {
        ...
        case TC_ENUM:
            return checkResolve(readEnum(unshared));
        ...
}
```

我们看到,在 readObject0()中调用了 readEnum()方法,来看 readEnum()方法的代码实现:

```
private Enum<?> readEnum(boolean unshared) throws IOException {
    if (bin.readByte() != TC_ENUM) {
        throw new InternalError();
    }

    ObjectStreamClass desc = readClassDesc(false);
    if (!desc.isEnum()) {
        throw new InvalidClassException("non-enum class: " + desc);
    }

    int enumHandle = handles.assign(unshared ? unsharedMarker : null);
    ClassNotFoundException resolveEx = desc.getResolveException();
    if (resolveEx != null) {
        handles.markException(enumHandle, resolveEx);
    }

    String name = readString(false);
    Enum<?> result = null;
    Class<?> cl = desc.forClass();
```

```
    if (cl != null) {
        try {
            @SuppressWarnings("unchecked")
            Enum<?> en = Enum.valueOf((Class)cl, name);
            result = en;
        } catch (IllegalArgumentException ex) {
            throw (IOException) new InvalidObjectException(
                "enum constant " + name + " does not exist in " +
                cl).initCause(ex);
        }
        if (!unshared) {
            handles.setObject(enumHandle, result);
        }
    }
    handles.finish(enumHandle);
    passHandle = enumHandle;
    return result;
}
```

我们发现，枚举类型其实通过类名和类对象类找到一个唯一的枚举对象。因此，枚举对象不可能被类加载器加载多次。那么反射是否能破坏枚举式单例模式呢？来看一段测试代码：

```java
public static void main(String[] args) {
    try {
        Class clazz = EnumSingleton.class;
        Constructor c = clazz.getDeclaredConstructor();
        c.newInstance();
    }catch (Exception e){
        e.printStackTrace();
    }
}
```

运行结果如下图所示。

```
Run: EnumSingletonTest
java.lang.NoSuchMethodException: com.gupaoedu.vip.pattern.singleton.register.EnumSingleton.<init>()
    at java.lang.Class.getConstructor0(Class.java:3082)
    at java.lang.Class.getDeclaredConstructor(Class.java:2178)
    at com.gupaoedu.vip.pattern.singleton.test.EnumSingletonTest.main(EnumSingletonTest.java:46)

Process finished with exit code 0
```

结果中报的是 java.lang.NoSuchMethodException 异常，意思是没找到无参的构造方法。这时候，我们打开 java.lang.Enum 的源码，查看它的构造方法，只有一个 protected 类型的构造方法，代码如下：

```java
protected Enum(String name, int ordinal) {
    this.name = name;
    this.ordinal = ordinal;
}
```

我们再来做一个下面这样的测试：

```java
public static void main(String[] args) {
    try {
        Class clazz = EnumSingleton.class;
        Constructor c = clazz.getDeclaredConstructor(String.class,int.class);
        c.setAccessible(true);
        EnumSingleton enumSingleton = (EnumSingleton)c.newInstance("Tom",666);

    }catch (Exception e){
        e.printStackTrace();
    }
}
```

运行结果如下图所示。

```
Run  EnumSingletonTest
  java.lang.IllegalArgumentException: Cannot reflectively create enum objects <1 internal calls>
      at com.gupaoedu.vip.pattern.singleton.test.EnumSingletonTest.main(EnumSingletonTest.java:59)

  Process finished with exit code 0
```

这时错误已经非常明显了，"Cannot reflectively create enum objects"，即不能用反射来创建枚举类型。还是习惯性地想来看看 JDK 源码，进入 Constructor 的 newInstance()方法：

```java
public T newInstance(Object ... initargs)
    throws InstantiationException, IllegalAccessException,
        IllegalArgumentException, InvocationTargetException
{
    if (!override) {
        if (!Reflection.quickCheckMemberAccess(clazz, modifiers)) {
            Class<?> caller = Reflection.getCallerClass();
            checkAccess(caller, clazz, null, modifiers);
        }
    }
    if ((clazz.getModifiers() & Modifier.ENUM) != 0)
        throw new IllegalArgumentException("Cannot reflectively create enum objects");
```

```
        ConstructorAccessor ca = constructorAccessor;
        if (ca == null) {
            ca = acquireConstructorAccessor();
        }
        @SuppressWarnings("unchecked")
        T inst = (T) ca.newInstance(initargs);
        return inst;
}
```

从上述代码可以看到，在 newInstance()方法中做了强制性的判断，如果修饰符是 Modifier.ENUM 枚举类型，则直接抛出异常。

到此为止，我们是不是已经非常清晰明了呢？枚举式单例模式也是 *Effective Java* 书中推荐的一种单例模式实现写法。JDK 枚举的语法特殊性及反射也为枚举保驾护航，让枚举式单例模式成为一种比较优雅的实现。

2. 容器式单例

接下来看注册式单例模式的另一种写法，即容器式单例模式，创建 ContainerSingleton 类：

```
public class ContainerSingleton {
    private ContainerSingleton(){}
    private static Map<String,Object> ioc = new ConcurrentHashMap<String,Object>();
    public static Object getBean(String className){
        synchronized (ioc) {
            if (!ioc.containsKey(className)) {
                Object obj = null;
                try {
                    obj = Class.forName(className).newInstance();
                    ioc.put(className, obj);
                } catch (Exception e) {
                    e.printStackTrace();
                }
                return obj;
            } else {
                return ioc.get(className);
            }
        }
    }
}
```

容器式单例模式适用于实例非常多的情况，便于管理。但它是非线程安全的。到此，注册式单例模式介绍完毕。我们再来看看 Spring 中的容器式单例模式的实现代码：

```java
public abstract class AbstractAutowireCapableBeanFactory extends AbstractBeanFactory
    implements AutowireCapableBeanFactory {
  /** Cache of unfinished FactoryBean instances: FactoryBean name --> BeanWrapper */
  private final Map<String, BeanWrapper> factoryBeanInstanceCache = new ConcurrentHashMap<>(16);
...
}
```

2.3.7 线程单例实现 ThreadLocal

最后赠送给大家一个彩蛋，讲讲线程单例实现 ThreadLocal。ThreadLocal 不能保证其创建的对象是全局唯一的，但是能保证在单个线程中是唯一的，天生是线程安全的。下面来看代码：

```java
public class ThreadLocalSingleton {
    private static final ThreadLocal<ThreadLocalSingleton> threadLocalInstance =
            new ThreadLocal<ThreadLocalSingleton>(){
                @Override
                protected ThreadLocalSingleton initialValue() {
                    return new ThreadLocalSingleton();
                }
            };
    private ThreadLocalSingleton(){}

    public static ThreadLocalSingleton getInstance(){
        return threadLocalInstance.get();
    }
}
```

写一下测试代码：

```java
public static void main(String[] args) {

    System.out.println(ThreadLocalSingleton.getInstance());
    System.out.println(ThreadLocalSingleton.getInstance());
    System.out.println(ThreadLocalSingleton.getInstance());
    System.out.println(ThreadLocalSingleton.getInstance());
    System.out.println(ThreadLocalSingleton.getInstance());

    Thread t1 = new Thread(new ExectorThread());
    Thread t2 = new Thread(new ExectorThread());
    t1.start();
    t2.start();
    System.out.println("End");
}
```

运行结果如下图所示。

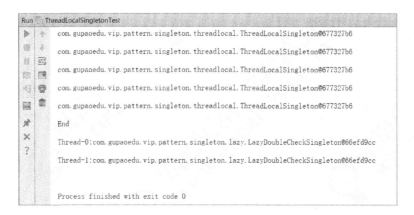

我们发现，在主线程中无论调用多少次，获取到的实例都是同一个，都在两个子线程中分别获取到了不同的实例。那么 ThreadLocal 是如何实现这样的效果的呢？我们知道，单例模式为了达到线程安全的目的，会给方法上锁，以时间换空间。ThreadLocal 将所有的对象全部放在 ThreadLocalMap 中，为每个线程都提供一个对象，实际上是以空间换时间来实现线程隔离的。

2.3.8 单例模式小结

单例模式可以保证内存里只有一个实例，减少了内存的开销，还可以避免对资源的多重占用。单例模式看起来非常简单，实现起来其实也非常简单，但是在面试中却是一个高频面试点。希望"小伙伴们"通过本章的学习，对单例模式有了非常深刻的认识，在面试中彰显技术深度，提升核心竞争力，给面试加分，顺利拿到录取通知（Offer）。

2.4 原型模式详解

2.4.1 原型模式的应用场景

你一定遇到过大篇幅使用 get 或 set 赋值的场景，例如下面这样的代码：

```
public void setParam(ExamPaperVo vo){
    ExamPaper examPaper = new ExamPaper();
    //试卷主键
    examPaper.setExaminationPaperId(vo.getExaminationPaperId());
    //剩余时间
    curForm.setLeavTime(examPaper.getLeavTime());
    //单位主键
    curForm.setOrganizationId(examPaper.getOrganizationId());
    //考试主键
```

```
        curForm.setId(examPaper.getId());
        //考场主键
        curForm.setExamroomId(examPaper.getExamroomId());
        //用户主键
        curForm.setUserId(examPaper.getUserId());
        //专业代码
        curForm.setSpecialtyCode(examPaper.getSpecialtyCode());
        //报考岗位
        curForm.setPostionCode(examPaper.getPostionCode());
        //报考等级
        curForm.setGradeCode(examPaper.getGradeCode());
        //考试开始时间
        curForm.setExamStartTime(examPaper.getExamStartTime());
        //考试结束时间
        curForm.setJudgementImpCount(examPaper.getJudgementImpCount());
        //考试时长
        curForm.setExamTime(examPaper.getExamTime());
        //总分
        curForm.setFullScore(examPaper.getFullScore());
        //及格分
        curForm.setPassScore(examPaper.getPassScore());
        //学员姓名
        curForm.setUserName(examPaper.getUserName());
        //考试得分
        curForm.setScore(examPaper.getScore());
        //是否及格
        curForm.setResult(examPaper.getResult());
        curForm.setIsPassed(examPaper.getIsPassed());
        //单选题答对数量
        curForm.setSingleOkCount(examPaper.getSingleOkCount());
        //多选题答对数量
        curForm.setMultiOkCount(examPaper.getMultiOkCount());
        //判断题答对数量
        curForm.setJudgementOkCount(examPaper.getJudgementOkCount());

        //提交试卷
        service.submit(examPaper);
}
```

上述代码非常工整，命名非常规范，注释也写得很全面，大家觉得这样的代码优雅吗？我认为，这样的代码属于"纯体力劳动"。原型模式就能帮助我们解决这样的问题。

原型模式（Prototype Pattern）是指原型实例指定创建对象的种类，并且通过复制这些原型创建新的对象。

原型模式主要适用于以下场景：

（1）类初始化消耗资源较多。

（2）使用 new 生成一个对象需要非常烦琐的过程（数据准备、访问权限等）。

（3）构造函数比较复杂。

（4）在循环体中产生大量对象。

在 Spring 中，原型模式应用得非常广泛。例如 scope="prototype"，我们经常用的 JSON.parseObject()也是一种原型模式。下面我们来看看原型模式的类结构图，如下图所示。

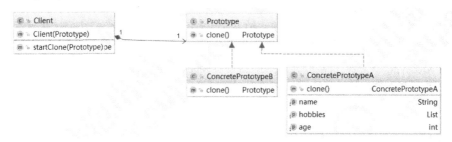

2.4.2　浅克隆

一个标准的原型模式代码应该是这样设计的，先创建原型 Prototype 接口：

```
public interface Prototype{
    Prototype clone();
}
```

创建具体需要克隆的类 ConcretePrototypeA：

```
import java.util.List;

public class ConcretePrototypeA implements Prototype {
    private int age;
    private String name;
    private List hobbies;

    public int getAge() { return age; }
    public void setAge(int age) { this.age = age; }
    public String getName() { return name; }
    public void setName(String name) { this.name = name; }
    public List getHobbies() { return hobbies; }
    public void setHobbies(List hobbies) { this.hobbies = hobbies; }

    @Override
    public ConcretePrototypeA clone() {
```

```java
        ConcretePrototypeA concretePrototype = new ConcretePrototypeA();
        concretePrototype.setAge(this.age);
        concretePrototype.setName(this.name);
        concretePrototype.setHobbies(this.hobbies);
        return concretePrototype;
    }
}
```

创建 Client 类：

```java
public class Client {
    private Prototype prototype;
    public Client(Prototype prototype){
        this.prototype = prototype;
    }
    public Prototype startClone(Prototype concretePrototype){
        return (Prototype)concretePrototype.clone();
    }
}
```

测试代码如下：

```java
import java.util.ArrayList;
import java.util.List;
public class PrototypeTest {
    public static void main(String[] args) {
        // 创建一个具体的需要克隆的对象
        ConcretePrototypeA concretePrototype = new ConcretePrototypeA();
        // 填充属性，方便测试
        concretePrototype.setAge(18);
        concretePrototype.setName("prototype");
        List hobbies = new ArrayList<String>();
        concretePrototype.setHobbies(hobbies);
        System.out.println(concretePrototype);

        // 创建 Client 对象，准备开始克隆
        Client client = new Client(concretePrototype);
        ConcretePrototypeA concretePrototypeClone = (ConcretePrototypeA) client.startClone(concretePrototype);
        System.out.println(concretePrototypeClone);

        System.out.println("克隆对象中的引用类型地址值：" + concretePrototypeClone.getHobbies());
        System.out.println("原对象中的引用类型地址值：" + concretePrototype.getHobbies());
        System.out.println("对象地址比较："+(concretePrototypeClone.getHobbies() == concretePrototype.getHobbies()));
    }
}
```

运行结果如下图所示。

从测试结果可以看出，hobbies 的引用地址是相同的，意味着复制的不是值，而是引用的地址。这样的话，如果我们修改任意一个对象的属性值，则 concretePrototype 和 concretePrototypeClone 的 hobbies 值都会改变，这就是我们常说的浅克隆。浅克隆只是完整复制了值类型数据，没有赋值引用对象。换言之，所有的引用对象仍然指向原来的对象，显然这不是我们想要的结果。

2.4.3 深克隆

我们换一个场景，大家都知道齐天大圣，首先它是一只猴子，有七十二般变化，拔一根毫毛就可以吹出千万个猴子，手里还拿着金箍棒，金箍棒可以变大或变小。这就是我们耳熟能详的原型模式的经典体现。

创建原型猴子类 Monkey：

```java
import java.util.Date;

public class Monkey {
    public int height;
    public int weight;
    public Date birthday;
}
```

创建引用对象金箍棒类 Jingubang：

```java
import java.io.Serializable;

public class JinGuBang implements Serializable {
    public float h = 100;
    public float d = 10;
    public void big(){
        this.d *= 2;
        this.h *= 2;
```

```java
    }
    public void small(){
        this.d /= 2;
        this.h /= 2;
    }
}
```

创建具体的对象齐天大圣类 QiTianDaSheng：

```java
import java.io.*;
import java.util.Date;

public class QiTianDaSheng extends Monkey implements Cloneable,Serializable {

    public JinGuBang jinGuBang;

    public  QiTianDaSheng(){
        //只是初始化
        this.birthday = new Date();
        this.jinGuBang = new JinGuBang();
    }

    @Override
    protected Object clone() throws CloneNotSupportedException {
        return this.deepClone();
    }

    public Object deepClone(){
        try{

            ByteArrayOutputStream bos = new ByteArrayOutputStream();
            ObjectOutputStream oos = new ObjectOutputStream(bos);
            oos.writeObject(this);

            ByteArrayInputStream bis = new ByteArrayInputStream(bos.toByteArray());
            ObjectInputStream ois = new ObjectInputStream(bis);

            QiTianDaSheng copy = (QiTianDaSheng)ois.readObject();
            copy.birthday = new Date();
            return copy;

        }catch (Exception e){
            e.printStackTrace();
            return null;
        }
    }

    public QiTianDaSheng shallowClone(QiTianDaSheng target){
```

```java
        QiTianDaSheng qiTianDaSheng = new QiTianDaSheng();
        qiTianDaSheng.height = target.height;
        qiTianDaSheng.weight = target.height;

        qiTianDaSheng.jinGuBang = target.jinGuBang;
        qiTianDaSheng.birthday = new Date();

        return  qiTianDaSheng;
    }
}
```

测试代码如下:

```java
public class DeepCloneTest {

    public static void main(String[] args) {
        QiTianDaSheng qiTianDaSheng = new QiTianDaSheng();
        try {
            QiTianDaSheng clone = (QiTianDaSheng)qiTianDaSheng.clone();
            System.out.println("深克隆: " + (qiTianDaSheng.jinGuBang == clone.jinGuBang));
        } catch (Exception e) {
            e.printStackTrace();
        }

        QiTianDaSheng q = new QiTianDaSheng();
        QiTianDaSheng n = q.shallowClone(q);
        System.out.println("浅克隆: " + (q.jinGuBang == n.jinGuBang));
    }
}
```

运行结果如下图所示。

2.4.4 克隆破坏单例模式

如果我们克隆的目标对象是单例对象,那么意味着深克隆会破坏单例模式。实际上防止克隆破坏单例模式的解决思路非常简单,禁止深克隆便可。要么我们的单例类不实现 Cloneable 接口,要么我们重写 clone() 方法,在 clone() 方法中返回单例对象即可,具体代码如下:

```java
@Override
protected Object clone() throws CloneNotSupportedException {
```

```
    return INSTANCE;
}
```

2.4.5　clone()方法的源码

我们常用的 ArrayList 实现了 Cloneable 接口，来看 clone()方法的源码：

```
public Object clone() {
    try {
        ArrayList<?> v = (ArrayList<?>) super.clone();
        v.elementData = Arrays.copyOf(elementData, size);
        v.modCount = 0;
        return v;
    } catch (CloneNotSupportedException e) {
        throw new InternalError(e);
    }
}
```

2.5　代理模式详解

2.5.1　代理模式的应用场景

生活中的租房中介、售票黄牛、婚介、经纪人、快递、事务代理、非侵入式日志监听等，都是代理模式的实际体现。代理模式（Proxy Pattern）的定义也非常简单，是指为其他对象提供一种代理，以控制对这个对象的访问。代理对象在客户端和目标对象之间起到中介作用，代理模式属于结构型设计模式。使用代理模式主要有两个目的：一是保护目标对象，二是增强目标对象。下面我们来看一下代理模式的类结构图，如下图所示。

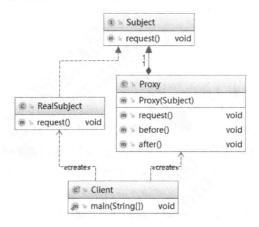

Subject 是顶层接口，RealSubject 是真实对象（被代理对象），Proxy 是代理对象，代理对象持有被代理对象的引用，客户端调用代理对象的方法，同时也调用被代理对象的方法，但是会在代理对象前后增加一些处理代码。在代码中，一般代理会被理解为代码增强，实际上就是在原代码逻辑前后增加一些代码逻辑，而使调用者无感知。代理模式属于结构型模式，分为静态代理和动态代理。

2.5.2 静态代理

举个例子，有些人到了适婚年龄，其父母总是迫不及待地希望早点抱孙子。而现在在各种压力之下，很多人都选择晚婚晚育。于是着急的父母就开始到处为自己的子女相亲，比子女自己还着急。这个相亲的过程就是一种"人人都有份"的代理。下面来看代码实现。

顶层接口 Person 的代码如下：

```java
/**
 * 人有很多行为，要谈恋爱，要住房子，要购物，要工作
 */
public interface Person {
    public void findLove();
    ...
}
```

儿子要找对象，实现 Son 类：

```java
public class Son implements Person{

    public void findLove(){
        //我没有时间，工作忙
        System.out.println("儿子要求：肤白貌美大长腿");
    }
}
```

父亲要帮儿子相亲，实现 Father 类：

```java
public class Father {
    private Son son;
    //没办法扩展
    public Father(Son son){
        this.son = son;
    }
    //获取目标对象的引用
    public void findLove(){
        System.out.println("父亲物色对象");
        this.son.findLove();
```

```
        System.out.println("双方同意交往,确立关系");
    }
}
```

来看测试代码:

```
public static void main(String[] args) {
    //只能帮儿子找对象,不能帮表妹、不能帮陌生人
    Father father = new Father(new Son());
    father.findLove();
}
```

运行结果如下图所示。

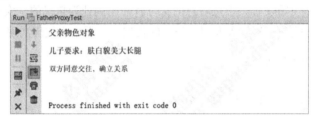

这里"小伙伴们"可能会觉得还是不知道如何将代理模式应用到业务场景中,我们来看一个实际的业务场景。在分布式业务场景中,通常会对数据库进行分库分表,分库分表之后使用 Java 操作时就可能需要配置多个数据源,我们通过设置数据源路由来动态切换数据源。先创建 Order 订单类:

```
public class Order {
    private Object orderInfo;
    private Long createTime;
    private String id;

    public Object getOrderInfo() {
        return orderInfo;
    }
    public void setOrderInfo(Object orderInfo) {
        this.orderInfo = orderInfo;
    }
    public Long getCreateTime() {
        return createTime;
    }
    public void setCreateTime(Long createTime) {
        this.createTime = createTime;
    }
    public String getId() {
        return id;
    }
}
```

```java
    public void setId(String id) {
        this.id = id;
    }
}
```

创建 OrderDao 持久层操作类：

```java
public class OrderDao {
    public int insert(Order order){
        System.out.println("OrderDao 创建 Order 成功!");
        return 1;
    }
}
```

创建 IOrderService 接口：

```java
public interface IOrderService {
    int createOrder(Order order);
}
```

创建 OrderService 实现类：

```java
public class OrderService implements IOrderService {
    private OrderDao orderDao;

    public OrderService(){
        //如果使用 Spring 应该是自动注入的
        //为了使用方便，我们在构造方法中将 orderDao 直接初始化
        orderDao = new OrderDao();
    }

    @Override
    public int createOrder(Order order) {
        System.out.println("OrderService 调用 orderDao 创建订单");
        return orderDao.insert(order);
    }
}
```

接下来使用静态代理，主要完成的功能是：根据订单创建时间自动按年进行分库。根据开闭原则，我们修改原来写好的代码逻辑，通过代理对象来完成。先创建数据源路由对象，使用 ThreadLocal 的单例实现 DynamicDataSourceEntry 类：

```java
package com.gupaoedu.vip.pattern.proxy.staticproxy.dbroute.db;

//动态切换数据源
public class DynamicDataSourceEntry {
```

```java
//默认数据源
public final static String DEFAULT_SOURCE = null;

private final static ThreadLocal<String> local = new ThreadLocal<String>();

private DynamicDataSourceEntry(){}

//清空数据源
public static void clear() {
    local.remove();
}

//获取当前正在使用的数据源名字
public static String get() {
    return local.get();
}

//还原当前切换的数据源
public static void restore() {
    local.set(DEFAULT_SOURCE);
}

//设置已知名字的数据源
public static void set(String source) {
    local.set(source);
}

//根据年份动态设置数据源
public static void set(int year) {
    local.set("DB_" + year);
}
}
```

创建切换数据源的代理类 OrderServiceSaticProxy：

```java
package com.gupaoedu.vip.pattern.proxy.staticproxy.dbroute.proxy;

import com.gupaoedu.vip.pattern.proxy.staticproxy.dbroute.IOrderService;
import com.gupaoedu.vip.pattern.proxy.staticproxy.dbroute.Order;
import com.gupaoedu.vip.pattern.proxy.staticproxy.dbroute.db.DynamicDataSourceEntry;

import java.text.SimpleDateFormat;
import java.util.Date;

public class OrderServiceStaticProxy implements IOrderService {

    private SimpleDateFormat yearFormat = new SimpleDateFormat("yyyy");
```

```java
    private IOrderService orderService;
    public OrderServiceStaticProxy(IOrderService orderService){
        this.orderService = orderService;
    }

    public int createOrder(Order order) {
        before();
        Long time = order.getCreateTime();
        Integer dbRouter = Integer.valueOf(yearFormat.format(new Date(time)));
        System.out.println("静态代理类自动分配到【DB_" + dbRouter + "】数据源处理数据");
        DynamicDataSourceEntry.set(dbRouter);
        orderService.createOrder(order);
        after();
        return 0;
    }

    private void before(){
        System.out.println("Proxy before method.");
    }

    private void after(){
        System.out.println("Proxy after method.");
    }
}
```

来看测试代码：

```java
public static void main(String[] args) {

    try {

        Order order = new Order();
        SimpleDateFormat sdf = new SimpleDateFormat("yyyy/MM/dd");
        Date date = sdf.parse("2017/02/01");
        order.setCreateTime(date.getTime());

        IOrderService orderService = new OrderServiceStaticProxy(new OrderService());
        orderService.createOrder(order);
    }catch (Exception e){
        e.printStackTrace();;
    }

}
```

运行结果如下图所示。

结果符合我们的预期。现在再来回顾一下类图，看是不是和我们最先画的一致，如下图所示。

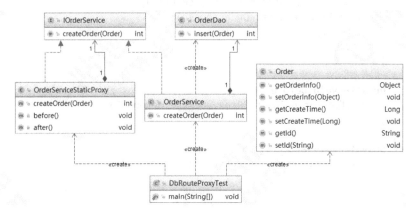

2.5.3 动态代理

动态代理和静态代理的基本思路是一致的，只不过动态代理功能更加强大，随着业务的扩展适应性更强。如果还以找对象为例，那么使用动态代理相当于能够适应复杂的业务场景。不仅包括父亲给儿子找对象，如果找对象这项业务发展成了一个产业，出现了媒婆、婚介所等，那么用静态代理成本太高了，需要一个更加通用的解决方案，满足任何单身人士找对象的需求。下面我们升级一下代码。

1. JDK 实现方式

创建媒婆（婚介所）类 JDKMeipo：

```
package com.gupaoedu.vip.pattern.proxy.dynamicproxy.jdkproxy;

import com.gupaoedu.vip.pattern.proxy.staticproxy.Person;
import java.lang.reflect.InvocationHandler;
import java.lang.reflect.Method;
import java.lang.reflect.Proxy;

public class JDKMeipo implements InvocationHandler{
```

```java
//被代理的对象,把引用保存下来
private Object target;
public Object getInstance(Object target) throws Exception{
    this.target = target;

    Class<?> clazz = target.getClass();
    return Proxy.newProxyInstance(clazz.getClassLoader(),clazz.getInterfaces(),this);
}

public Object invoke(Object proxy, Method method, Object[] args) throws Throwable {
    before();
    Object obj = method.invoke(this.target,args);
    after();
    return  obj;
}

private void before(){
    System.out.println("我是媒婆:我要给你找对象,现在已经确认你的需求");
    System.out.println("开始物色");
}

private void after(){
    System.out.println("如果合适的话,就准备办事");
}
}
```

创建单身客户类 Customer:

```java
package com.gupaoedu.vip.pattern.proxy.dynamicproxy.jdkproxy;

import com.gupaoedu.vip.pattern.proxy.Person;

public class Customer implements Person{
    public void findLove(){
        System.out.println("高富帅");
        System.out.println("身高180cm");
        System.out.println("有6块腹肌");

    }
}
```

测试代码如下:

```java
public static void main(String[] args) {
    try {
        Person obj = (Person)new JDKMeipo().getInstance(new Customer());
        obj.findLove();
    } catch (Exception e) {
        e.printStackTrace();
    }
}
```

运行效果如下图所示。

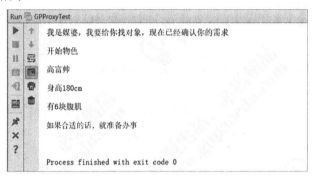

上面的案例理解了,我们再来看数据源动态路由业务,帮助"小伙伴们"加深对动态代理的印象。创建动态代理的类 OrderServiceDynamicProxy:

```
package com.gupaoedu.vip.pattern.proxy.dbroute.proxy;

import com.gupaoedu.vip.pattern.proxy.dbroute.db.DynamicDataSourceEntry;

import java.lang.reflect.InvocationHandler;
import java.lang.reflect.Method;
import java.lang.reflect.Proxy;
import java.text.SimpleDateFormat;
import java.util.Date;

public class OrderServiceDynamicProxy implements InvocationHandler {

    private SimpleDateFormat yearFormat = new SimpleDateFormat("yyyy");
    private Object target;

    public Object getInstance(Object target){
        this.target = target;
        Class<?> clazz = target.getClass();
        return Proxy.newProxyInstance(clazz.getClassLoader(),clazz.getInterfaces(),this);
    }

    public Object invoke(Object proxy, Method method, Object[] args) throws Throwable {
        before(args[0]);
        Object object = method.invoke(target,args);
        after();
        return object;
    }

    private void before(Object target){
```

```java
    try {
        System.out.println("Proxy before method.");
        Long time = (Long) target.getClass().getMethod("getCreateTime").invoke(target);
        Integer dbRouter = Integer.valueOf(yearFormat.format(new Date(time)));
        System.out.println("静态代理类自动分配到【DB_" + dbRouter + "】数据源处理数据");
        DynamicDataSourceEntry.set(dbRouter);
    }catch (Exception e){
        e.printStackTrace();
    }
}

private void after(){
    System.out.println("Proxy after method.");
}
```

测试代码如下：

```java
public static void main(String[] args) {

    try {

        Order order = new Order();

        SimpleDateFormat sdf = new SimpleDateFormat("yyyy/MM/dd");
        Date date = sdf.parse("2018/02/01");
        order.setCreateTime(date.getTime());

        IOrderService orderService = (IOrderService)new OrderServiceDynamicProxy().getInstance(new OrderService());
        orderService.createOrder(order);
    }catch (Exception e){
        e.printStackTrace();
    }

}
```

依然能够达到相同运行效果。但是，使用动态代理实现之后，我们不仅能实现 Order 的数据源动态路由，还可以实现其他任何类的数据源路由。当然，有个比较重要的约定，必须实现 getCreateTime()方法，因为路由规则是根据时间来运算的。我们可以通过接口规范来达到约束的目的，在此就不再举例。

2. 手写实现 JDK 动态代理

不仅知其然，还得知其所以然。既然 JDK 动态代理功能如此强大，那么它是如何实现的呢？我们现在来探究一下原理，并模仿 JDK 动态代理动手写一个属于自己的动态代理。

我们都知道 JDK 动态代理采用字节重组，重新生成对象来替代原始对象，以达到动态代理的目的。JDK 动态代理生成对象的步骤如下：

（1）获取被代理对象的引用，并且获取它的所有接口，反射获取。

（2）JDK 动态代理类重新生成一个新的类，同时新的类要实现被代理类实现的所有接口。

（3）动态生成 Java 代码，新加的业务逻辑方法由一定的逻辑代码调用（在代码中体现）。

（4）编译新生成的 Java 代码.class 文件。

（5）重新加载到 JVM 中运行。

以上过程就叫字节码重组。JDK 中有一个规范，在 ClassPath 下只要是$开头的.class 文件，一般都是自动生成的。那么我们有没有办法看到代替后的对象的"真容"呢？做一个这样测试，我们将内存中的对象字节码通过文件流输出到一个新的.class 文件，然后利用反编译工具查看其源代码。

```java
package com.gupaoedu.vip.pattern.proxy.dynamicproxy.jdkproxy;
import com.gupaoedu.vip.pattern.proxy.Person;
import sun.misc.ProxyGenerator;
import java.io.FileOutputStream;
public class JDKProxyTest {
    public static void main(String[] args) {
        try {
            Person obj = (Person)new JDKMeipo().getInstance(new Customer());
            obj.findLove();

            //通过反编译工具可以查看源代码
            byte [] bytes = ProxyGenerator.generateProxyClass("$Proxy0", new Class[]{Person.class});
            FileOutputStream os = new FileOutputStream("E://$Proxy0.class");
            os.write(bytes);
            os.close();
        } catch (Exception e) {
            e.printStackTrace();
        }
    }
}
```

运行以上代码，我们能在 E 盘下找到一个$Proxy0.class 文件。使用 Jad 反编译，得到$Proxy0.jad 文件，打开它可以看到如下内容：

```java
import com.gupaoedu.vip.pattern.proxy.Person;
import java.lang.reflect.*;
```

```java
public final class $Proxy0 extends Proxy
    implements Person
{

    public $Proxy0(InvocationHandler invocationhandler)
    {
        super(invocationhandler);
    }

    public final boolean equals(Object obj)
    {
        try
        {
            return ((Boolean)super.h.invoke(this, m1, new Object[] {
                obj
            })).booleanValue();
        }
        catch(Error _ex) { }
        catch(Throwable throwable)
        {
            throw new UndeclaredThrowableException(throwable);
        }
    }

    public final void findLove()
    {
        try
        {
            super.h.invoke(this, m3, null);
            return;
        }
        catch(Error _ex) { }
        catch(Throwable throwable)
        {
            throw new UndeclaredThrowableException(throwable);
        }
    }

    public final String toString()
    {
        try
        {
            return (String)super.h.invoke(this, m2, null);
        }
        catch(Error _ex) { }
        catch(Throwable throwable)
```

```
        throw new UndeclaredThrowableException(throwable);
    }
}

public final int hashCode()
{
    try
    {
        return ((Integer)super.h.invoke(this, m0, null)).intValue();
    }
    catch(Error _ex) { }
    catch(Throwable throwable)
    {
        throw new UndeclaredThrowableException(throwable);
    }
}

private static Method m1;
private static Method m3;
private static Method m2;
private static Method m0;

static
{
    try
    {
        m1 = Class.forName("java.lang.Object").getMethod("equals", new Class[] {
            Class.forName("java.lang.Object")
        });
        m3 = Class.forName("com.gupaoedu.vip.pattern.proxy.Person").getMethod("findLove", new Class[0]);
        m2 = Class.forName("java.lang.Object").getMethod("toString", new Class[0]);
        m0 = Class.forName("java.lang.Object").getMethod("hashCode", new Class[0]);
    }
    catch(NoSuchMethodException nosuchmethodexception)
    {
        throw new NoSuchMethodError(nosuchmethodexception.getMessage());
    }
    catch(ClassNotFoundException classnotfoundexception)
    {
        throw new NoClassDefFoundError(classnotfoundexception.getMessage());
    }
}
```

我们发现，$Proxy0 继承了 Proxy 类，同时还实现了 Person 接口，而且重写了 findLove()等方法。在静态块中用反射查找到了目标对象的所有方法，而且保存了所有方法的引用，重写的方法用反射调用目标对象的方法。"小伙伴们"此时一定会好奇：这些代码是哪里来的呢？其实是 JDK 帮我们自动生成的。现在我们不依赖 JDK，自己来动态生成源代码、动态完成编译，然后替代目标对象并执行。

创建 GPInvocationHandler 接口：

```java
package com.gupaoedu.vip.pattern.proxy.dynamicproxy.gpproxy;

import java.lang.reflect.Method;

public interface GPInvocationHandler {
    public Object invoke(Object proxy, Method method, Object[] args) throws Throwable;
}
```

创建 GPProxy 类：

```java
package com.gupaoedu.vip.pattern.proxy.dynamicproxy.gpproxy;

import javax.tools.JavaCompiler;
import javax.tools.StandardJavaFileManager;
import javax.tools.ToolProvider;
import java.io.File;
import java.io.FileWriter;
import java.lang.reflect.Constructor;
import java.lang.reflect.Method;
import java.util.HashMap;
import java.util.Map;

//用来生成源代码的工具类
public class GPProxy {

    public static final String ln = "\r\n";

    public static Object newProxyInstance(GPClassLoader classLoader, Class<?> [] interfaces, GPInvocationHandler h){
        try {
            //动态生成源代码.java 文件
            String src = generateSrc(interfaces);

            //Java 文件输出磁盘
            String filePath = GPProxy.class.getResource("").getPath();
            File f = new File(filePath + "$Proxy0.java");
            FileWriter fw = new FileWriter(f);
            fw.write(src);
            fw.flush();
            fw.close();
```

```java
        //把生成的.java文件编译成.class文件
        JavaCompiler compiler = ToolProvider.getSystemJavaCompiler();
        StandardJavaFileManager manage = compiler.getStandardFileManager(null,null,null);
        Iterable iterable = manage.getJavaFileObjects(f);

        JavaCompiler.CompilationTask task = compiler.getTask(null,manage,null,null,null,iterable);
        task.call();
        manage.close();

        //把编译生成的.class文件加载到JVM中
        Class proxyClass = classLoader.findClass("$Proxy0");
        Constructor c = proxyClass.getConstructor(GPInvocationHandler.class);
        f.delete();

        //返回字节码重组以后的新的代理对象
        return c.newInstance(h);
    }catch (Exception e){
        e.printStackTrace();
    }
    return null;
}

private static String generateSrc(Class<?>[] interfaces){
    StringBuffer sb = new StringBuffer();
    sb.append("package com.gupaoedu.vip.pattern.proxy.dynamicproxy.gpproxy;" + ln);
    sb.append("import com.gupaoedu.vip.pattern.proxy.Person;" + ln);
    sb.append("import java.lang.reflect.*;" + ln);
    sb.append("public class $Proxy0 implements " + interfaces[0].getName() + "{" + ln);
        sb.append("GPInvocationHandler h;" + ln);
        sb.append("public $Proxy0(GPInvocationHandler h) { " + ln);
            sb.append("this.h = h;");
        sb.append("}" + ln);
        for (Method m : interfaces[0].getMethods()){
            Class<?>[] params = m.getParameterTypes();

            StringBuffer paramNames = new StringBuffer();
            StringBuffer paramValues = new StringBuffer();
            StringBuffer paramClasses = new StringBuffer();

            for (int i = 0; i < params.length; i++) {
                Class clazz = params[i];
                String type = clazz.getName();
                String paramName = toLowerFirstCase(clazz.getSimpleName());
                paramNames.append(type + " " + paramName);
                paramValues.append(paramName);
                paramClasses.append(clazz.getName() + ".class");
```

```java
                    if(i > 0 && i < params.length-1){
                        paramNames.append(",");
                        paramClasses.append(",");
                        paramValues.append(",");
                    }
                }
                sb.append("public " + m.getReturnType().getName() + " " + m.getName() + "("
+ paramNames.toString() + ") {" + ln);
                sb.append("try{" + ln);
                sb.append("Method m = " + interfaces[0].getName() + ".class.getMethod
(\"" + m.getName() + "\",new Class[]{" + paramClasses.toString() + "});" + ln);
                sb.append((hasReturnValue(m.getReturnType()) ? "return " : "") +
getCaseCode("this.h.invoke(this,m,new Object[]{" + paramValues + "})",m.getReturnType()) + ";"
+ ln);
                sb.append("}catch(Error _ex) { }");
                sb.append("catch(Throwable e){" + ln);
                sb.append("throw new UndeclaredThrowableException(e);" + ln);
                sb.append("}");
                sb.append(getReturnEmptyCode(m.getReturnType()));
            sb.append("}");
        }
        sb.append("}" + ln);
        return sb.toString();
    }


    private static Map<Class,Class> mappings = new HashMap<Class, Class>();
    static {
        mappings.put(int.class,Integer.class);
    }

    private static String getReturnEmptyCode(Class<?> returnClass){
        if(mappings.containsKey(returnClass)){
            return "return 0;";
        }else if(returnClass == void.class){
            return "";
        }else {
            return "return null;";
        }
    }

    private static String getCaseCode(String code,Class<?> returnClass){
        if(mappings.containsKey(returnClass)){
            return "((" + mappings.get(returnClass).getName() +    ")" + code + ")." +
returnClass.getSimpleName() + "Value()";
        }
```

```java
        return code;
    }

    private static boolean hasReturnValue(Class<?> clazz){
        return clazz != void.class;
    }

    private static String toLowerFirstCase(String src){
        char [] chars = src.toCharArray();
        chars[0] += 32;
        return String.valueOf(chars);
    }
}
```

创建 GPClassLoader 类：

```java
package com.gupaoedu.vip.pattern.proxy.dynamicproxy.gpproxy;

import java.io.ByteArrayOutputStream;
import java.io.File;
import java.io.FileInputStream;
import java.io.IOException;

public class GPClassLoader extends ClassLoader{

    private File classPathFile;

    public GPClassLoader(){
        String classPath = GPClassLoader.class.getResource("").getPath();
        this.classPathFile = new File(classPath);
    }

    protected Class<?> findClass(String name) throws ClassNotFoundException {

        String className = GPClassLoader.class.getPackage().getName() + "." + name;

        if(classPathFile != null){
            File classFile = new File(classPathFile,name.replaceAll("\\.","/") + ".class");
            if(classFile.exists()){
                FileInputStream in = null;
                ByteArrayOutputStream out = null;

                try{
                    in = new FileInputStream(classFile);
                    out = new ByteArrayOutputStream();
                    byte [] buff = new byte[1024];
                    int len;
```

```java
                    while ((len = in.read(buff)) != -1){
                        out.write(buff,0,len);
                    }
                    return  defineClass(className,out.toByteArray(),0,out.size());
                }catch (Exception e){
                    e.printStackTrace();
                }finally {
                    if(null != in){
                        try {
                            in.close();
                        } catch (IOException e) {
                            e.printStackTrace();
                        }
                    }

                    if(out != null){
                        try {
                            out.close();
                        } catch (IOException e) {
                            e.printStackTrace();
                        }
                    }
                }
            }

            return null;
        }
    }
```

创建 GPMeipo 类：

```java
package com.gupaoedu.vip.pattern.proxy.dynamicproxy.gpproxy;

import com.gupaoedu.vip.pattern.proxy.Person;
import java.lang.reflect.Method;

public class GPMeipo implements GPInvocationHandler {

    //被代理的对象，把引用保存下来
    private Object target;

    public Object getInstance(Object target) throws Exception{
        this.target = target;

        Class<?> clazz = target.getClass();
```

```java
        return GPProxy.newProxyInstance(new GPClassLoader(),clazz.getInterfaces(),this);
    }

    public Object invoke(Object proxy, Method method, Object[] args) throws Throwable {
        before();
        method.invoke(this.target,args);
        after();
        return null;
    }

    private void before(){
        System.out.println("我是媒婆：我要给你找对象，现在已经确认你的需求");
        System.out.println("开始物色");
    }

    private void after(){
        System.out.println("如果合适的话，就准备办事");
    }
}
```

客户端测试代码如下：

```java
public static void main(String[] args) {

    try {
        Person obj = (Person)new GPMeipo().getInstance(new Customer());
        System.out.println(obj.getClass());
        obj.findLove();

    } catch (Exception e) {
        e.printStackTrace();
    }
}
```

到此，手写 JDK 动态代理就完成了。"小伙伴们"是不是又多了一个面试用的"撒手锏"呢？

3. CGLib 代理调用 API 及原理分析

简单看一下 CGLib 代理的使用，还是以媒婆为例，创建 CglibMeipo 类：

```java
package com.gupaoedu.vip.pattern.proxy.dynamicproxy.cglibproxy;
import net.sf.cglib.proxy.Enhancer;
import net.sf.cglib.proxy.MethodInterceptor;
import net.sf.cglib.proxy.MethodProxy;
import java.lang.reflect.Method;
public class CglibMeipo implements MethodInterceptor{
    public Object getInstance(Class<?> clazz) throws Exception{
```

```java
        Enhancer enhancer = new Enhancer();
        //要把哪个设置为即将生成的新类的父类
        enhancer.setSuperclass(clazz);
        enhancer.setCallback(this);

        return enhancer.create();
    }

    public Object intercept(Object o, Method method, Object[] objects, MethodProxy methodProxy) throws Throwable {
        //业务的增强
        before();
        Object obj = methodProxy.invokeSuper(o,objects);
        after();
        return obj;
    }

    private void before(){
        System.out.println("我是媒婆：我要给你找对象，现在已经确认你的需求");
        System.out.println("开始物色");
    }
    private void after(){
        System.out.println("如果合适的话，就准备办事");
    }
}
```

创建单身客户类 Customer：

```java
package com.gupaoedu.vip.pattern.proxy.dynamicproxy.cglibproxy;

public class Customer {

    public void findLove(){
        System.out.println("肤白貌美大长腿");
    }
}
```

有个小细节，CGLib 代理的目标对象不需要实现任何接口，它是通过动态继承目标对象实现动态代理的。来看测试代码：

```java
package com.gupaoedu.vip.pattern.proxy.dynamicproxy.cglibproxy;

public class CglibTest {
    public static void main(String[] args) {
        try {
            Customer obj = (Customer)new CglibMeipo().getInstance(Customer.class);
```

```
        obj.findLove();
    } catch (Exception e) {
        e.printStackTrace();
    }
  }
}
```

CGLib 代理的实现原理又是怎样的呢？我们可以在测试代码中加上一句代码，将 CGLib 代理后的.class 文件写入磁盘，然后反编译来一探究竟，代码如下：

```
public static void main(String[] args) {
    try {

        //利用 CGlib 的代理类可以将内存中的.class 文件写入本地磁盘
        System.setProperty(DebuggingClassWriter.DEBUG_LOCATION_PROPERTY,
"E://cglib_proxy_class/");

        Customer obj = (Customer)new CglibMeipo().getInstance(Customer.class);
        obj.findLove();

    } catch (Exception e) {
        e.printStackTrace();
    }
  }
}
```

重新执行代码，我们会发现在 E://cglib_proxy_class 目录下多了三个.class 文件，如下图所示。

通过调试跟踪发现，Customer$$EnhancerByCGLIB$$3feeb52a.class 就是 CGLib 代理生成的代理类，继承了 Customer 类。小伙伴们可自行反编译看看源代码。

```
package com.gupaoedu.vip.pattern.proxy.dynamicproxy.cglibproxy;

import java.lang.reflect.Method;
import net.sf.cglib.core.ReflectUtils;
import net.sf.cglib.core.Signature;
import net.sf.cglib.proxy.*;

public class Customer$$EnhancerByCGLIB$$3feeb52a extends Customer
    implements Factory
{
```

```
    ...
    final void CGLIB$findLove$0()
    {
        super.findLove();
    }

    public final void findLove()
    {
        CGLIB$CALLBACK_0;
        if(CGLIB$CALLBACK_0 != null) goto _L2; else goto _L1
_L1:
        JVM INSTR pop ;
        CGLIB$BIND_CALLBACKS(this);
        CGLIB$CALLBACK_0;
_L2:
        JVM INSTR dup ;
        JVM INSTR ifnull 37;
           goto _L3 _L4
_L3:
        break MISSING_BLOCK_LABEL_21;
_L4:
        break MISSING_BLOCK_LABEL_37;
        this;
        CGLIB$findLove$0$Method;
        CGLIB$emptyArgs;
        CGLIB$findLove$0$Proxy;
        intercept();
        return;
        super.findLove();
        return;
    }
    ...
}
```

我们重写了 Customer 类的所有方法,通过代理类的源码可以看到,代理类会获得所有从父类继承来的方法,并且会有 MethodProxy 与之对应,比如 Method CGLIB$findLove$0$Method、MethodProxy CGLIB$findLove0Proxy 这些方法在代理类的 findLove()方法中都有调用。

```
//代理方法（methodProxy.invokeSuper()方法会调用）
    final void CGLIB$findLove$0()
    {
        super.findLove();
    }
```

//被代理方法（methodProxy.invoke()方法会调用，这就是为什么在拦截器中调用 methodProxy.invoke 会发生死循环，一直在调用拦截器）
```java
public final void findLove()
{
    ...
    //调用拦截器
    intercept();
    return;
    super.findLove();
    return;
}
```

调用过程为：代理对象调用 this.findLove()方法→调用拦截器→methodProxy.invokeSuper→CGLIB$findLove$0→被代理对象 findLove()方法。

此时，我们发现拦截器 MethodInterceptor 中就是由 MethodProxy 的 invokeSuper()方法调用代理方法的，MethodProxy 非常关键，我们分析一下它具体做了什么。

```java
package net.sf.cglib.proxy;

import java.lang.reflect.InvocationTargetException;
import java.lang.reflect.Method;
import net.sf.cglib.core.AbstractClassGenerator;
import net.sf.cglib.core.CodeGenerationException;
import net.sf.cglib.core.GeneratorStrategy;
import net.sf.cglib.core.NamingPolicy;
import net.sf.cglib.core.Signature;
import net.sf.cglib.reflect.FastClass;
import net.sf.cglib.reflect.FastClass.Generator;

public class MethodProxy {
    private Signature sig1;
    private Signature sig2;
    private MethodProxy.CreateInfo createInfo;
    private final Object initLock = new Object();
    private volatile MethodProxy.FastClassInfo fastClassInfo;

    public static MethodProxy create(Class c1, Class c2, String desc, String name1, String name2) {
        MethodProxy proxy = new MethodProxy();
        proxy.sig1 = new Signature(name1, desc);
        proxy.sig2 = new Signature(name2, desc);
        proxy.createInfo = new MethodProxy.CreateInfo(c1, c2);
        return proxy;
    }

    ...
```

```java
    private static class CreateInfo {
        Class c1;
        Class c2;
        NamingPolicy namingPolicy;
        GeneratorStrategy strategy;
        boolean attemptLoad;

        public CreateInfo(Class c1, Class c2) {
            this.c1 = c1;
            this.c2 = c2;
            AbstractClassGenerator fromEnhancer = AbstractClassGenerator.getCurrent();
            if(fromEnhancer != null) {
                this.namingPolicy = fromEnhancer.getNamingPolicy();
                this.strategy = fromEnhancer.getStrategy();
                this.attemptLoad = fromEnhancer.getAttemptLoad();
            }

        }
    }
    ...
}
```

继续看 invokeSuper() 方法：

```java
public Object invokeSuper(Object obj, Object[] args) throws Throwable {
    try {
        this.init();
        MethodProxy.FastClassInfo fci = this.fastClassInfo;
        return fci.f2.invoke(fci.i2, obj, args);
    } catch (InvocationTargetException var4) {
        throw var4.getTargetException();
    }
}

...

private static class FastClassInfo {
    FastClass f1;
    FastClass f2;
    int i1;
    int i2;

    private FastClassInfo() {
    }
}
```

上面的代码调用就是获取代理类对应的 FastClass，并执行代理方法。还记得之前生成的三个.class 文件吗？Customer$$EnhancerByCGLIB$$3feeb52a$$FastClassByCGLIB$$6aad62f1.class 就是代理类的 FastClass，Customer$$FastClassByCGLIB$$2669574a.class 就是被代理类的 FastClass。

CGLib 代理执行代理方法的效率之所以比 JDK 的高，是因为 CGlib 采用了 FastClass 机制，它的原理简单来说就是：为代理类和被代理类各生成一个类，这个类会为代理类或被代理类的方法分配一个 index（int 类型）；这个 index 当作一个入参，FastClass 就可以直接定位要调用的方法并直接进行调用，省去了反射调用，所以调用效率比 JDK 代理通过反射调用高。下面我们反编译一个 FastClass 看看：

```
public int getIndex(Signature signature)
    {
        String s = signature.toString();
        s;
        s.hashCode();
        JVM INSTR lookupswitch 11: default 223
        ...
        JVM INSTR pop ;
        return -1;
    }

//部分代码省略

    //根据 index 直接定位执行方法
    public Object invoke(int i, Object obj, Object aobj[])
        throws InvocationTargetException
    {
        (Customer)obj;
        i;
        JVM INSTR tableswitch 0 10: default 161
           goto _L1 _L2 _L3 _L4 _L5 _L6 _L7 _L8 _L9 _L10 _L11 _L12
_L2:
        eat();
        return null;
_L3:
        findLove();
        return null;
        ...
        throw new IllegalArgumentException("Cannot find matching method/constructor");
    }
```

FastClass 并不是跟代理类一起生成的，而是在第一次执行 MethodProxy 的 invoke() 或 invokeSuper() 方法时生成的，并放在了缓存中。

```
//MethodProxy 的 invoke() 或 invokeSuper()方法都调用了 init()方法
private void init() {
```

```
if(this.fastClassInfo == null) {
    Object var1 = this.initLock;
    synchronized(this.initLock) {
        if(this.fastClassInfo == null) {
            MethodProxy.CreateInfo ci = this.createInfo;
            MethodProxy.FastClassInfo fci = new MethodProxy.FastClassInfo();
            fci.f1 = helper(ci, ci.c1);//如果在缓存中就取出,没有没在缓存中就生成新的 FastClass
            fci.f2 = helper(ci, ci.c2);
            fci.i1 = fci.f1.getIndex(this.sig1);//获取方法的 index
            fci.i2 = fci.f2.getIndex(this.sig2);
            this.fastClassInfo = fci;
        }
    }
}
```

至此,CGLib 代理的原理我们就基本搞清楚了,对代码细节有兴趣的"小伙伴"可以自行深入研究。

4. CGLib 和 JDK 动态代理对比

(1) JDK 动态代理实现了被代理对象的接口,CGLib 代理继承了被代理对象。

(2) JDK 动态代理和 CGLib 代理都在运行期生成字节码,JDK 动态代理直接写 Class 字节码,CGLib 代理使用 ASM 框架写 Class 字节码,CGlib 代理实现更复杂,生成代理类比 JDK 动态代理效率低。

(3) JDK 动态代理调用代理方法是通过反射机制调用的,CGLib 代理是通过 FastClass 机制直接调用方法的,CGLib 代理的执行效率更高。

2.5.4 代理模式与 Spring

1. 代理模式在 Spring 源码中的应用

先看 ProxyFactoryBean 核心方法 getObject(),源码如下:

```
public Object getObject() throws BeansException {
  initializeAdvisorChain();
  if (isSingleton()) {
    return getSingletonInstance();
  }
  else {
    if (this.targetName == null) {
      logger.warn("Using non-singleton proxies with singleton targets is often undesirable. " +
          "Enable prototype proxies by setting the 'targetName' property.");
```

```
        }
        return newPrototypeInstance();
    }
}
```

在 getObject()方法中，主要调用 getSingletonInstance()和 newPrototypeInstance()。在 Spring 的配置中如果不做任何设置，那么 Spring 代理生成的 Bean 都是单例对象。如果修改 scope，则每次创建一个新的原型对象。newPrototypeInstance()里面的逻辑比较复杂，我们后面再做深入研究，这里先做简单了解。

Spring 利用动态代理实现 AOP 时有两个非常重要的类：JdkDynamicAopProxy 类和 CglibAopProxy 类，来看一下类图，如下图所示。

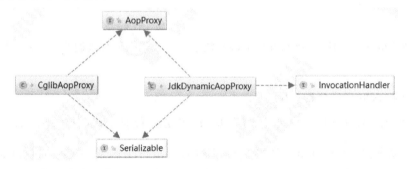

2. Spring 中的代理选择原则

（1）当 Bean 有实现接口时，Spring 就会用 JDK 动态代理。

（2）当 Bean 没有实现接口时，Spring 会选择 CGLib 代理。

（3）Spring 可以通过配置强制使用 CGLib 代理，只需在 Spring 的配置文件中加入如下代码：

```
<aop:aspectj-autoproxy proxy-target-class="true"/>
```

2.5.5 静态代理和动态代理的本质区别

（1）静态代理只能通过手动完成代理操作，如果被代理类增加了新的方法，代理类需要同步增加，违背开闭原则。

（2）动态代理采用在运行时动态生成代码的方式，取消了对被代理类的扩展限制，遵循开闭原则。

（3）若动态代理要对目标类的增强逻辑进行扩展，结合策略模式，只需要新增策略类便可完成，无须修改代理类的代码。

2.5.6　代理模式的优缺点

代理模式具有以下优点：

（1）代理模式能将代理对象与真实被调用目标对象分离。
（2）在一定程度上降低了系统的耦合性，扩展性好。
（3）可以起到保护目标对象的作用。
（4）可以增强目标对象的功能。

当然，代理模式也有缺点：

（1）代理模式会造成系统设计中类的数量增加。
（2）在客户端和目标对象中增加一个代理对象，会导致请求处理速度变慢。
（3）增加了系统的复杂度。

2.6　委派模式详解

2.6.1　委派模式的定义及应用场景

委派模式（Delegate Pattern）不属于GoF 23种设计模式。委派模式的基本作用就是负责任务的调用和分配，跟代理模式很像，可以看作一种特殊情况下的静态的全权代理，但是代理模式注重过程，而委派模式注重结果。委派模式在Spring中应用得非常多，大家常用的DispatcherServlet就用到了委派模式。现实生活中也常有委派的场景发生，例如老板（Boss）给项目经理（Leader）下达任务，项目经理会根据实际情况给每个员工派发任务，待员工把任务完成后，再由项目经理向老板汇报结果。我们用代码来模拟一下这个业务场景，先来看一下类图，如下图所示。

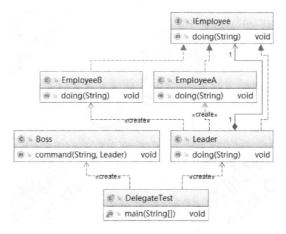

创建 IEmployee 员工接口：

```java
package com.gupaoedu.vip.pattern.delegate.simple;

public interface IEmployee {

    public void doing(String command);

}
```

创建员工类 EmployeeA：

```java
package com.gupaoedu.vip.pattern.delegate.simple;

public class EmployeeA implements IEmployee {
    @Override
    public void doing(String command) {
        System.out.println("我是员工A，我现在开始干" + command + "工作");
    }
}
```

创建员工类 EmployeeB：

```java
package com.gupaoedu.vip.pattern.delegate.simple;

public class EmployeeB implements IEmployee {
    @Override
    public void doing(String command) {
        System.out.println("我是员工B，我现在开始干" + command + "工作");
    }
}
```

创建项目经理类 Leader：

```java
package com.gupaoedu.vip.pattern.delegate.simple;

import java.util.HashMap;
import java.util.Map;

public class Leader implements IEmployee {

    private Map<String,IEmployee> targets = new HashMap<String,IEmployee>();

    public Leader() {
        targets.put("加密",new EmployeeA());
        targets.put("登录",new EmployeeB());
    }

    //项目经理自己不干活
```

```java
    public void doing(String command){
        targets.get(command).doing(command);
    }
}
```

创建 Boss 类下达命令：

```java
package com.gupaoedu.vip.pattern.delegate.simple;
public class Boss {

    public void command(String command,Leader leader){
        leader.doing(command);
    }
}
```

测试代码如下：

```java
package com.gupaoedu.vip.pattern.delegate.simple;

public class DelegateTest {

    public static void main(String[] args) {

        //代理模式注重的是过程，委派模式注重的是结果
        //策略模式注重可扩展性（外部扩展性），委派模式注重内部的灵活性和可复用性
        //委派模式的核心就是分发、调度、派遣，委派模式是静态代理和策略模式的一种特殊组合
        new Boss().command("登录",new Leader());

    }
}
```

上面的代码生动地还原了项目经理分配任务的业务场景，也是委派模式的生动体现。

2.6.2 委派模式在源码中的体现

下面我们再来还原一下 Spring MVC 的 DispatcherServlet 是如何实现委派模式的。创建业务类 MemberController：

```java
package com.gupaoedu.vip.pattern.delegate.mvc.controllers;

public class MemberController {

    public void getMemberById(String mid){

    }
}
```

OrderController 类如下：

```java
package com.gupaoedu.vip.pattern.delegate.mvc.controllers;

public class OrderController {

    public void getOrderById(String mid){

    }

}
```

SystemController 类如下：

```java
package com.gupaoedu.vip.pattern.delegate.mvc.controllers;

public class SystemController {

    public void logout(){

    }

}
```

创建 DispatcherServlet 类：

```java
package com.gupaoedu.vip.pattern.delegate.mvc;

import com.gupaoedu.vip.pattern.delegate.mvc.controllers.MemberController;
import com.gupaoedu.vip.pattern.delegate.mvc.controllers.OrderController;
import com.gupaoedu.vip.pattern.delegate.mvc.controllers.SystemController;

import javax.servlet.ServletException;
import javax.servlet.http.HttpServlet;
import javax.servlet.http.HttpServletRequest;
import javax.servlet.http.HttpServletResponse;
import java.io.IOException;
import java.lang.reflect.InvocationTargetException;
import java.lang.reflect.Method;
import java.util.ArrayList;
import java.util.List;

//相当于项目经理的角色
public class DispatcherServlet extends HttpServlet{

    private void doDispatch(HttpServletRequest request, HttpServletResponse response) throws Exception{

        String uri = request.getRequestURI();
```

```java
        String mid = request.getParameter("mid");

    if("getMemberById".equals(uri)){
        new MemberController().getMemberById(mid);
    }else if("getOrderById".equals(uri)){
        new OrderController().getOrderById(mid);
    }else if("logout".equals(uri)){
        new SystemController().logout();
    }else {
        response.getWriter().write("404 Not Found!!");
    }
}

    protected void service(HttpServletRequest req, HttpServletResponse resp) throws ServletException, IOException {
    try {
        doDispatch(req,resp);
    } catch (Exception e) {
        e.printStackTrace();
    }
}

}
```

配置 web.xml 文件：

```xml
<?xml version="1.0" encoding="UTF-8"?>
<web-app xmlns:xsi="http://www.w3.org/2001/XMLSchema-instance"
   xmlns="http://java.sun.com/xml/ns/j2ee" xmlns:javaee="http://java.sun.com/xml/ns/javaee"
   xmlns:web="http://java.sun.com/xml/ns/javaee/web-app_2_5.xsd"
   xsi:schemaLocation="http://java.sun.com/xml/ns/j2ee
http://java.sun.com/xml/ns/j2ee/web-app_2_4.xsd"
   version="2.4">
   <display-name>Gupao Web Application</display-name>


   <servlet>
      <servlet-name>delegateServlet</servlet-name>
      <servlet-class>com.gupaoedu.vip.pattern.delegate.mvc.DispatcherServlet</servlet-class>
      <load-on-startup>1</load-on-startup>
   </servlet>

   <servlet-mapping>
      <servlet-name>delegateServlet</servlet-name>
      <url-pattern>/*</url-pattern>
   </servlet-mapping>
```

```
</web-app>
```

一个完整的委派模式就实现了。当然，在 Spring 中运用到委派模式的地方还有很多，"小伙伴们"通过命名就可以识别出来。在 Spring 源码中，以 Delegate 结尾的地方都实现了委派模式，例如 BeanDefinitionParserDelegate 根据不同类型委派不同的逻辑解析 BeanDefinition。

2.7 策略模式详解

策略模式（Strategy Pattern）是指定义了算法家族并分别封装起来，让它们之间可以互相替换，此模式使得算法的变化不会影响使用算法的用户。

2.7.1 策略模式的应用场景

策略模式的应用场景如下。

（1）系统中有很多类，而它们的区别仅仅在于行为不同。

（2）一个系统需要动态地在几种算法中选择一种。

2.7.2 用策略模式实现选择支付方式的业务场景

大家都知道，咕泡学院的架构师课程经常有优惠活动，优惠策略有很多种，如优惠券抵扣、返现促销、拼团。下面我们用代码来模拟这个场景，首先创建一个促销策略的接口 PromotionStrategy：

```java
package com.gupaoedu.vip.pattern.strategy.promotion;

public interface PromotionStrategy {
    void doPromotion();
}
```

然后分别创建优惠券抵扣策略类 CouponStrategy、返现促销策略类 CashbackStrategy、拼团策略类 GroupbuyStrategy 和无优惠策略类 EmptyStrategy。

CouponStrategy 类如下：

```java
package com.gupaoedu.vip.pattern.strategy.promotion;
public class CouponStrategy implements PromotionStrategy {
    public void doPromotion() {
        System.out.println("领取优惠券，课程的价格直接减优惠券面值抵扣");
    }
}
```

CashbackStrategy 类如下：

```java
package com.gupaoedu.vip.pattern.strategy.promotion;
public class CashbackStrategy implements PromotionStrategy {
    public void doPromotion() {
        System.out.println("返现促销，返回的金额转到支付宝账号");
    }
}
```

GroupbuyStrategy 类如下：

```java
package com.gupaoedu.vip.pattern.strategy.promotion;
public class GroupbuyStrategy implements PromotionStrategy{
    public void doPromotion() {
        System.out.println("拼团，满 20 人成团，全团享受团购价");
    }
}
```

EmptyStrategy 类如下：

```java
package com.gupaoedu.vip.pattern.strategy.promotion;
public class EmptyStrategy implements PromotionStrategy {
    public void doPromotion() {
        System.out.println("无促销活动");
    }
}
```

创建促销活动方案类 PromotionActivity：

```java
package com.gupaoedu.vip.pattern.strategy.promotion;
public class PromotionActivity {
    private PromotionStrategy promotionStrategy;

    public PromotionActivity(PromotionStrategy promotionStrategy) {
        this.promotionStrategy = promotionStrategy;
    }
    public void execute(){
        promotionStrategy.doPromotion();
    }
}
```

编写客户端测试类：

```java
public static void main(String[] args) {
    PromotionActivity activity618 = new PromotionActivity(new CouponStrategy());
    PromotionActivity activity1111 = new PromotionActivity(new CashbackStrategy());

    activity618.execute();
```

```
        activity1111.execute();
}
```

此时,"小伙伴们"会发现,把上面这段测试代码放到实际的业务场景并不实用。因为我们做活动的时候往往要根据不同的需求对促销策略进行动态选择,并不会一次性执行多种优惠,我们的代码通常会这样写:

```java
public static void main(String[] args) {
    PromotionActivity promotionActivity = null;

    String promotionKey = "COUPON";

    if(StringUtils.equals(promotionKey,"COUPON")){
        promotionActivity = new PromotionActivity(new CouponStrategy());
    }else if(StringUtils.equals(promotionKey,"CASHBACK")){
        promotionActivity = new PromotionActivity(new CashbackStrategy());
    }
    promotionActivity.execute();
}
```

这样改造之后,满足了业务需求,客户可根据自己的需求选择不同的优惠策略。但是,经过一段时间的业务积累,促销活动会越来越多,于是我们的程序员"小哥"就忙不停了,每次上活动之前都要通宵改代码,而且要做重复测试,判断逻辑可能也变得越来越复杂。这时候,我们是不是需要考虑重构代码了?回顾之前学过的设计模式,应该如何来优化这段代码呢?其实,我们可以结合单例模式和工厂模式来进行优化。创建 PromotionStrategyFactory 类:

```java
package com.gupaoedu.vip.pattern.strategy.promotion;

import java.util.HashMap;
import java.util.Map;

public class PromotionStrategyFactory {
    private static Map<String,PromotionStrategy> PROMOTION_STRATEGY_MAP = new HashMap<String,PromotionStrategy>();
    static {
        PROMOTION_STRATEGY_MAP.put(PromotionKey.COUPON,new CouponStrategy());
        PROMOTION_STRATEGY_MAP.put(PromotionKey.CASHBACK,new CashbackStrategy());
        PROMOTION_STRATEGY_MAP.put(PromotionKey.GROUPBUY,new GroupbuyStrategy());
    }

    private static final PromotionStrategy NON_PROMOTION = new EmptyStrategy();

    private PromotionStrategyFactory(){}

    public static PromotionStrategy getPromotionStrategy(String promotionKey){
```

```java
        PromotionStrategy promotionStrategy = PROMOTION_STRATEGY_MAP.get(promotionKey);
        return promotionStrategy == null ? NON_PROMOTION : promotionStrategy;
    }

    private interface PromotionKey{
        String COUPON = "COUPON";
        String CASHBACK = "CASHBACK";
        String GROUPBUY = "GROUPBUY";
    }
}
```

这时候客户端代码就应该这样写了：

```java
public static void main(String[] args) {
    String promotionKey = "GROUPBUY";
    PromotionActivity promotionActivity = new PromotionActivity(PromotionStrategyFactory.getPromotionStrategy(promotionKey));
    promotionActivity.execute();
}
```

优化之后，程序员"小哥"的维护工作是不是就轻松了？每次上新活动，不影响原来的代码逻辑。为了加深大家对策略模式的理解，再来举一个例子。相信"小伙伴们"都用过支付宝支付、微信支付、银联支付及京东白条支付。一个常见的应用场景就是大家在支付时会提示选择支付方式，如果用户未选，系统也会使用默认的支付方式进行结算。来看一下类图，如下图所示。

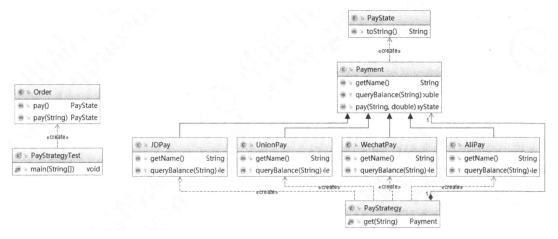

下面我们用策略模式来模拟此业务场景。

创建 Payment 抽象类，定义支付规范和支付逻辑：

```
package com.gupaoedu.vip.pattern.strategy.pay.payport;
```

```java
import com.gupaoedu.vip.pattern.strategy.pay.PayState;

public abstract class Payment {

    //支付类型
    public abstract String getName();

    //查询余额
    protected abstract double queryBalance(String uid);

    //扣款支付
    public PayState pay(String uid,double amount) {
        if(queryBalance(uid) < amount){
            return new PayState(500,"支付失败","余额不足");
        }
        return new PayState(200,"支付成功","支付金额: " + amount);
    }
}
```

分别创建具体的支付方式,支付宝支付类 AliPay:

```java
package com.gupaoedu.vip.pattern.strategy.pay.payport;

public class AliPay extends Payment {

    public String getName() {
        return "支付宝";
    }

    protected double queryBalance(String uid) {
        return 900;
    }
}
```

微信支付类 WechatPay:

```java
package com.gupaoedu.vip.pattern.strategy.pay.payport;
public class WechatPay extends Payment {

    public String getName() {
        return "微信支付";
    }

    protected double queryBalance(String uid) {
        return 256;
    }
}
```

银联支付类 UnionPay：

```java
package com.gupaoedu.vip.pattern.strategy.pay.payport;
public class UnionPay extends Payment {
    public String getName() {
        return "银联支付";
    }

    protected double queryBalance(String uid) {
        return 120;
    }
}
```

京东白条支付类 JDPay：

```java
package com.gupaoedu.vip.pattern.strategy.pay.payport;
public class JDPay extends Payment {

    public String getName() {
        return "京东白条";
    }

    protected double queryBalance(String uid) {
        return 500;
    }
}
```

创建支付状态的包装类 PayState：

```java
package com.gupaoedu.vip.pattern.strategy.pay;

public class PayState {
    private int code;
    private Object data;
    private String msg;

    public PayState(int code, String msg,Object data) {
        this.code = code;
        this.data = data;
        this.msg = msg;
    }

    public String toString(){
        return ("支付状态：[" + code + "]," + msg + ",交易详情: " + data);
    }
}
```

创建支付策略管理类 PayStrategy：

```java
package com.gupaoedu.vip.pattern.strategy.pay.payport;

import java.util.HashMap;
import java.util.Map;

public class PayStrategy {
    public static final String ALI_PAY = "AliPay";
    public static final String JD_PAY = "JdPay";
    public static final String UNION_PAY = "UnionPay";
    public static final String WECHAT_PAY = "WechatPay";
    public static final String DEFAULT_PAY = ALI_PAY;

    private static Map<String,Payment> payStrategy = new HashMap<String,Payment>();
    static {
        payStrategy.put(ALI_PAY,new AliPay());
        payStrategy.put(WECHAT_PAY,new WechatPay());
        payStrategy.put(UNION_PAY,new UnionPay());
        payStrategy.put(JD_PAY,new JDPay());
    }

    public static Payment get(String payKey){
        if(!payStrategy.containsKey(payKey)){
            return payStrategy.get(DEFAULT_PAY);
        }
        return payStrategy.get(payKey);
    }
}
```

创建订单类 Order：

```java
package com.gupaoedu.vip.pattern.strategy.pay;

import com.gupaoedu.vip.pattern.strategy.pay.payport.PayStrategy;
import com.gupaoedu.vip.pattern.strategy.pay.payport.Payment;

public class Order {
    private String uid;
    private String orderId;
    private double amount;

    public Order(String uid,String orderId,double amount){
        this.uid = uid;
        this.orderId = orderId;
        this.amount = amount;
    }

    //完美地解决了 switch 的过程，不需要在代码中写 switch 了
```

```java
    //更不需要写if...else if
    public PayState pay(){
        return pay(PayStrategy.DEFAULT_PAY);
    }

    public PayState pay(String payKey){
        Payment payment = PayStrategy.get(payKey);
        System.out.println("欢迎使用" + payment.getName());
        System.out.println("本次交易金额为: " + amount + "，开始扣款...");
        return payment.pay(uid,amount);
    }
}
```

测试代码如下：

```java
package com.gupaoedu.vip.pattern.strategy.pay;

import com.gupaoedu.vip.pattern.strategy.pay.payport.PayStrategy;

public class PayStrategyTest {

    public static void main(String[] args) {

        //省略把商品添加到购物车再从购物车下单，直接从订单开始
        Order order = new Order("1","201803110010000009",324.45);

        //开始支付，选择支付宝支付、微信支付、银联支付或京东白条支付
        //每个渠道支付的具体算法是不一样的，基本算法是固定的

        //在支付的时候才决定这个值用哪个
        System.out.println(order.pay(PayStrategy.ALI_PAY));

    }
}
```

运行结果如下图所示。

```
Run  PayStrategyTest
    欢迎使用支付宝
    本次交易金额为: 324.45，开始扣款...
    支付状态：[200],支付成功,交易详情：支付金额：324.45

    Process finished with exit code 0
```

这里通过大家耳熟能详的业务场景来举例，希望"小伙伴们"能更深刻地理解策略模式。同

时也希望"小伙伴们"在面试和工作中都能体现出自己的优势。

2.7.3 策略模式在 JDK 源码中的体现

首先来看一个比较常用的比较器——Comparator 接口，大家常用的 compare() 方法就是一个策略模式的抽象实现：

```java
public interface Comparator<T> {
    int compare(T o1, T o2);
...
}
```

Comparator 接口下面有非常多的实现类，我们经常会把 Comparator 接口作为传入参数实现排序策略，例如 Arrays 类的 parallelSort() 方法等：

```java
public class Arrays {
    ...
    public static <T> void parallelSort(T[] a, int fromIndex, int toIndex,
                                        Comparator<? super T> cmp) {
    ...
    }
    ...
}
```

还有 TreeMap 类的构造方法：

```java
public class TreeMap<K,V>
    extends AbstractMap<K,V>
    implements NavigableMap<K,V>, Cloneable, java.io.Serializable
{
    ...
    public TreeMap(Comparator<? super K> comparator) {
        this.comparator = comparator;
    }
...
}
```

这就是策略模式在 JDK 源码中的应用。下面我们来看策略模式在 Spring 源码中的应用，来看 Resource 接口：

```java
package org.springframework.core.io;

import java.io.File;
import java.io.IOException;
```

```java
import java.net.URI;
import java.net.URL;
import java.nio.channels.Channels;
import java.nio.channels.ReadableByteChannel;
import org.springframework.lang.Nullable;

public interface Resource extends InputStreamSource {
    boolean exists();

    default boolean isReadable() {
        return true;
    }

    default boolean isOpen() {
        return false;
    }

    default boolean isFile() {
        return false;
    }

    URL getURL() throws IOException;

    URI getURI() throws IOException;

    File getFile() throws IOException;

    default ReadableByteChannel readableChannel() throws IOException {
        return Channels.newChannel(this.getInputStream());
    }

    long contentLength() throws IOException;

    long lastModified() throws IOException;

    Resource createRelative(String var1) throws IOException;

    @Nullable
    String getFilename();

    String getDescription();
}
```

我们虽然不经常直接使用 Resource 接口，但是经常使用它的子类，如下图所示。

```
public interface Resource extends InputStreamSource {
                                        Choose Implementation of Resource (38 fo
    /**                 ⊙ AbstractFileResolvingResource (org.springframework.core.io)
     * Determine        ⊙ AbstractResource (org.springframework.core.io)
     * <p>This me       ⊙ Anonymous in readInternal() in ResourceHttpMessageConverter (org.springf
     * existence        ⊙ Anonymous in readInternal() in ResourceHttpMessageConverter (org.springf
     * descriptor       ⊙ Anonymous in testAbstractResourceExceptions() in ResourceTests (org.spri
     */                 ⊙ Anonymous in testContentLength() in ResourceTests (org.springframework.c
    boolean exist       ⊙ Anonymous in testFirstFound() in YamlMapFactoryBeanTests (org.springfram
                        ⊙ Anonymous in testSignificantLoad() in CachingMetadataReaderLeakTests (or
                        ⊙ Anonymous in writeMultipart() in FormHttpMessageConverterTests (org.spri
                        ⊙ Anonymous in writeMultipart() in MultipartHttpMessageWriterTests (org.sp
                        ⊙ BaseResource in CollectionToCollectionConverterTests (org.springframewor
    /**                 ⊙ BeanDefinitionResource (org.springframework.beans.factory.support)
     * Indicate w       ⊙ ByteArrayResource (org.springframework.core.io)
     * {@link #ge       ⊙ ClassPathContextResource in DefaultResourceLoader (org.springframework.c
     * <p>Will be       ⊙ ClassPathResource (org.springframework.core.io)
                        ⊙ ClassRelativeContextResource in ClassRelativeResourceLoader (org.springf
```

还有一个非常典型的场景，Spring 的初始化也采用了策略模式，即不同类型的类采用不同的初始化策略。有一个 InstantiationStrategy 接口，我们来看一下源码：

```java
package org.springframework.beans.factory.support;

import java.lang.reflect.Constructor;
import java.lang.reflect.Method;
import org.springframework.beans.BeansException;
import org.springframework.beans.factory.BeanFactory;
import org.springframework.lang.Nullable;

public interface InstantiationStrategy {
    Object instantiate(RootBeanDefinition var1, @Nullable String var2, BeanFactory var3) throws BeansException;

    Object instantiate(RootBeanDefinition var1, @Nullable String var2, BeanFactory var3, Constructor<?> var4, @Nullable Object... var5) throws BeansException;

    Object instantiate(RootBeanDefinition var1, @Nullable String var2, BeanFactory var3, @Nullable Object var4, Method var5, @Nullable Object... var6) throws BeansException;
}
```

顶层的策略抽象非常简单，它下面有两种策略：SimpleInstantiationStrategy 和 CglibSubclassing-InstantiationStrategy。我们看一下类图，如下图所示。

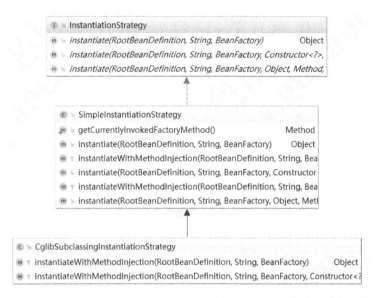

从类图我们发现，CglibSubclassingInstantiationStrategy 类继承了 SimpleInstantiationStrategy 类，说明在实际应用中多种策略之间可以继承使用。"小伙伴们"可以将其作为一个参考，在实际业务场景中根据需要来设计。

2.7.4 策略模式的优缺点

策略模式的优点如下：

（1）策略模式符合开闭原则。

（2）策略模式可避免使用多重条件语句，如 if...else 语句、switch 语句。

（3）使用策略模式可以提高算法的保密性和安全性。

策略模式的缺点如下：

（1）客户端必须知道所有的策略，并且自行决定使用哪一个策略类。

（2）代码中会产生非常多的策略类，增加了代码的维护难度。

2.7.5 委派模式与策略模式综合应用

在前面的代码中我们列举了几个业务场景，相信"小伙伴们"对委派模式和策略模式都有了非常深刻的理解。我们再来回顾一下 DispatcherServlet：

```
package com.gupaoedu.vip.pattern.delegate.mvc;
...
public class DispatcherServlet extends HttpServlet{

    private void doDispatch(HttpServletRequest request, HttpServletResponse response) throws Exception{

        String uri = request.getRequestURI();
        String mid = request.getParameter("mid");

        if("getMemberById".equals(uri)){
            new MemberController().getMemberById(mid);
        }else if("getOrderById".equals(uri)){
            new OrderController().getOrderById(mid);
        }else if("logout".equals(uri)){
            new SystemController().logout();
        }else {
            response.getWriter().write("404 Not Found!!");
        }
    }
...
}
```

这样的代码扩展性不太好，也不现实，因为在实际项目中一定不只有这几个 Controller，往往有成千上万个 Controller，显然我们不能写成千上万个 if...else。那么我们如何来改造呢？"小伙伴们"一定想到了策略模式，来看一下我是怎么优化的：

```
package com.gupaoedu.vip.pattern.delegate.mvc;

import com.gupaoedu.vip.pattern.delegate.mvc.controllers.MemberController;
import com.gupaoedu.vip.pattern.delegate.mvc.controllers.OrderController;
import com.gupaoedu.vip.pattern.delegate.mvc.controllers.SystemController;

import javax.servlet.ServletException;
import javax.servlet.http.HttpServlet;
import javax.servlet.http.HttpServletRequest;
import javax.servlet.http.HttpServletResponse;
import java.io.IOException;
import java.lang.reflect.InvocationTargetException;
import java.lang.reflect.Method;
import java.util.ArrayList;
import java.util.List;
```

```java
public class DispatcherServlet extends HttpServlet{

    private List<Handler> handlerMapping = new ArrayList<Handler>();

    public void init() throws ServletException {
        try {
            Class<?> memberControllerClass = MemberController.class;
            handlerMapping.add(new Handler()
                    .setController(memberControllerClass.newInstance())
                    .setMethod(memberControllerClass.getMethod("getMemberById", new Class[]{String.class}))
                    .setUrl("/web/getMemberById.json"));
        }catch(Exception e){

        }
    }

    private void doDispatch(HttpServletRequest request, HttpServletResponse response){

        //获取用户请求的 URL
        //如果按照 J2EE 的标准，每个 URL 对应一个 Serlvet，URL 从浏览器输入
        String uri = request.getRequestURI();

        //Servlet 拿到 URL 以后，要做权衡（要做判断，要做选择）
        //根据用户请求的 URL，找到这个 URL 对应的某个 Java 类的方法

        //通过获取的 URL 去做 handlerMapping（我们认为它是策略常量）
        Handler handle = null;
        for (Handler h: handlerMapping) {
            if(uri.equals(h.getUrl())){
                handle = h;
                break;
            }
        }

        //将具体的任务分发给 Method（通过反射调用对应的方法）
        Object object = null;
        try {
            object = handle.getMethod().invoke(handle.getController(),request.getParameter("mid"));
        } catch (IllegalAccessException e) {
            e.printStackTrace();
        } catch (InvocationTargetException e) {
            e.printStackTrace();
        }

    }
}
```

```java
protected void service(HttpServletRequest req, HttpServletResponse resp) throws
ServletException, IOException {
    try {
        doDispatch(req,resp);
    } catch (Exception e) {
        e.printStackTrace();
    }
}

class Handler{

    private Object controller;
    private Method method;
    private String url;

    public Object getController() {
        return controller;
    }

    public Handler setController(Object controller) {
        this.controller = controller;
        return this;
    }

    public Method getMethod() {
        return method;
    }

    public Handler setMethod(Method method) {
        this.method = method;
        return this;
    }

    public String getUrl() {
        return url;
    }

    public Handler setUrl(String url) {
        this.url = url;
        return this;
    }
}
```

上面的代码结合了策略模式、工厂模式、单例模式。当然，我的优化方案不一定是最完美的，仅代表个人观点。感兴趣的"小伙伴"可以继续思考，如何让这段代码变得更优雅。我们后面在讲 Spring 源码时还会讲到 DispatcherServlet 的相关内容。

2.8 模板模式详解

2.8.1 模板模式的应用场景

我们平时办理入职的流程是：填写入职登记表→打印简历→复印学历→复印身份证→签订劳动合同→建立花名册→办理工牌→安排工位等。我平时在家里炒菜的流程是：洗锅→点火→热锅→上油→下原料→翻炒→放调料→出锅。赵本山问宋丹丹："如何把大象放进冰箱？"宋丹丹回答："第一步：打开冰箱门；第二步：把大象塞进冰箱；第三步：关闭冰箱门。"如下图所示。赵本山再问："怎么把长颈鹿放进冰箱？"宋丹丹答："第一步：打开冰箱门；第二步：把大象拿出来；第三步：把长颈鹿塞进去；第四步：关闭冰箱门。"以上这些都是模板模式的体现。

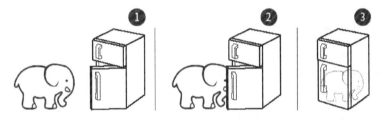

模板模式又叫模板方法模式（Template Method Pattern），是指定义一个算法的骨架，并允许子类为一个或者多个步骤提供实现。模板模式使得子类可以在不改变算法结构的情况下，重新定义算法的某些步骤，属于行为型设计模式。模板模式适用于以下场景：

（1）一次性实现一个算法的不变部分，并将可变的行为留给子类来实现。

（2）各子类中公共的行为被提取出来并集中到一个公共的父类中，从而避免代码重复。

以咕泡学院的课程创建流程为例：发布预习资料→制作课件 PPT→在线直播→提交课堂笔记→提交源码→布置作业→检查作业。首先创建 NetworkCourse 抽象类：

```
package com.gupaoedu.vip.pattern.template.course;
//模板会有一个或者多个未现实方法，而且这几个未实现方法有固定的执行顺序
public abstract class NetworkCourse {

    protected final void createCourse(){
        //发布预习资料
        this.postPreResource();
```

```
    //制作课件 PPT
    this.createPPT();

    //在线直播
    this.liveVideo();

    //提交课堂笔记
    this.postNote();

    //提交源码
    this.postSource();

    //布置作业,有些课是没有作业的,有些课是有作业的
    //如果有作业,检查作业,如果没有作业,流程结束
    if(needHomework()){
        checkHomework();
    }
}

abstract void checkHomework();

//钩子方法:实现流程的微调
protected boolean needHomework(){return false;}

final void postSource(){
    System.out.println("提交源代码");
}

final void postNote(){
    System.out.println("提交课件和笔记");
}

final void liveVideo(){
    System.out.println("直播授课");
}

final void createPPT(){
    System.out.println("创建备课 PPT");
}

final void postPreResource(){
    System.out.println("分发预习资料");
}
}
```

在上面的代码中有个钩子方法,可能有些"小伙伴"还不太理解,在此稍做解释。设计钩子方法的主要目的是干预执行流程,使得控制行为更加灵活,更符合实际业务的需求。钩子方法的

返回值一般为适合条件分支语句的返回值。"小伙伴们"可以根据自己的业务场景来决定是否使用钩子方法。接下来创建 JavaCourse 类：

```java
package com.gupaoedu.vip.pattern.template.course;

public class JavaCourse extends NetworkCourse {
    void checkHomework() {
        System.out.println("检查 Java 的架构课件");
    }
}
```

创建 BigDataCourse 类：

```java
package com.gupaoedu.vip.pattern.template.course;

public class BigDataCourse extends NetworkCourse {

    private boolean needHomeworkFlag = false;

    public BigDataCourse(boolean needHomeworkFlag) {
        this.needHomeworkFlag = needHomeworkFlag;
    }

    void checkHomework() {
        System.out.println("检查大数据的课后作业");
    }

    @Override
    protected boolean needHomework() {
        return this.needHomeworkFlag;
    }
}
```

客户端测试代码如下：

```java
package com.gupaoedu.vip.pattern.template.course;

public class NetworkCourseTest {
    public static void main(String[] args) {

        System.out.println("---Java 架构师课程---");
        NetworkCourse javaCourse = new JavaCourse();
        javaCourse.createCourse();

        System.out.println("---大数据课程---");
        NetworkCourse bigDataCourse = new BigDataCourse(true);
        bigDataCourse.createCourse();
```

```
    }
}
```

通过这样一个案例，相信"小伙伴们"对模板模式已经有了一个基本的印象。为了加深理解，下面我们来看一个常见的业务场景。

2.8.2 利用模板模式重构 JDBC 操作业务场景

创建一个模板类 JdbcTemplate，封装所有的 JDBC 操作。以查询为例，每次查询的表不同，返回的数据结构也就不一样。我们针对不同的数据，都要将其封装成不同的实体对象。而每个实体对象的封装逻辑是不一样的，但封装前和封装后的处理流程是不变的，因此可以使用模板模式来进行设计。先创建约束 ORM 逻辑的接口 RowMapper：

```java
package com.gupaoedu.vip.pattern.template;

import java.sql.ResultSet;

public interface RowMapper<T> {

    T mapRow(ResultSet rs, int rowNum) throws Exception;

}
```

再创建封装了所有处理流程的抽象类 JdbcTemplate：

```java
package com.gupaoedu.vip.pattern.template.jdbc;

import com.sun.org.apache.regexp.internal.RE;
import com.sun.org.apache.xerces.internal.xs.datatypes.ObjectList;

import javax.sql.DataSource;
import java.sql.Connection;
import java.sql.PreparedStatement;
import java.sql.ResultSet;
import java.sql.SQLException;
import java.util.ArrayList;
import java.util.List;

public abstract class JdbcTemplate {
    private DataSource dataSource;

    public JdbcTemplate(DataSource dataSource) {
        this.dataSource = dataSource;
    }
```

```java
public List<?> executeQuery(String sql, RowMapper<?> rowMapper, Object[] values){
    try {
        //获取连接
        Connection conn = this.getConnection();
        //创建语句集
        PreparedStatement pstm = this.createPrepareStatement(conn,sql);
        //执行语句集
        ResultSet rs = this.executeQuery(pstm,values);
        //处理结果集
        List<?> result = this.paresResultSet(rs,rowMapper);
        //关闭结果集
        this.closeResultSet(rs);
        //关闭语句集
        this.closeStatement(pstm);
        //关闭连接
        this.closeConnection(conn);
        return result;
    }catch (Exception e){
        e.printStackTrace();
    }
    return null;
}

protected void closeConnection(Connection conn) throws Exception {
    //我们不关闭数据库连接池
    conn.close();
}

protected void closeStatement(PreparedStatement pstm) throws Exception {
    pstm.close();
}

protected void closeResultSet(ResultSet rs) throws Exception {
    rs.close();
}

protected List<?> paresResultSet(ResultSet rs, RowMapper<?> rowMapper) throws Exception {
    List<Object> result = new ArrayList<Object>();
    int rowNum = 1;
    while (rs.next()){
        result.add(rowMapper.mapRow(rs,rowNum ++));
    }
    return result;
}

protected ResultSet executeQuery(PreparedStatement pstm, Object[] values) throws Exception {
    for (int i = 0; i < values.length; i++) {
```

```java
            pstm.setObject(i,values[i]);
        }
        return pstm.executeQuery();
    }

    protected PreparedStatement createPrepareStatement(Connection conn, String sql) throws Exception {
        return conn.prepareStatement(sql);
    }

    public Connection getConnection() throws Exception {
        return this.dataSource.getConnection();
    }
}
```

创建实体对象类 Member：

```java
package com.gupaoedu.vip.pattern.template.entity;

public class Member {

    private String username;
    private String password;
    private String nickName;

    private int age;
    private String addr;

    public String getUsername() {
        return username;
    }

    public void setUsername(String username) {
        this.username = username;
    }

    public String getPassword() {
        return password;
    }

    public void setPassword(String password) {
        this.password = password;
    }

    public String getNickName() {
        return nickName;
    }
```

```java
    public void setNickName(String nickName) {
        this.nickName = nickName;
    }

    public int getAge() {
        return age;
    }

    public void setAge(int age) {
        this.age = age;
    }

    public String getAddr() {
        return addr;
    }

    public void setAddr(String addr) {
        this.addr = addr;
    }
}
```

创建数据库操作类 MemberDao：

```java
package com.gupaoedu.vip.pattern.template.jdbc.dao;

import com.gupaoedu.vip.pattern.template.jdbc.JdbcTemplate;
import com.gupaoedu.vip.pattern.template.jdbc.Member;
import com.gupaoedu.vip.pattern.template.jdbc.RowMapper;

import javax.sql.DataSource;
import java.sql.ResultSet;
import java.util.List;

public class MemberDao extends JdbcTemplate {
    public MemberDao(DataSource dataSource) {
        super(dataSource);
    }

    public List<?> selectAll(){
        String sql = "select * from t_member";
        return super.executeQuery(sql, new RowMapper<Member>() {
            public Member mapRow(ResultSet rs, int rowNum) throws Exception {
                Member member = new Member();
                //字段过多，原型模式
                member.setUsername(rs.getString("username"));
                member.setPassword(rs.getString("password"));
```

```
                member.setAge(rs.getInt("age"));
                member.setAddr(rs.getString("addr"));
                return member;
            }
        },null);
    }
}
```

客户端测试代码如下:

```java
package com.gupaoedu.vip.pattern.template.jdbc;

import com.gupaoedu.vip.pattern.template.jdbc.dao.MemberDao;
import java.util.List;

public class MemberDaoTest {

    public static void main(String[] args) {
        MemberDao memberDao = new MemberDao(null);
        List<?> result = memberDao.selectAll();
        System.out.println(result);
    }
}
```

通过这两个案例的业务场景分析,"小伙伴们"对模板模式应该有了更深的理解。

2.8.3 模板模式在源码中的体现

先来看 JDK 中的 AbstractList 类的代码:

```java
package java.util;

public abstract class AbstractList<E> extends AbstractCollection<E> implements List<E> {
    ...
    abstract public E get(int index);
    ...
}
```

我们看到,get()是一个抽象方法,交给子类来实现,大家所熟知的 ArrayList 就是 AbstractList 的子类。同理,有 AbstractList 就有 AbstractSet 和 AbstractMap,有兴趣的"小伙伴"可以去看看它们的源码。还有很多人每天都在用的 HttpServlet,有三个方法:service()、doGet()和 doPost(),都是模板方法的抽象实现。

在 MyBatis 框架中也有一些经典的应用,我们来一下 BaseExecutor 类。它是一个基础的 SQL 执行类,实现了大部分 SQL 的执行逻辑,然后把几个方法交给子类定制化完成,源码如下:

```
...
public abstract class BaseExecutor implements Executor {
    ...
    protected abstract int doUpdate(MappedStatement var1, Object var2) throws SQLException;

    protected abstract List<BatchResult> doFlushStatements(boolean var1) throws SQLException;

    protected abstract <E> List<E> doQuery(MappedStatement var1, Object var2, RowBounds var3,
ResultHandler var4, BoundSql var5) throws SQLException;

    protected abstract <E> Cursor<E> doQueryCursor(MappedStatement var1, Object var2, RowBounds
var3, BoundSql var4) throws SQLException;
...
}
```

doUpdate()、doFlushStatements()、doQuery()和 doQueryCursor()方法就是由子类来实现的。BaseExecutor 有哪些子类呢？我们来看一下它的类图，如下图所示。

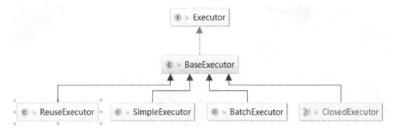

我们一起来看一下 SimpleExecutor 的 doUpdate()实现：

```
public int doUpdate(MappedStatement ms, Object parameter) throws SQLException {
    Statement stmt = null;

    int var6;
    try {
        Configuration configuration = ms.getConfiguration();
        StatementHandler handler = configuration.newStatementHandler(this, ms, parameter,
RowBounds.DEFAULT, (ResultHandler)null, (BoundSql)null);
        stmt = this.prepareStatement(handler, ms.getStatementLog());
        var6 = handler.update(stmt);
    } finally {
        this.closeStatement(stmt);
    }

    return var6;
}
```

再来对比一下 BatchExecutor 的 doUpate()实现：

```java
public int doUpdate(MappedStatement ms, Object parameterObject) throws SQLException {
    Configuration configuration = ms.getConfiguration();
    StatementHandler handler = configuration.newStatementHandler(this, ms, parameterObject,
RowBounds.DEFAULT, (ResultHandler)null, (BoundSql)null);
    BoundSql boundSql = handler.getBoundSql();
    String sql = boundSql.getSql();
    Statement stmt;
    if(sql.equals(this.currentSql) && ms.equals(this.currentStatement)) {
        int last = this.statementList.size() - 1;
        stmt = (Statement)this.statementList.get(last);
        this.applyTransactionTimeout(stmt);
        handler.parameterize(stmt);
        BatchResult batchResult = (BatchResult)this.batchResultList.get(last);
        batchResult.addParameterObject(parameterObject);
    } else {
        Connection connection = this.getConnection(ms.getStatementLog());
        stmt = handler.prepare(connection, this.transaction.getTimeout());
        handler.parameterize(stmt);
        this.currentSql = sql;
        this.currentStatement = ms;
        this.statementList.add(stmt);
        this.batchResultList.add(new BatchResult(ms, sql, parameterObject));
    }

    handler.batch(stmt);
    return -2147482646;
}
```

细心的"小伙伴"一定看出差异了。当然，这里暂时不对 MyBatis 源码进行深入分析，感兴趣的"小伙伴"可以关注我们的相关课程。

2.8.4 模板模式的优缺点

模板模式的优点如下：

（1）利用模板模式将相同处理逻辑的代码放到抽象父类中，可以提高代码的复用性。

（2）将不同的代码放到不同的子类中，通过对子类的扩展增加新的行为，可以提高代码的扩展性。

（3）把不变的行为写在父类中，去除子类的重复代码，提供了一个很好的代码复用平台，符合开闭原则。

模板模式的缺点如下：

（1）每个抽象类都需要一个子类来实现，导致了类的数量增加。

(2)类数量的增加间接地增加了系统的复杂性。

(3)因为继承关系自身的缺点,如果父类添加新的抽象方法,所有子类都要改一遍。

模板模式比较简单,相信"小伙伴们"肯定能学会,也肯定能理解好!只要勤加练习,多结合业务场景思考问题,就能够把模板模式运用好。

2.9 适配器模式详解

2.9.1 适配器模式的应用场景

适配器模式(Adapter Pattern)是指将一个类的接口转换成用户期望的另一个接口,使原本接口不兼容的类可以一起工作,属于结构型设计模式。

适配器模式适用于以下几种业务场景:

(1)已经存在的类的方法和需求不匹配(方法结果相同或相似)的情况。

(2)适配器模式不是软件初始阶段考虑的设计模式,是随着软件的发展,由于不同产品、不同厂家造成功能类似而接口不同的问题的解决方案,有点亡羊补牢的感觉。

生活中也有类似的应用场景,例如电源插座转换头、手机充电转换头、显示器转接头,如下图所示。

两脚插座转三角插座　　　　　　手机充电转换头　　　　　　显示器转接头

在中国,民用电都是220V交流电,但手机锂电池使用的是5V直流电。因此,我们给手机充电时就需要使用电源适配器来进行转换。下面用代码来还原这个生活场景,创建AC220类,表示220V交流电:

```
package com.gupaoedu.vip.pattern.adapter.objectadapter;

public class AC220 {
```

```java
    public int outputAC220V(){
        int output = 220;
        System.out.println("输出交流电"+output+"V");
        return output;
    }
}
```

创建 DC5 接口，表示 5V 直流电：

```java
package com.gupaoedu.vip.pattern.adapter.objectadapter;

public interface DC5 {
    int outputDC5V();
}
```

创建电源适配器类 PowerAdapter：

```java
package com.gupaoedu.vip.pattern.adapter.objectadapter;

public class PowerAdapter implements DC5{
    private AC220 ac220;
    public PowerAdapter(AC220 ac220){
        this.ac220 = ac220;
    }
    public int outputDC5V() {
        int adapterInput = ac220.outputAC220V();
        //变压器...
        int adapterOutput = adapterInput/44;
        System.out.println("使用 PowerAdapter 输入 AC:"+adapterInput+"V"+"输出 DC:"+adapterOutput+"V");
        return adapterOutput;
    }
}
```

客户端测试代码如下：

```java
package com.gupaoedu.vip.pattern.adapter.objectadapter;

public class ObjectAdapterTest {
    public static void main(String[] args) {
        DC5 dc5 = new PowerAdapter(new AC220());
        dc5.outputDC5V();
    }
}
```

在上面的案例中，通过增加电源适配器类 PowerAdapter 实现了二者的兼容。

2.9.2 重构第三方登录自由适配的业务场景

下面我们来看一个实际的业务场景，利用适配模式来解决实际问题。年纪稍微大一点的"小伙伴"一定经历过这样一个过程。我们很早以前开发的老系统都有登录接口，但是随着业务的发展和社会的进步，单纯地依赖用户名和密码登录显然不能满足用户的需求了。现在，大部分系统都已经支持多种登录方式，如 QQ 登录、微信登录、手机登录、微博登录等，同时保留用户名和密码的登录方式。虽然登录形式丰富了，但是登录后台的处理逻辑可以不改，同样是将登录状态保存到 session，遵循开闭原则。

创建统一的返回结果类 ResultMsg：

```java
package com.gupaoedu.vip.pattern.adapter.loginadapter;

public class ResultMsg {

    private int code;
    private String msg;
    private Object data;

    public ResultMsg(int code, String msg, Object data) {
        this.code = code;
        this.msg = msg;
        this.data = data;
    }

    public int getCode() {
        return code;
    }

    public void setCode(int code) {
        this.code = code;
    }

    public String getMsg() {
        return msg;
    }

    public void setMsg(String msg) {
        this.msg = msg;
    }

    public Object getData() {
        return data;
    }
}
```

```
    public void setData(Object data) {
        this.data = data;
    }
}
```

假设老系统的登录代码如下:

```
package com.gupaoedu.vip.pattern.adapter.loginadapter.v1.service;

import com.gupaoedu.vip.pattern.adapter.loginadapter.Member;
import com.gupaoedu.vip.pattern.adapter.loginadapter.ResultMsg;

public class SiginService {

    /**
     * 注册方法
     * @param username
     * @param password
     * @return
     */
    public ResultMsg regist(String username,String password){
        return new ResultMsg(200,"注册成功",new Member());
    }


    /**
     * 登录的方法
     * @param username
     * @param password
     * @return
     */
    public ResultMsg login(String username,String password){
        return null;
    }
}
```

为了遵循开闭原则,我们不修改老系统的代码。下面开始重构代码,先创建 Member 类:

```
package com.gupaoedu.vip.pattern.adapter.loginadapter;

public class Member {

    private String username;
    private String password;
    private String mid;
    private String info;
```

```java
    public String getUsername() {
        return username;
    }

    public void setUsername(String username) {
        this.username = username;
    }

    public String getPassword() {
        return password;
    }

    public void setPassword(String password) {
        this.password = password;
    }

    public String getMid() {
        return mid;
    }

    public void setMid(String mid) {
        this.mid = mid;
    }

    public String getInfo() {
        return info;
    }

    public void setInfo(String info) {
        this.info = info;
    }
}
```

再创建一个新的类继承原来的代码：

```java
package com.gupaoedu.vip.pattern.adapter.loginadapter.v1.service;

import com.gupaoedu.vip.pattern.adapter.loginadapter.ResultMsg;

//稳定的方法直接继承下来
public class SigninForThirdService extends SiginService {

    public ResultMsg loginForQQ(String openId){
        //1. openId 是全局唯一的，我们可以把它当作一个用户名（加长）
        //2. 密码默认为 QQ_EMPTY
        //3. 注册（在原有系统里面创建一个用户）
```

```java
        //4.调用原来的登录方法
        return loginForRegist(openId,null);
    }

    public ResultMsg loginForWechat(String openId){
        return null;
    }

    public ResultMsg loginForToken(String token){
        //通过Token获取用户信息,然后重新登录一次
        return null;
    }

    public ResultMsg loginForTelphone(String telphone,String code){

        return null;
    }

    public ResultMsg loginForRegist(String username,String password){
        super.regist(username,null);
        return super.login(username,null);
    }
}
```

客户端测试代码如下:

```java
package com.gupaoedu.vip.pattern.adapter.loginadapter.v1;

import com.gupaoedu.vip.pattern.adapter.loginadapter.v1.service.SigninForThirdService;

public class SigninForThirdServiceTest {

    public static void main(String[] args) {

        SigninForThirdService service = new SigninForThirdService();

        //不改变原来的代码,也能够兼容新的需求,还可以再加一层策略模式
        service.loginForQQ("sdfgdgfwresdf9123sdf");

    }
}
```

通过这么一个简单的过程,就完成了代码的兼容。当然,我们的代码还可以更加优雅,根据不同的登录方式创建不同的"Adapter"。

首先，创建 LoginAdapter 接口：

```java
package com.gupaoedu.vip.pattern.adapter.loginadapter.v2.adapters;

import com.gupaoedu.vip.pattern.adapter.loginadapter.ResultMsg;

public interface LoginAdapter {
    boolean support(Object adapter);
    ResultMsg login(String id,Object adapter);
}
```

然后，分别实现不同的登录方式，QQ 登录 LoginForQQAdapter 如下：

```java
package com.gupaoedu.vip.pattern.adapter.loginadapter.v2.adapters;

import com.gupaoedu.vip.pattern.adapter.loginadapter.ResultMsg;

public class LoginForQQAdapter implements LoginAdapter {
    public boolean support(Object adapter) {
        return adapter instanceof LoginForQQAdapter;
    }

    public ResultMsg login(String id, Object adapter) {

        return null;
    }
}
```

新浪微博登录 LoginForSinaAdapter 如下：

```java
package com.gupaoedu.vip.pattern.adapter.loginadapter.v2.adapters;

import com.gupaoedu.vip.pattern.adapter.loginadapter.ResultMsg;

public class LoginForSinaAdapter implements LoginAdapter {
    public boolean support(Object adapter) {
        return adapter instanceof LoginForSinaAdapter;
    }
    public ResultMsg login(String id, Object adapter) {
        return null;
    }
}
```

手机号登录 LoginForTelAdapter 如下：

```java
package com.gupaoedu.vip.pattern.adapter.loginadapter.v2.adapters;

import com.gupaoedu.vip.pattern.adapter.loginadapter.ResultMsg;
```

```java
public class LoginForTelAdapter implements LoginAdapter {
    public boolean support(Object adapter) {
        return adapter instanceof LoginForTelAdapter;
    }
    public ResultMsg login(String id, Object adapter) {
        return null;
    }
}
```

Token 自动登录 LoginForTokenAdapter 如下：

```java
package com.gupaoedu.vip.pattern.adapter.loginadapter.v2.adapters;

import com.gupaoedu.vip.pattern.adapter.loginadapter.ResultMsg;

public class LoginForTokenAdapter implements LoginAdapter {
    public boolean support(Object adapter) {
        return adapter instanceof LoginForTokenAdapter;
    }
    public ResultMsg login(String id, Object adapter) {
        return null;
    }
}
```

微信登录 LoginForWechatAdapter 如下：

```java
package com.gupaoedu.vip.pattern.adapter.loginadapter.v2.adapters;
import com.gupaoedu.vip.pattern.adapter.loginadapter.ResultMsg;

public class LoginForWechatAdapter implements LoginAdapter {
    public boolean support(Object adapter) {
        return adapter instanceof LoginForWechatAdapter;
    }
    public ResultMsg login(String id, Object adapter) {
        return null;
    }
}
```

接着，创建第三方登录兼容接口 IPassportForThird：

```java
package com.gupaoedu.vip.pattern.adapter.loginadapter.v2;

import com.gupaoedu.vip.pattern.adapter.loginadapter.ResultMsg;

public interface IPassportForThird {
    /**
     * QQ 登录
     * @param id
```

```java
     * @return
     */
    ResultMsg loginForQQ(String id);

    /**
     * 微信登录
     * @param id
     * @return
     */
    ResultMsg loginForWechat(String id);

    /**
     * 记住登录状态后自动登录
     * @param token
     * @return
     */
    ResultMsg loginForToken(String token);

    /**
     * 手机号登录
     * @param telphone
     * @param code
     * @return
     */
    ResultMsg loginForTelphone(String telphone, String code);

    /**
     * 注册后自动登录
     * @param username
     * @param passport
     * @return
     */
    ResultMsg loginForRegist(String username, String passport);
}
```

实现兼容 PassportForThirdAdapter：

```java
package com.gupaoedu.vip.pattern.adapter.loginadapter.v2;

import com.gupaoedu.vip.pattern.adapter.loginadapter.ResultMsg;
import com.gupaoedu.vip.pattern.adapter.loginadapter.v1.service.SiginService;
import com.gupaoedu.vip.pattern.adapter.loginadapter.v2.adapters.*;

//第三方登录自由适配
public class PassportForThirdAdapter extends SiginService implements IPassportForThird {

    public ResultMsg loginForQQ(String id) {
```

```java
        return processLogin(id,LoginForQQAdapter.class);
    }

    public ResultMsg loginForWechat(String id) {
        return processLogin(id,LoginForWechatAdapter.class);
    }

    public ResultMsg loginForToken(String token) {
        return processLogin(token,LoginForTokenAdapter.class);
    }

    public ResultMsg loginForTelphone(String telphone, String code) {
        return processLogin(telphone,LoginForTelAdapter.class);
    }

    public ResultMsg loginForRegist(String username, String passport) {
        super.regist(username,null);
        return super.login(username,null);
    }

    //这里用到了简单工厂模式及策略模式
    private ResultMsg processLogin(String key,Class<? extends LoginAdapter> clazz){
        try {
            LoginAdapter adapter = clazz.newInstance();
            if(adapter.support(adapter)) {
                return adapter.login(key, adapter);
            }else {
                return null;
            }
        }catch (Exception e){
            e.printStackTrace();;
        }
        return null;
    }
}
```

客户端测试代码如下：

```java
package com.gupaoedu.vip.pattern.adapter.loginadapter.v2;
public class PassportTest {
    public static void main(String[] args) {
        IPassportForThird passportForThird = new PassportForThirdAdapter();
        passportForThird.loginForQQ("");
    }
}
```

最后，来看一下类图，如下图所示。

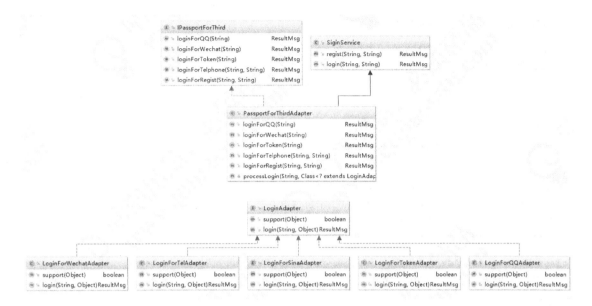

至此，我们在遵循开闭原则的前提下，完整地实现了一个兼容多平台登录方式的系统。当然，目前这个设计也并不完美，仅供参考，感兴趣的"小伙伴"可以继续完善代码。例如，适配器中的参数目前是固定为 String 的，改为 Object[]应该更合理。

学习到这里，"小伙伴们"会有一个疑问了：适配器模式跟策略模式好像区别不大？在这里要强调一下，适配器模式主要解决的是功能兼容问题，单场景适配时大家可能不会和策略模式对比，但多场景适配时大家就会产生联想和混淆。不知道大家有没有发现一个细节，前面给每个适配器都加上了一个 support()方法，用来判断是否兼容。support()方法的参数也是 Object 类型的，而 support()方法来自接口。适配器的实现并不依赖接口，我们完全可以将 LoginAdapter 接口去掉，加上它只是为了代码规范。上面的代码可以说是策略模式、简单工厂模式和适配器模式的综合运用。

2.9.3 适配器模式在源码中的体现

Spring 中适配器模式也应用得非常广泛，例如 Spring AOP 中的 AdvisorAdapter 类，它有三个实现类：MethodBeforeAdviceAdapter、AfterReturningAdviceAdapter 和 ThrowsAdviceAdapter。先来看顶层接口 AdvisorAdapter 的源代码：

```
package org.springframework.aop.framework.adapter;

import org.aopalliance.aop.Advice;
import org.aopalliance.intercept.MethodInterceptor;
```

```
import org.springframework.aop.Advisor;

public interface AdvisorAdapter {
    boolean supportsAdvice(Advice var1);
    MethodInterceptor getInterceptor(Advisor var1);
}
```

再看 MethodBeforeAdviceAdapter 类：

```
package org.springframework.aop.framework.adapter;

import java.io.Serializable;
import org.aopalliance.aop.Advice;
import org.aopalliance.intercept.MethodInterceptor;
import org.springframework.aop.Advisor;
import org.springframework.aop.MethodBeforeAdvice;

class MethodBeforeAdviceAdapter implements AdvisorAdapter, Serializable {
    MethodBeforeAdviceAdapter() {
    }

    public boolean supportsAdvice(Advice advice) {
        return advice instanceof MethodBeforeAdvice;
    }
    public MethodInterceptor getInterceptor(Advisor advisor) {
        MethodBeforeAdvice advice = (MethodBeforeAdvice)advisor.getAdvice();
        return new MethodBeforeAdviceInterceptor(advice);
    }
}
```

其他两个类这里就不看了。Spring 会根据不同的 AOP 配置来使用对应的"Advice"，与策略模式不同的是，一个方法可以同时拥有多个"Advice"。

下面再来看 Spring MVC 中的 HandlerAdapter 类，它也有多个子类，如下图所示。

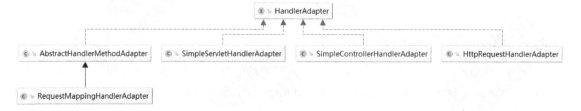

其适配器调用的关键代码也在 DispatcherServlet 的 doDispatch()方法中，下面我们来看源码：

```
protected void doDispatch(HttpServletRequest request, HttpServletResponse response) throws Exception {
```

```java
HttpServletRequest processedRequest = request;
HandlerExecutionChain mappedHandler = null;
boolean multipartRequestParsed = false;
WebAsyncManager asyncManager = WebAsyncUtils.getAsyncManager(request);

try {
    try {
        ModelAndView mv = null;
        Object dispatchException = null;

        try {
            processedRequest = this.checkMultipart(request);
            multipartRequestParsed = processedRequest != request;
            mappedHandler = this.getHandler(processedRequest);
            if(mappedHandler == null) {
                this.noHandlerFound(processedRequest, response);
                return;
            }

            HandlerAdapter ha = this.getHandlerAdapter(mappedHandler.getHandler());
            String method = request.getMethod();
            boolean isGet = "GET".equals(method);
            if(isGet || "HEAD".equals(method)) {
                long lastModified = ha.getLastModified(request, mappedHandler.getHandler());
                if(this.logger.isDebugEnabled()) {
                    this.logger.debug("Last-Modified value for [" + getRequestUri(request) + "] is: " + lastModified);
                }

                if((new ServletWebRequest(request, response)).checkNotModified(lastModified) && isGet) {
                    return;
                }
            }

            if(!mappedHandler.applyPreHandle(processedRequest, response)) {
                return;
            }

            mv = ha.handle(processedRequest, response, mappedHandler.getHandler());
            if(asyncManager.isConcurrentHandlingStarted()) {
                return;
            }

            this.applyDefaultViewName(processedRequest, mv);
            mappedHandler.applyPostHandle(processedRequest, response, mv);
        } catch (Exception var20) {
```

```
            dispatchException = var20;
        } catch (Throwable var21) {
            dispatchException = new NestedServletException("Handler dispatch failed", var21);
        }

        this.processDispatchResult(processedRequest, response, mappedHandler, mv,
(Exception)dispatchException);
    } catch (Exception var22) {
        this.triggerAfterCompletion(processedRequest, response, mappedHandler, var22);
    } catch (Throwable var23) {
        this.triggerAfterCompletion(processedRequest, response, mappedHandler, new
NestedServletException("Handler processing failed", var23));
    }

} finally {
    if(asyncManager.isConcurrentHandlingStarted()) {
        if(mappedHandler != null) {
            mappedHandler.applyAfterConcurrentHandlingStarted(processedRequest, response);
        }
    } else if(multipartRequestParsed) {
        this.cleanupMultipart(processedRequest);
    }
}
}
```

在 doDispatch()方法中调用了 getHandlerAdapter()方法,来看代码:

```
protected HandlerAdapter getHandlerAdapter(Object handler) throws ServletException {
    if(this.handlerAdapters != null) {
        Iterator var2 = this.handlerAdapters.iterator();

        while(var2.hasNext()) {
            HandlerAdapter ha = (HandlerAdapter)var2.next();
            if(this.logger.isTraceEnabled()) {
                this.logger.trace("Testing handler adapter [" + ha + "]");
            }

            if(ha.supports(handler)) {
                return ha;
            }
        }
    }

    throw new ServletException("No adapter for handler [" + handler + "]: The DispatcherServlet
configuration needs to include a HandlerAdapter that supports this handler");
}
```

在 getHandlerAdapter()方法中循环调用了 supports()方法判断是否兼容,循环迭代集合中的

"Adapter"在初始化时早已赋值，这里不再深入。

2.9.4　适配器模式的优缺点

适配器模式的优点如下：

（1）能提高类的透明性和复用性，现有的类会被复用但不需要改变。

（2）目标类和适配器类解耦，可以提高程序的扩展性。

（3）在很多业务场景中符合开闭原则。

适配器模式的缺点如下：

（1）在适配器代码编写过程中需要进行全面考虑，可能会增加系统的复杂性。

（2）增加了代码的阅读难度，降低了代码的可读性，过多使用适配器会使系统的代码变得凌乱。

2.10　装饰者模式详解

2.10.1　装饰者模式的应用场景

装饰者模式（Decorator Pattern）是指在不改变原有对象的基础上，将功能附加到对象上，提供了比继承更有弹性的方案（扩展原有对象的功能），属于结构型模式。装饰者模式在生活中的应用也比较多，如给煎饼加鸡蛋、给蛋糕加一些水果、给房子装修等，都是在为对象扩展一些额外的职责。装饰者模式适用于以下场景：

（1）扩展一个类的功能或给一个类添加附加职责。

（2）动态给一个对象添加功能，这些功能可以再动态地撤销。

来看一个这样的场景。上班族大多有睡懒觉的习惯，每天早上上班都"踩着点"，于是很多"小伙伴"为了多赖一会儿床都不吃早餐。也有些"小伙伴"可能在上班路上碰到卖煎饼的路边摊儿，会带一个到公司茶水间吃。卖煎饼的大姐可以给你的煎饼加鸡蛋，也可以加香肠（我买煎饼一般都要求不加生菜）。

下面我们用代码还原一下这个生活场景。首先创建一个煎饼类 Battercake：

```
package com.gupaoedu.vip.pattern.decorator.battercake.v1;

public class Battercake {
```

```java
    protected String getMsg(){
        return "煎饼";
    }

    public int getPrice(){
        return 5;
    }
}
```

然后创建一个加鸡蛋的煎饼类 BattercakeWithEgg：

```java
package com.gupaoedu.vip.pattern.decorator.battercake.v1;

public class BattercakeWithEgg extends Battercake{
    @Override
    protected String getMsg() {
        return super.getMsg() + "+1个鸡蛋";
    }

    @Override
    //加1个鸡蛋加1元钱
    public int getPrice() {
        return super.getPrice() + 1;
    }
}
```

再创建一个既加鸡蛋又加香肠的 BattercakeWithEggAndSausage 类：

```java
package com.gupaoedu.vip.pattern.decorator.battercake.v1;

public class BattercakeWithEggAndSausage extends BattercakeWithEgg{
    @Override
    protected String getMsg() {
        return super.getMsg() + "+1根香肠";
    }

    @Override
    //加1根香肠加2元钱
    public int getPrice() {
        return super.getPrice() + 2;
    }
}
```

编写客户端测试代码：

```java
package com.gupaoedu.vip.pattern.decorator.battercake.v1;

public class BattercakeTest {
```

```java
public static void main(String[] args) {

    Battercake battercake = new Battercake();
    System.out.println(battercake.getMsg() + ",总价格: " + battercake.getPrice());

    Battercake battercakeWithEgg = new BattercakeWithEgg();
    System.out.println(battercakeWithEgg.getMsg() + ",总价格: " + battercakeWithEgg.getPrice());

    Battercake battercakeWithEggAndSausage = new BattercakeWithEggAndSausage();
    System.out.println(battercakeWithEggAndSausage.getMsg() + ",总价格: " + battercakeWithEggAndSausage.getPrice());

}
}
```

运行结果如下图所示。

运行结果没有问题。但是，如果用户需要一个加 2 个鸡蛋、加 1 根香肠的煎饼，用我们现在的类结构是创建不出来的，也无法自动计算出价格，除非再创建一个类做定制。如果需求再变，一直做定制显然是不科学的。下面我们就用装饰者模式来解决上面的问题。首先，创建一个煎饼的抽象类 Battercake：

```java
package com.gupaoedu.vip.pattern.decorator.battercake.v2;
public abstract class Battercake {
    protected abstract String getMsg();
    protected abstract int getPrice();
}
```

然后，创建一个基本煎饼类（或者叫基础套餐）BaseBattercake：

```java
package com.gupaoedu.vip.pattern.decorator.battercake.v2;
public class BaseBattercake extends Battercake {
    protected String getMsg(){
        return "煎饼";
    }

    public int getPrice(){ return 5;  }
}
```

再创建一个扩展套餐的抽象装饰者类 BattercakeDecotator：

```
package com.gupaoedu.vip.pattern.decorator.battercake.v2;
public abstract class BattercakeDecorator extends Battercake {
    //静态代理，委派
    private Battercake battercake;

    public BattercakeDecorator(Battercake battercake) {
        this.battercake = battercake;
    }
    protected abstract void doSomething();

    @Override
    protected String getMsg() {
        return this.battercake.getMsg();
    }
    @Override
    protected int getPrice() {
        return this.battercake.getPrice();
    }
}
```

接下来，创建鸡蛋装饰者类 EggDecorator：

```
package com.gupaoedu.vip.pattern.decorator.battercake.v2;
public class EggDecorator extends BattercakeDecorator {
    public EggDecorator(Battercake battercake) {
        super(battercake);
    }

    protected void doSomething() {}

    @Override
    protected String getMsg() {
        return super.getMsg() + "+1 个鸡蛋";
    }

    @Override
    protected int getPrice() {
        return super.getPrice() + 1;
    }
}
```

最后，创建香肠装饰者类 SausageDecorator：

```
package com.gupaoedu.vip.pattern.decorator.battercake.v2;
public class SausageDecorator extends BattercakeDecorator {
    public SausageDecorator(Battercake battercake) {
```

```java
        super(battercake);
    }

    protected void doSomething() {}

    @Override
    protected String getMsg() {
        return super.getMsg() + "+1根香肠";
    }
    @Override
    protected int getPrice() {
        return super.getPrice() + 2;
    }
}
```

编写客户端测试代码：

```java
package com.gupaoedu.vip.pattern.decorator.battercake.v2;
public class BattercakeTest {
    public static void main(String[] args) {
        Battercake battercake;
        //路边摊儿买一个煎饼
        battercake = new BaseBattercake();
        //煎饼有点小，想再加1个鸡蛋
        battercake = new EggDecorator(battercake);
        //再加1个鸡蛋
        battercake = new EggDecorator(battercake);
        //很饿，再加1根香肠
        battercake = new SausageDecorator(battercake);

        //跟静态代理最大的区别就是职责不同
        //静态代理不一定满足is-a的关系
        //静态代理会做功能增强，使同一个职责变得不一样

        //装饰者模式更多考虑扩展
        System.out.println(battercake.getMsg() + ",总价： " + battercake.getPrice());
    }
}
```

运行结果如下图所示。

```
Run  BattercakeTest (1)
   煎饼+1个鸡蛋+1个鸡蛋+1根香肠,总价: 9

   Process finished with exit code 0
```

来看一下类图，如下图所示。

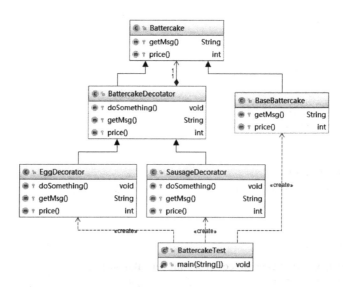

为了加深印象,我们再来看一个应用场景。在讲适配器模式时,为了实现新功能与老功能的兼容,创建了一个新的类继承已有的类,实现功能的扩展,遵循开闭原则。现在我们再用装饰者模式来升级一次代码,同时也做一个对比。先看一下 Member 类:

```java
package com.gupaoedu.vip.pattern.decorator.passport.old;

public class Member {

    private String username;
    private String password;
    private String mid;
    private String info;

    public String getUsername() {
        return username;
    }

    public void setUsername(String username) {
        this.username = username;
    }

    public String getPassword() {
        return password;
    }

    public void setPassword(String password) {
        this.password = password;
    }
```

```java
    public String getMid() {
        return mid;
    }

    public void setMid(String mid) {
        this.mid = mid;
    }

    public String getInfo() {
        return info;
    }

    public void setInfo(String info) {
        this.info = info;
    }
}
```

再看一下 ResultMsg 类：

```java
package com.gupaoedu.vip.pattern.decorator.passport.old;

public class ResultMsg {

    private int code;
    private String msg;
    private Object data;

    public ResultMsg(int code, String msg, Object data) {
        this.code = code;
        this.msg = msg;
        this.data = data;
    }

    public int getCode() {
        return code;
    }

    public void setCode(int code) {
        this.code = code;
    }

    public String getMsg() {
        return msg;
    }

    public void setMsg(String msg) {
```

```java
        this.msg = msg;
    }

    public Object getData() {
        return data;
    }

    public void setData(Object data) {
        this.data = data;
    }
}
```

接着看 ISigninService 接口：

```java
package com.gupaoedu.vip.pattern.decorator.passport.old;
public interface ISigninService {
    ResultMsg regist(String username, String password);
    /**
     * 登录的方法
     * @param username
     * @param password
     * @return
     */
    ResultMsg login(String username, String password);
}
```

接下来看 SigninService 实现类：

```java
package com.gupaoedu.vip.pattern.decorator.passport.old;

public class SigninService implements ISigninService {

    public ResultMsg regist(String username,String password){
        return new ResultMsg(200,"注册成功",new Member());
    }

    /**
     * 登录的方法
     * @param username
     * @param password
     * @return
     */
    public ResultMsg login(String username,String password){
        return null;
    }
}
```

下面进行升级，创建一个新的接口继承原来的接口：

```java
package com.gupaoedu.vip.pattern.decorator.passport.upgrade;

import com.gupaoedu.vip.pattern.decorator.passport.old.ISigninService;
import com.gupaoedu.vip.pattern.decorator.passport.old.ResultMsg;

public interface ISigninForThirdService extends ISigninService {

    /**
     * QQ 登录
     * @param id
     * @return
     */
    ResultMsg loginForQQ(String id);

    /**
     * 微信登录
     * @param id
     * @return
     */
    ResultMsg loginForWechat(String id);

    /**
     * 记住登录状态后自动登录
     * @param token
     * @return
     */
    ResultMsg loginForToken(String token);

    /**
     * 手机号登录
     * @param telphone
     * @param code
     * @return
     */
    ResultMsg loginForTelphone(String telphone, String code);

    /**
     * 注册后自动登录
     * @param username
     * @param passport
     * @return
     */
    ResultMsg loginForRegist(String username, String passport);

}
```

创建新的逻辑处理类 SigninForThirdService，实现新创建的接口：

```java
package com.gupaoedu.vip.pattern.decorator.passport.upgrade;

import com.gupaoedu.vip.pattern.decorator.passport.old.ISigninService;
import com.gupaoedu.vip.pattern.decorator.passport.old.ResultMsg;

/**
 * 第三方登录自由适配
 */
public class SigninForThirdService implements ISigninForThirdService {

    private ISigninService signin;

    public SigninForThirdService(ISigninService iSignin){
        this.signin = signin;
    }

    public ResultMsg regist(String username, String passport){
        return signin.regist(username,passport);
    }

    public ResultMsg login(String username,String passport){
        return  signin.login(username,passport);
    }

    public ResultMsg loginForQQ(String id) {
        return null;

    }

    public ResultMsg loginForWechat(String id) {
        return null;
    }

    public ResultMsg loginForToken(String token) {
        return null;
    }

    public ResultMsg loginForTelphone(String telphone, String code) {
        return null;
    }

    public ResultMsg loginForRegist(String username, String passport) {
        return null;
    }
}
```

客户端测试代码如下：

```
package com.gupaoedu.vip.pattern.decorator.passport;
import com.gupaoedu.vip.pattern.decorator.passport.old.SigninService;
import com.gupaoedu.vip.pattern.decorator.passport.upgrade.ISigninForThirdService;
import com.gupaoedu.vip.pattern.decorator.passport.upgrade.SigninForThirdService;

public class DecoratorTest {

    public static void main(String[] args) {

        ISigninForThirdService signinForThirdService = new SigninForThirdService(new SigninService());

        signinForThirdService.loginForQQ("xdcdfswrwsdfssdfqsdf");

        //动态增加或者覆盖原有方法时，采用装饰者模式

    }

}
```

装饰者模式最本质的特征是将原有类的附加功能抽离出来，简化原有类的逻辑。通过案例可以总结出，其实抽象的装饰者是可有可无的，具体可以根据业务模型来选择。

2.10.2 装饰者模式和适配器模式对比

装饰者模式和适配器模式都是包装模式（Wrapper Pattern），装饰者模式也是一种特殊的代理模式，二者对比如下表所示。

	装饰者模式	适配器模式
形式	是一种非常特别的适配器模式	没有层级关系，装饰者模式有层级关系
定义	装饰者和被装饰者实现同一个接口，主要目的是扩展之后依旧保留OOP关系	适配器和被适配者没有必然的联系，通常采用继承或代理的形式进行包装
关系	满足is-a的关系	满足has-a的关系
功能	注重覆盖、扩展	注重兼容、转换
设计	前置考虑	后置考虑

2.10.3 装饰者模式在源码中的应用

装饰者模式在源码中应用得也非常多，在 JDK 中体现最明显的类就是 I/O 相关的类，如 BufferedReader、InputStream、OutputStream，看一下常用的 InputStream 类的结构图，如下图所示。

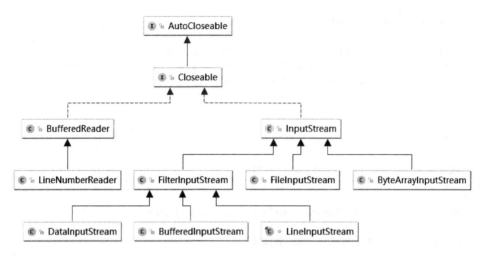

我们尝试理解一下 Spring 中的 TransactionAwareCacheDecorator 类，这个类主要是用来处理事务缓存的，来看代码：

```
public class TransactionAwareCacheDecorator implements Cache {
    private final Cache targetCache;
    public TransactionAwareCacheDecorator(Cache targetCache) {
        Assert.notNull(targetCache, "Target Cache must not be null");
        this.targetCache = targetCache;
    }
    public Cache getTargetCache() {
        return this.targetCache;
    }
    ...
}
```

TransactionAwareCacheDecorator 就是对 Cache 的一个包装。再来看一个 MVC 中的装饰者模式的 HttpHeadResponseDecorator 类：

```
public class HttpHeadResponseDecorator extends ServerHttpResponseDecorator {
    public HttpHeadResponseDecorator(ServerHttpResponse delegate) {
        super(delegate);
    }
    ...
}
```

最后看 MyBatis 中的一段处理缓存的代码，即 org.apache.ibatis.cache.Cache 类，找到它的包定位，如下图所示。

从名字来看更容易理解。比如 FifoCache 是先入先出算法的缓存，LruCache 是最近最少使用的缓存，TransactionlCache 是事务相关的缓存，它们都采用了装饰者模式。

2.10.4 装饰者模式的优缺点

装饰者模式的优点如下：

（1）装饰者模式是继承的有力补充，且比继承灵活，可以在不改变原有对象的情况下动态地给一个对象扩展功能，即插即用。

（2）使用不同的装饰类及这些装饰类的排列组合，可以实现不同的效果。

（3）装饰者模式完全符合开闭原则。

装饰者模式的缺点如下：

（1）会出现更多的代码、更多的类，增加程序的复杂性。

（2）动态装饰时，多层装饰会更复杂。

装饰者模式就讲解到这里，希望"小伙伴们"认真体会，加深理解。

2.11 观察者模式详解

2.11.1 观察者模式的应用场景

观察者模式（Observer Pattern）定义了对象之间的一对多依赖，让多个观察者对象同时监听一个主体对象，当主体对象发生变化时，它的所有依赖者（观察者）都会收到通知并更新，属于行为型模式。观察者模式有时也叫作发布订阅模式。观察者模式主要用于在关联行为之间建立一套触发机制的场景。观察者模式在现实生活应用得也非常广泛，比如微信朋友圈动态通知（如下左图所示）、GPer 生态圈消息通知（如下右图所示）、邮件通知、广播通知、桌面程序的事件响应等。

当"小伙伴们"在 GPer 生态圈中提问的时候，如果设置了指定老师回答，对应的老师就会收到邮件通知，这就是观察者模式的一种应用场景。有些"小伙伴"可能会想到 MQ、异步队列等，其实 JDK 本身就提供这样的 API。我们用代码来还原一下这个应用场景，创建 GPer 类：

```java
package com.gupaoedu.vip.pattern.observer.gperadvice;
import java.util.Observable;
public class GPer extends Observable{
    private String name = "GPer 生态圈";
    private static GPer gper = null;
    private GPer(){}

    public static GPer getInstance(){
        if(null == gper){
            gper = new GPer();
        }
        return gper;
    }
    public String getName() {
        return name;
    }
    public void publishQuestion(Question question){
        System.out.println(question.getUserName() + "在" + this.name + "上提交了一个问题。");
        setChanged();
        notifyObservers(question);
    }
}
```

创建问题类 Question：

```java
package com.gupaoedu.vip.pattern.observer.gperadvice;
public class Question {
    private String userName;
    private String content;

    public String getUserName() {
        return userName;
    }
    public void setUserName(String userName) {
        this.userName = userName;
    }
    public String getContent() {
        return content;
    }
    public void setContent(String content) {
        this.content = content;
    }
}
```

创建老师类 Teacher：

```java
package com.gupaoedu.vip.pattern.observer.gperadvice;
import java.util.Observable;
import java.util.Observer;
/**
 * 观察者
 */
public class Teacher implements Observer {
    private String name;
    public Teacher(String name){
        this.name = name;
    }
    public void update(Observable o, Object arg) {
        GPer gper = (GPer)o;
        Question question = (Question)arg;
        System.out.println("==============================");
        System.out.println(name + "老师，你好！\n" +
        "您收到了一个来自""+ gper.getName() + ""的提问，希望您解答，问题内容如下：\n" +
        question.getContent() + "\n" +
        "提问者：" + question.getUserName());
    }
}
```

客户端测试代码如下：

```java
package com.gupaoedu.vip.pattern.observer.gperadvice;
public class ObserverTest {
    public static void main(String[] args) {
```

```
        GPer gper = GPer.getInstance();
        Teacher tom = new Teacher("Tom");
        Teacher mic = new Teacher("Mic");

        gper.addObserver(tom);
        gper.addObserver(mic);

        //业务逻辑代码
        Question question = new Question();
        question.setUserName("小明");
        question.setContent("观察者设计模式适用于哪些场景？");

        gper.publishQuestion(gper,question);
    }
}
```

运行结果如下图所示。

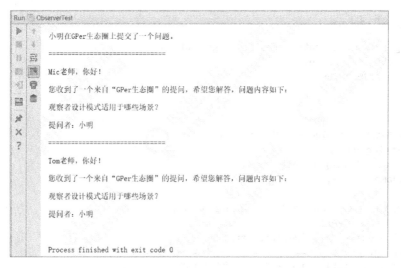

下面再来设计一个业务场景，帮助"小伙伴们"更好地理解观察者模式。在 JDK 源码中，观察者模式也应用得非常多。例如 java.awt.Event 就是观察者模式的一种，只不过 Java 很少被用来写桌面程序。我们用代码来实现一下，以帮助"小伙伴们"更深刻地了解观察者模式的实现原理。

首先，创建 Event 类：

```
package com.gupaoedu.vip.pattern.observer.events.core;

import java.lang.reflect.Method;
```

```java
/**
 * 监听器的一种包装，标准事件源格式的定义
 */
public class Event {
    //事件源，事件是由谁发起的，保存起来
    private Object source;
    //事件触发，要通知谁
    private Object target;
    //事件触发，要做什么动作，回调
    private Method callback;
    //事件的名称，触发的是什么事件
    private String trigger;
    //事件触发的时间
    private long time;

    public Event(Object target, Method callback) {
        this.target = target;
        this.callback = callback;
    }

    public Event setSource(Object source) {
        this.source = source;
        return this;
    }

    public Event setTime(long time) {
        this.time = time;
        return this;
    }

    public Object getSource() {
        return source;
    }

    public Event setTrigger(String trigger) {
        this.trigger = trigger;
        return this;
    }

    public long getTime() {
        return time;
    }

    public Object getTarget() {
        return target;
    }
```

```java
    public Method getCallback() {
        return callback;
    }

    @Override
    public String toString() {
        return "Event{" + "\n" +
                "\tsource=" + source.getClass() + ",\n" +
                "\ttarget=" + target.getClass() + ",\n" +
                "\tcallback=" + callback + ",\n" +
                "\ttrigger='" + trigger + "',\n" +
                "\ttime=" + time + "'\n" +
                '}';
    }
}
```

创建 EventLisenter 类：

```java
package com.gupaoedu.vip.pattern.observer.events.core;

import java.lang.reflect.Method;
import java.util.HashMap;
import java.util.Map;

/**
 * 监听器，它就是观察者的桥梁
 */
public class EventLisenter {

    //JDK 底层的 Lisenter 通常也是这样来设计的
    protected Map<String,Event> events = new HashMap<String,Event>();

    //通过事件名称和一个目标对象来触发事件
    public void addLisenter(String eventType,Object target){
        try {
            this.addLisenter(
                    eventType,
                    target,
                    target.getClass().getMethod("on" + toUpperFirstCase(eventType),Event.class));
        }catch (Exception e){
            e.printStackTrace();
        }
    }

    public void addLisenter(String eventType,Object target,Method callback){
        //注册事件
        events.put(eventType, new Event(target, callback));
```

```java
}

//触发,只要有动作就触发
private void trigger(Event event) {
    event.setSource(this);
    event.setTime(System.currentTimeMillis());

    try {
        //发起回调
        if(event.getCallback() != null){
            //用反射调用它的回调函数
            event.getCallback().invoke(event.getTarget(),event);
        }
    } catch (Exception e) {
        e.printStackTrace();
    }
}

//事件名称触发
protected void trigger(String trigger){
    if(!this.events.containsKey(trigger)){return;}
    trigger(this.events.get(trigger).setTrigger(trigger));
}

//逻辑处理的私有方法,首字母大写
private String toUpperFirstCase(String str){
    char[] chars = str.toCharArray();
    chars[0] -= 32;
    return String.valueOf(chars);
}

}
```

创建 MouseEventType 接口:

```java
package com.gupaoedu.vip.pattern.observer.events.mouseevent;
public interface MouseEventType {
    //单击
    String ON_CLICK = "click";

    //双击
    String ON_DOUBLE_CLICK = "doubleClick";

    //弹起
    String ON_UP = "up";
```

```java
    //按下
    String ON_DOWN = "down";

    //移动
    String ON_MOVE = "move";

    //滚动
    String ON_WHEEL = "wheel";

    //悬停
    String ON_OVER = "over";

    //失焦
    String ON_BLUR = "blur";

    //获焦
    String ON_FOCUS = "focus";
}
```

创建 Mouse 类：

```java
package com.gupaoedu.vip.pattern.observer.events.mouseevent;

import com.gupaoedu.vip.pattern.observer.events.core.EventLisenter;

public class Mouse extends EventLisenter {

    public void click(){
        System.out.println("调用单击方法");
        this.trigger(MouseEventType.ON_CLICK);
    }

    public void doubleClick(){
        System.out.println("调用双击方法");
        this.trigger(MouseEventType.ON_DOUBLE_CLICK);
    }

    public void up(){
        System.out.println("调用弹起方法");
        this.trigger(MouseEventType.ON_UP);
    }

    public void down(){
        System.out.println("调用按下方法");
        this.trigger(MouseEventType.ON_DOWN);
    }
```

```java
    public void move(){
        System.out.println("调用移动方法");
        this.trigger(MouseEventType.ON_MOVE);
    }

    public void wheel(){
        System.out.println("调用滚动方法");
        this.trigger(MouseEventType.ON_WHEEL);
    }

    public void over(){
        System.out.println("调用悬停方法");
        this.trigger(MouseEventType.ON_OVER);
    }

    public void blur(){
        System.out.println("调用获焦方法");
        this.trigger(MouseEventType.ON_BLUR);
    }

    public void focus(){
        System.out.println("调用失焦方法");
        this.trigger(MouseEventType.ON_FOCUS);
    }
}
```

创建 MouseEventCallback 类：

```java
package com.gupaoedu.vip.pattern.observer.events.mouseevent;

import com.gupaoedu.vip.pattern.observer.events.core.Event;

public class MouseEventCallback {

    public void onClick(Event e){
        System.out.println("===========触发鼠标单击事件==========" + "\n" + e);
    }

    public void onDoubleClick(Event e){
        System.out.println("===========触发鼠标双击事件==========" + "\n" + e);
    }

    public void onUp(Event e){
        System.out.println("===========触发鼠标弹起事件==========" + "\n" + e);
    }

    public void onDown(Event e){
```

```java
            System.out.println("==========触发鼠标按下事件==========" + "\n" + e);
        }

        public void onMove(Event e){
            System.out.println("==========触发鼠标移动事件==========" + "\n" + e);
        }

        public void onWheel(Event e){
            System.out.println("==========触发鼠标滚动事件==========" + "\n" + e);
        }

        public void onOver(Event e){
            System.out.println("==========触发鼠标悬停事件==========" + "\n" + e);
        }

        public void onBlur(Event e){
            System.out.println("==========触发鼠标失焦事件==========" + "\n" + e);
        }

        public void onFocus(Event e){
            System.out.println("==========触发鼠标获焦事件==========" + "\n" + e);
        }
}
```

客户端测试代码如下：

```java
package com.gupaoedu.vip.pattern.observer.events;

import com.gupaoedu.vip.pattern.observer.events.mouseevent.Mouse;
import com.gupaoedu.vip.pattern.observer.events.mouseevent.MouseEventCallback;
import com.gupaoedu.vip.pattern.observer.events.mouseevent.MouseEventType;
public class MouseEventTest {
    public static void main(String[] args) {

        try {
            MouseEventCallback callback = new MouseEventCallback();

            //注册事件
            Mouse mouse = new Mouse();
            mouse.addLisenter(MouseEventType.ON_CLICK, callback);
            mouse.addLisenter(MouseEventType.ON_MOVE, callback);
            mouse.addLisenter(MouseEventType.ON_WHEEL, callback);
            mouse.addLisenter(MouseEventType.ON_OVER, callback);

            //调用方法
            mouse.click();
```

```
            //失焦事件
            mouse.blur();

        }catch(Exception e){
            e.printStackTrace();
        }
    }
}
```

2.11.2 观察者模式在源码中的应用

Spring 中的 ContextLoaderListener 类实现了 ServletContextListener 接口，ServletContextListener 接口又继承了 EventListener 接口，在 JDK 中 EventListener 接口有非常广泛的应用。我们看一下 ContextLoaderListener 类：

```java
package org.springframework.web.context;

import javax.servlet.ServletContextEvent;
import javax.servlet.ServletContextListener;

public class ContextLoaderListener extends ContextLoader implements ServletContextListener {
    public ContextLoaderListener() {
    }

    public ContextLoaderListener(WebApplicationContext context) {
        super(context);
    }

    public void contextInitialized(ServletContextEvent event) {
        this.initWebApplicationContext(event.getServletContext());
    }

    public void contextDestroyed(ServletContextEvent event) {
        this.closeWebApplicationContext(event.getServletContext());
        ContextCleanupListener.cleanupAttributes(event.getServletContext());
    }
}
```

ServletContextListener 接口如下：

```java
package javax.servlet;
import java.util.EventListener;
public interface ServletContextListener extends EventListener {
    public void contextInitialized(ServletContextEvent sce);
    public void contextDestroyed(ServletContextEvent sce);
}
```

EventListener 接口如下：

```
package java.util;
public interface EventListener {
}
```

2.11.3　基于 Guava API 轻松落地观察者模式

这里给大家推荐一个非常好用的实现观察者模式的框架，API 使用也非常简单。举个例子，先引入 maven 依赖包：

```
<dependency>
    <groupId>com.google.guava</groupId>
    <artifactId>guava</artifactId>
    <version>20.0</version>
</dependency>
```

创建监听事件 GuavaEvent：

```
package com.gupaoedu.vip.pattern.observer.guava;

import com.google.common.eventbus.Subscribe;

public class GuavaEvent {
    @Subscribe
    public void subscribe(String str){
        //业务逻辑
        System.out.println("执行 subscribe 方法,传入的参数是:" + str);
    }
}
```

客户端测试代码如下：

```
package com.gupaoedu.vip.pattern.observer.guava;

import com.google.common.eventbus.EventBus;

public class GuavaEventTest {
    public static void main(String[] args) {
        EventBus eventbus = new EventBus();
        GuavaEvent guavaEvent = new GuavaEvent();
        eventbus.register(guavaEvent);
        eventbus.post("Tom");
    }
}
```

2.11.4 观察者模式的优缺点

观察者模式的优点如下：

（1）在观察者和被观察者之间建立了一个抽象的耦合。

（2）观察者模式支持广播通信。

观察者模式的缺点如下：

（1）观察者之间有过多的细节依赖、时间消耗多，程序的复杂性更高。

（2）使用不当会出现循环调用。

2.12 各设计模式的总结与对比

2.12.1 GoF 23 种设计模式简介

设计模式其实是一门艺术。设计模式来源于生活，不要为了套用设计模式而使用设计模式。设计模式是在我们迷茫时提供的一种解决问题的方案，或者说用好设计模式可以防范于未然。自古以来，在人生迷茫时，人们往往都会寻求帮助，或上门咨询，或查经问典。

中国人都知道：出生元婴、二十加冕、三十而立、四十不惑、五十知天命、六十花甲、七十古稀不逾矩、八九十耄耋……这就是在用模板模式，当然有些人不会选择这套模板。

设计模式是经验之谈，总结的是前人的经验，提供给后人借鉴使用，正所谓："前人栽树，后人乘凉。"设计模式可以帮助我们提升代码的可读性、可扩展性，降低维护成本，解决复杂的业务问题，但是千万不要死记硬背，生搬硬套。

GoF 23 种设计模式的归纳和总结如下表所示。

分类	设计模式
创建型	工厂方法模式、抽象工厂模式、建造者模式、原型模式、单例模式
结构型	适配器模式、桥接模式、组合模式、装饰者模式、门面模式、享元模式、代理模式
行为型	解释器模式、模板方法模式（模板模式）、责任链模式、命令模式、迭代器模式、调解者模式、备忘录模式、观察者模式、状态模式、策略模式、访问者模式

各设计模式的关联关系如下图所示。

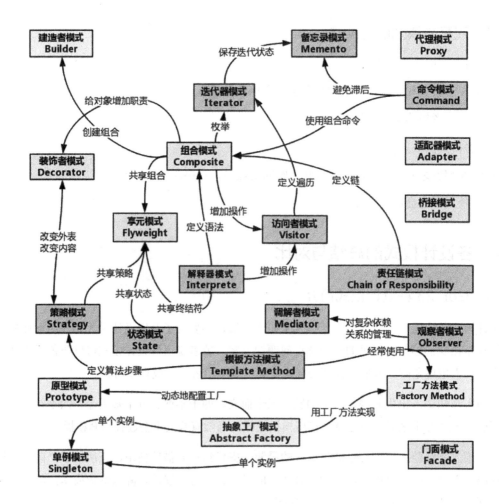

2.12.2 设计模式之间的关联关系

1. 单例模式和工厂模式

在实际业务代码中,通常会把工厂类设计为单例模式。

2. 策略模式和工厂模式

(1)工厂模式包含工厂方法模式和抽象工厂模式,是创建型模式,策略模式属于行为型模式。

(2)工厂模式的主要目的是封装好创建逻辑,策略模式接收工厂创建好的对象,从而实现不同的行为。

3. 策略模式和委派模式

（1）策略模式是委派模式内部的一种实现形式，策略模式关注结果是否能相互替代。

（2）委派模式更关注分发和调度的过程。

4. 模板方法模式和工厂方法模式

工厂方法模式是模板方法模式的一种特殊实现，二者举例如下图所示。

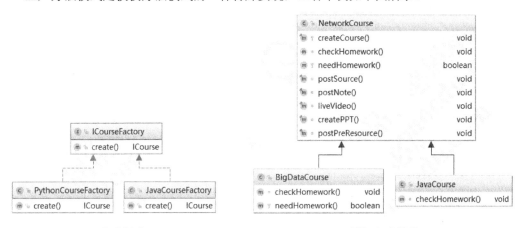

工厂方法模式　　　　　　　　　　模板方法模式

对于工厂方法模式的 create()方法而言，工厂方法模式相当于只有一个步骤的模板方法模式，这个步骤交给子类实现。而模板方法模式将 needHomework()方法和 checkHomework()方法交给子类实现，needHomework()方法和 checkHomework()方法又属于父类的某个步骤，且不可变更。

5. 模板方法模式和策略模式

（1）模板方法模式和策略模式都有封装算法。

（2）策略模式使不同算法可以相互替换，且不影响客户端应用层的使用。

（3）模板方法模式针对定义一个算法的流程，将一些有细微差异的部分交给子类实现。

（4）模板方法模式不能改变算法流程，策略模式可以改变算法流程且可替换。策略模式通常用来代替 if...else 等条件分支语句。

二者举例如下图所示。

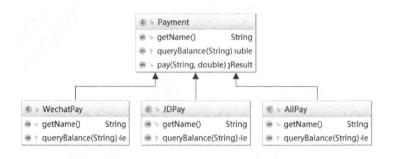

策略模式

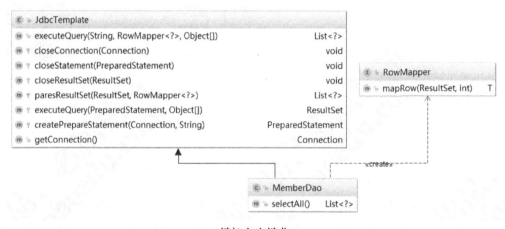

模板方法模式

上图说明如下。

（1）WechatPay、JDPay、AliPay 是交给用户选择且相互替代的解决方案，而 JdbcTemplate 下面的子类是不能相互代替的。

（2）策略模式中的 queryBalance()方法虽然在 pay()方法中也有调用，但是这只是出于程序健壮性考虑，用户完全可以自主调用 queryBalance()方法。而模板方法模式中的 mapRow()方法一定要在获得 ResultSet 值之后调用，否则没有意义。

6. 装饰者模式和代理模式

（1）装饰者模式的关注点在于给对象动态添加方法，而代理模式更加注重控制对象的访问。

（2）代理模式通常会在代理类中创建被代理对象的实例，而装饰者模式通常会把被装饰者作为构造参数。

二者举例如下图所示。

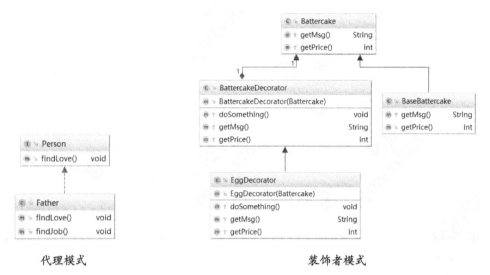

代理模式　　　　　　　　　　　装饰者模式

装饰者和代理者虽然都持有对方的引用，但处理重心是不一样的。

7．装饰者模式和适配器模式

（1）装饰者模式和适配器模式都属于包装器模式（Wrapper Pattern）。

（2）装饰者模式可以实现与被装饰者相同的接口，或者继承被装饰者作为它的子类，而适配器和被适配者可以实现不同的接口。

二者举例如下图所示。

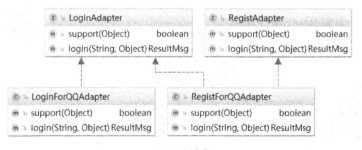

适配器模式

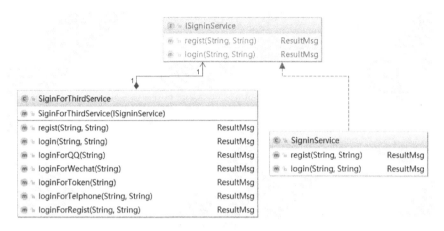

装饰者模式

装饰者和适配器都是对 SiginService 类的包装和扩展,属于装饰器模式的实现形式。但是装饰者需要满足 OOP 的 is-a 的关系,我们讲过煎饼的例子,不管如何包装都有共同的父类。适配器主要解决兼容问题,不一定要统一父类,上图适配器模式中的 LoginAdapter 类和 RegistAdapter 类就是兼容不同功能的两个类,但 RegistForQQAdapter 类需要注册后自动登录,因此既继承了 RegistAdpter 类又继承了 LoginAdapter 类。

8. 适配器模式和静态代理模式

适配器模式可以结合静态代理模式来实现,保存被适配对象的引用,但不是唯一的实现方式。

9. 适配器模式和策略模式

在业务复杂的情况下,可利用策略模式优化适配器模式。

2.12.3 Spring 中常用的设计模式

Spring 中常用的设计模式如下表所示。

设计模式	一句话归纳	举例
工厂模式	只对结果负责,封装创建过程	BeanFactory、Calender
单例模式	保证独一无二	ApplicationContext、Calender
原型模式	拔一根猴毛,吹出千万个猴子	ArrayList、PrototypeBean
代理模式	找人办事,增强职责	ProxyFactoryBean、JdkDynamicAopProxy、CglibAopProxy
委派模式	干活算你的(普通员工),功劳算我的(一些项目经理)	DispatcherServlet、BeanDefinitionParserDelegate

续表

设计模式	一句话归纳	举例
策略模式	用户选择，结果统一	InstantiationStrategy
模板模式	流程标准化，自己实现定制	JdbcTemplate、HttpServlet
适配器模式	兼容转换头	AdvisorAdapter、HandlerAdapter
装饰者模式	包装，同宗同源	BufferedReader、InputStream、OutputStream、HttpHeadResponseDecorator
观察者模式	在任务完成时通知	ContextLoaderListener

2.13 Spring 中的编程思想总结

Spring 中的编程思想总结如下表所示。

Spring思想	应用场景（特点）	一句话归纳
OOP	Object Oriented Programming（面向对象编程），用程序归纳总结生活中的一切事物	封装、继承、多态
BOP	Bean Oriented Programming（面向Bean编程），面向Bean（普通的Java类）设计程序，解放程序员	一切从Bean开始
AOP	Aspect Oriented Programming（面向切面编程），找出多个类中有一定规律的代码，开发时拆开，运行时再合并。面向切面编程即面向规则编程	解耦，专人做专事
IoC	Inversion of Control（控制反转），将new对象的动作交给Spring管理，并由Spring保存已创建的对象（IoC容器）	转交控制权（即控制权反转）
DI/DL	Dependency Injection（依赖注入）或者Dependency Lookup（依赖查找），Spring不仅保存自己创建的对象，而且保存对象与对象之间的关系。注入即赋值，主要有三种方式：构造方法、set方法、直接赋值	赋值

希望"小伙伴们"通过设计模式的系统学习，修炼好内功，在以后的源码生涯中不再"晕车"。

第 2 篇
Spring 环境预热

第 3 章　Spring 的前世今生
第 4 章　Spring 5 系统架构
第 5 章　Spring 版本命名规则
第 6 章　Spring 源码下载及构建技巧

第 3 章
Spring 的前世今生

相信经历过不使用框架开发 Web 程序的 70 后、80 后都会有如此感触：如今的程序员开发程序太轻松了，基本只需要关心业务如何实现，通用技术问题集成框架便可。早在 2007 年，一个基于 Java 语言的开源框架正式发布，取了一个非常有活力且美好的名字，叫作 Spring。它是一个开源的轻量级 Java SE（Java 标准版本）/Java EE（Java 企业版本）开发应用框架，其目的是简化企业级应用程序开发。在传统应用程序开发中，一个完整的应用程序是由一组相互协作的对象组成的。所以开发一个应用程序除了要开发业务逻辑，还要关注这些对象如何协作来实现所需功能，而且要低耦合、高聚合。业务逻辑开发是不可避免的，如果有个框架来帮我们创建对象及管理这些对象之间的依赖关系就好了。可能有人说了，抽象工厂模式、工厂方法模式可以帮我们创建对象，建造者模式可以帮我们处理对象之间的依赖关系。可是这些又需要创建另一些工厂类、建造者类，我们又要管理这些类，增加了我们的负担。如果能通过配置的方式来创建对象、管理对象之间依赖关系，我们不需要通过工厂和建造者来创建及管理对象之间的依赖关系，可以减少许多工作，加速开发，节省时间。Spring 框架刚出来时主要就是干这个的。

Spring 框架除了帮我们管理对象及其依赖关系，还提供通用日志记录、性能统计、安全控制、异常处理等面向切面的能力，能帮我们管理最头疼的数据库事务，它本身提供了一套简单的 JDBC 访问实现，可与第三方数据访问框架集成（如 Hibernate、JPA），与各种 Java EE 技术整合（如

Java Mail、任务调度等），提供一套自己的 Web 层框架 Spring MVC，还能非常简单地与第三方 Web 框架集成。Spring 是一个超级黏合大平台，除了自己提供功能，还提供黏合其他技术和框架的能力，从而使我们可以更自由地选择使用什么技术进行开发。而且不管是 Java SE（C/S 架构）应用程序还是 Java EE（B/S 架构）应用程序，都可以使用这个平台进行开发。如今的 Spring 已经不再是一个框架，早已成为一个生态，Spring Boot 的便捷式开发实现了零配置，Spring Cloud 全家桶提供了非常方便的解决方案。接下来，让我们来深入探讨 Spring 到底能给我们带来什么。

3.1 一切从 Bean 开始

说到 Bean，还得从 Java 的起源说起。早在 1996 年，Java 还只是一门新兴的、初出茅庐的编程语言。人们之所以关注它，仅仅是因为可以使用 Java 的 Applet 来开发 Web 应用，并作为浏览器组件。但开发者很快就发现这种新兴的语言还能做更多的事情。与之前的所有语言不同，Java 让模块化构建复杂的系统成为可能（当时的软件行业虽然在业务上突飞猛进，但用的是传统的面向过程开发思想，软件的开发效率一直不高。伴随着业务复杂性的不断增加，开发也变得越发困难。其实，当时也是面向对象思想飞速发展的时期，它在 20 世纪 80 年代末被提出，成熟于 20 世纪 90 年代，现今大多数编程语言都已经是面向对象的）。

同年 12 月，Sun 公司发布了当时还名不见经传但后来人尽皆知的 JavaBean 1.00-A 规范。早期的 JavaBean 规范针对 Java 定义了软件组件模型。这个规范规定了一整套编码策略，使简单的 Java 对象不仅可以被重用，而且还可以轻松地构建更为复杂的应用。尽管 JavaBean 最初是为重用应用组件而设计的，但当时却主要用作构建窗体控件，毕竟在 PC 时代那才是主流。但相比当时正如日中天的 Delphi、VB 和 C++，它看起来还是太简易了，以至于无法胜任任何"实际的"工作。

复杂的应用通常需要事务、安全、分布式等的支持，但 JavaBean 并未直接提供这些支持。所以到了 1998 年 3 月，Sun 公司发布了 EJB 1.0 规范，该规范把 Java 组件的设计理念延伸到了服务器端，并提供了许多企业级服务，但它也不再像早期的 JavaBean 那么简单了。实际上，除了名字叫 EJB（企业级 JavaBean），其他的和 JavaBean 关系不大。

尽管现实中有很多系统是基于 EJB 构建的，但 EJB 从来没有实现它最初的目标：简化开发。EJB 的声明式编程模型的确简化了一些基础架构层面的开发，例如事务和安全，但另一方面 EJB 在部署描述符和配套代码实现等方面变得异常复杂。随着时间的推移，很多开发者对 EJB 不再抱有幻想，开始寻求更简洁的方法。

然后 Java 组件开发理念重新回归正轨。新的编程技术 AOP 和 DI 的出现，为 JavaBean 提供了之前 EJB 拥有的强大功能。这些技术为 POJO 提供了类似 EJB 的声明式编程模型，而没有引入任何 EJB 的复杂性。当简单的 JavaBean 足以胜任时，人们便不愿编写笨重的 EJB 组件了。

客观地讲，EJB 的发展甚至促进了基于 POJO 的编程模型的发展。引入新的理念，最新的 EJB 规范相比之前的规范有了前所未有的改变，但对很多开发者而言，这一切的一切都来得太迟了。到了 EJB 3.0 规范发布时，其他基于 POJO 的开发框架已经成为事实的标准了，而 Spring 框架也就是在这样的大环境下出现的。

3.2 Spring 的设计初衷

Spring 是为降低企业级应用开发的复杂性而设计的，它可以做很多事。但归根到底支撑 Spring 的仅仅是少许的基本理念，所有的这些基本理念都可以追溯到一个最根本的使命：简化开发。这是一个郑重的承诺，其实许多框架都声称在某些方面做了简化，而 Spring 则立志于全方位地简化 Java 开发。对此，它主要采取了 4 个关键策略：

（1）基于 POJO 的轻量级和最小侵入性编程。

（2）通过依赖注入和面向接口实现松耦合。

（3）基于切面和惯性进行声明式编程。

（4）通过切面和模板减少样板式代码。

以上策略主要是通过面向 Bean（BOP）、依赖注入（DI）及面向切面（AOP）这三种方式来实现的。

3.3 BOP 编程伊始

Spring 是面向 Bean 的编程（Bean Oriented Programming，BOP），Bean 在 Spring 中才是真正的主角。Bean 对于 Spring 的意义就像 Object 对于 OOP 的意义一样，Spring 中没有 Bean 也就没有 Spring 存在的意义。Spring IoC 容器通过配置文件或者注解的方式来管理对象之间的依赖关系。

控制反转（其中最常见的实现方式叫作依赖注入，还有一种方式叫依赖查找，在 C++、Java、PHP 及.NET 中都有运用。在最早的 Spring 中是包含依赖注入和依赖查找的，但因为依赖查找使用频率过低，不久就被 Spring 移除了，所以在 Spring 中控制反转也被直接称作依赖注入）的基本概念是：不创建对象，但是描述创建它们的方式。在代码中不直接与对象和服务连接，但在配置文件中描述哪一个组件需要哪一项服务。容器（在 Spring 框架中是 IoC 容器）负责将这些联系在一起。

在典型的 IoC 场景中，容器创建了所有对象，并设置必要的属性将它们连接在一起，决定什么时间调用方法。

3.4 理解 BeanFactory

Spring 的设计核心 org.springframework.beans 包（架构核心是 org.springframework.core 包）的设计目标是与 JavaBean 组件一起使用。这个包通常不由用户直接使用，而是由服务器将其用作其他功能的底层中介。Spring 的最高级抽象是 BeanFactory 接口，它是工厂模式的实现，允许通过名称创建和检索对象。BeanFactory 也可以管理对象之间的关系。

BeanFactory 底层支持两个对象模型。

（1）单例模型：提供了具有特定名称的全局共享实例对象，可以在查询时对其进行检索。Singleton 是默认的、也是最常用的单例模型。

（2）原型模型：确保每次检索都会创建单独的实例对象。在每个用户都需要自己的对象时，采用原型模型。

BeanFactory（Bean 工厂）是 Spring 作为 IoC 容器的基础。IoC 则将处理事情的责任从应用程序代码转移到框架。

3.5 AOP 编程理念

AOP 即面向切面编程，是一种编程思想，它允许程序员对横切关注点或横切典型的职责分界线的行为（例如日志和事务管理）进行模块化。AOP 的核心构造是切面，它将那些影响多个类的行为封装到可重用的模块中。

AOP 和 IoC 是补充性的技术，它们都运用模块化方式解决企业应用程序开发中的复杂问题。在典型的面向对象开发方式中，可能要将日志记录语句放在所有方法和 Java 类中才能实现日志功能。在 AOP 方式中，可以反过来将日志服务模块化，并以声明的方式将它们应用到需要日志的组件上。当然，优势就是 Java 类不需要知道日志服务的存在，也不需要考虑相关的代码。所以，用 Spring AOP 编写的应用程序代码是松耦合的。

AOP 的功能完全集成到了 Spring 事务、日志和其他各种特性的上下文中。

AOP 编程的常用场景有：Authentication（权限认证）、Auto Caching（自动缓存）、Error Handling（错误处理）、Debugging（调试）、Logging（日志）、Transaction（事务）等。

第 4 章
Spring 5 系统架构

Spring 大约有 20 个模块，由 1300 多个不同的文件构成。这些模块可以分为核心容器、AOP 和设备支持、数据访问与集成、Web 组件、通信报文和集成测试、集成兼容等类。Spring 5 的模块结构如下图所示。

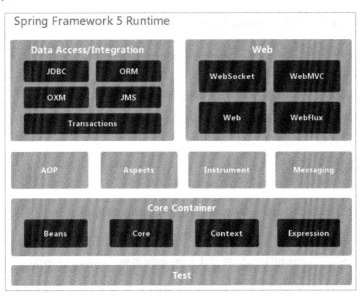

组成 Spring 框架的每个模块都可以单独存在，也可以将一个或多个模块联合实现。下面分别介绍每类模块的组成和功能。

4.1 核心容器

核心容器由 spring-beans、spring-core、spring-context 和 spring-expression（Spring Expression Language，SpEL）4 个模块组成。

spring-beans 和 spring-core 模块是 Spring 框架的核心模块，包含了控制反转（Inversion of Control，IOC）和依赖注入（Dependency Injection，DI）。BeanFactory 使用控制反转对应用程序的配置和依赖性规范与实际的应用程序代码进行了分离。但 BeanFactory 实例化后并不会自动实例化 Bean，只有当 Bean 被使用时，BeanFactory 才会对该 Bean 进行实例化与依赖关系的装配。

spring-context 模块构架于核心模块之上，扩展了 BeanFactory，为它添加了 Bean 生命周期控制、框架事件体系及资源加载透明化等功能。此外，该模块还提供了许多企业级支持，如邮件访问、远程访问、任务调度等，ApplicationContext 是该模块的核心接口，它的超类是 BeanFactory。与 BeanFactory 不同，ApplicationContext 实例化后会自动对所有的单实例 Bean 进行实例化与依赖关系的装配，使之处于待用状态。

spring-context-support 模块是对 Spring IoC 容器及 IoC 子容器的扩展支持。

spring-context-indexer 模块是 Spring 的类管理组件和 Classpath 扫描组件。

spring-expression 模块是统一表达式语言（EL）的扩展模块，可以查询、管理运行中的对象，同时也可以方便地调用对象方法，以及操作数组、集合等。它的语法类似于传统 EL，但提供了额外的功能，最出色的要数函数调用和简单字符串的模板函数。EL 的特性是基于 Spring 产品的需求而设计的，可以非常方便地同 Spring IoC 进行交互。

4.2 AOP 和设备支持

AOP 和设备支持由 spring-aop、spring-aspects 和 spring-instrument 3 个模块组成。

spring-aop 是 Spring 的另一个核心模块，是 AOP 主要的实现模块。作为继 OOP 后对程序员影响最大的编程思想之一，AOP 极大地拓展了人们的编程思路。Spring 以 JVM 的动态代理技术为基础，设计出了一系列的 AOP 横切实现，比如前置通知、返回通知、异常通知等。同时，Pointcut

接口可以匹配切入点，可以使用现有的切入点来设计横切面，也可以扩展相关方法根据需求进行切入。

spring-aspects 模块集成自 AspectJ 框架，主要是为 Spring 提供多种 AOP 实现方法。

spring-instrument 模块是基于 Java SE 中的 java.lang.instrument 进行设计的，应该算 AOP 的一个支援模块，主要作用是在 JVM 启用时生成一个代理类，程序员通过代理类在运行时修改类的字节，从而改变一个类的功能，实现 AOP。

4.3 数据访问与集成

数据访问与集成由 spring-jdbc、spring-tx、spring-orm、spring-oxm 和 spring-jms 5 个模块组成。

spring-jdbc 模块是 Spring 提供的 JDBC 抽象框架的主要实现模块，用于简化 Spring JDBC 操作。主要提供 JDBC 模板方式、关系数据库对象化方式、SimpleJdbc 方式、事务管理来简化 JDBC 编程，主要实现类有 JdbcTemplate、SimpleJdbcTemplate 及 NamedParameterJdbcTemplate。

spring-tx 模块是 Spring JDBC 事务控制实现模块。Spring 对事务做了很好的封装，通过它的 AOP 配置，可以灵活地在任何一层配置。但是在很多需求和应用中，直接使用 JDBC 事务控制还是有优势的。事务是以业务逻辑为基础的，一个完整的业务应该对应业务层里的一个方法，如果业务操作失败，则整个事务回滚，所以事务控制是应该放在业务层的。持久层的设计则应该遵循一个很重要的原则：保证操作的原子性，即持久层里的每个方法都应该是不可分割的。在使用 Spring JDBC 控制事务时，应该注意其特殊性。

spring-orm 模块是 ORM 框架支持模块，主要集成 Hibernate，Java Persistence API（JPA）和 Java Data Objects（JDO）用于资源管理、数据访问对象（DAO）的实现和事务策略。

spring-oxm 模块主要提供一个抽象层以支撑 OXM（OXM 是 Object-to-XML-Mapping 的缩写，它是一个 O/M-mapper，将 Java 对象映射成 XML 数据，或者将 XML 数据映射成 Java 对象），例如 JAXB、Castor、XMLBeans、JiBX 和 XStream 等。

spring-jms 模块能够发送和接收信息，自 Spring 4.1 开始，它还提供了对 spring-messaging 模块的支撑。

4.4 Web 组件

Web 组件由 spring-web、spring-webmvc、spring-websocket 和 spring-webflux 4 个模块组成。

spring-web 模块为 Spring 提供了最基础的 Web 支持，主要建立在核心容器之上，通过 Servlet 或者 Listeners 来初始化 IoC 容器，也包含一些与 Web 相关的支持。

众所周知，spring-webmvc 模块是一个 Web-Servlet 模块，实现了 Spring MVC（Model-View-Controller）的 Web 应用。

spring-websocket 模块是与 Web 前端进行全双工通信的协议。

spring-webflux 是一个新的非堵塞函数式 Reactive Web 框架，可以用来建立异步的、非阻塞的、事件驱动的服务，并且扩展性非常好。

4.5 通信报文

通信报文即 spring-messaging 模块，它是 Spring 4 新加入的一个模块，主要职责是为 Spring 框架集成一些基础的报文传送应用。

4.6 集成测试

集成测试即 spring-test 模块，主要为测试提供支持，使得在不需要将程序发布到应用服务器或者连接到其他设施的情况下能够进行一些集成测试或者其他测试，这对于任何企业都是非常重要的。

4.7 集成兼容

集成兼容即 spring-framework-bom 模块，主要解决 Spring 的不同模块依赖版本不同的问题。

4.8 各模块之间的依赖关系

Spring 官网对 Spring 5 各模块之间的关系做了详细说明，如下图所示。

第 4 章 Spring 5 系统架构

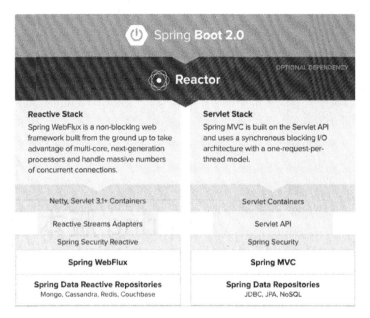

下图对 Spring 5 各模块做了一次系统的总结，描述了模块之间的依赖关系，希望能对"小伙伴们"有所帮助。

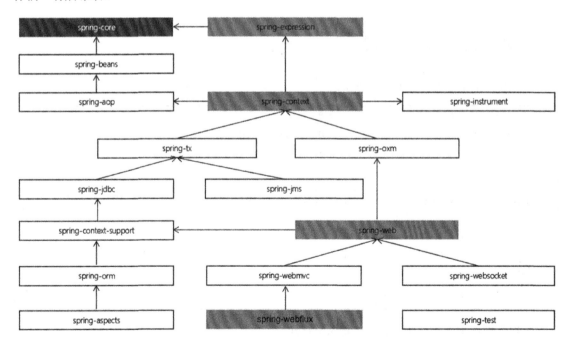

第 5 章 Spring 版本命名规则

5.1 常见软件的版本命名

常见软件的版本命名举例如下表所示。

软件	升级过程	说明
Linux Kernel	0.0.1 1.0.0 2.6.32 3.0.18	若用 X.Y.Z 表示，则偶数 Y 表示稳定版本，奇数 Y 表示开发版本
Windows	Windows 98 Windows 2000 Windows XP Windows 7	最大的特点是杂乱无章，毫无规律
SSH Client	0.9.8	
OpenStack	2014.1.3 2015.1.1.dev8	

可以看到，不同的软件的版本命名风格各异。系统的规模越大，依赖的软件越多，如果这些

软件没有遵循一套规范的命名风格，容易造成"Dependency Hell"。所以当我们发布版本时，命名需要遵循某种规则，Semantic Versioning 2.0.0 定义了一套简单的规则及条件来约束版本号的配置和增长。本书根据 Semantic Versionning 2.0.0 和 Semantic Versioning 3.0.0 选择性地整理出一些版本命名规则。

5.2 语义化版本命名通行规则

语义化版本命名通行规则对版本的迭代顺序做了很好的规范，其版本号的格式为 $X.Y.Z$（又称 Major.Minor.Patch），递增的规则如下表所示。

序号	格式要求	说明
X	非负整数	表示主版本号（Major），当API的兼容性发生变化时，X 必须递增
Y	非负整数	表示次版本号（Minor），当增加功能时（不影响API的兼容性），Y 必须递增
Z	非负整数	表示修订号（Patch），当修复漏洞时（不影响API的兼容性），Z 必须递增

详细的使用规则如下：

- X、Y、Z 必须为非负整数，且不得包含前导零，必须按数值递增，如 1.9.0→1.10.0→1.11.0。
- 0.Y.Z 表明软件处于初始开发阶段，意味着 API 可能不稳定；1.0.0 表明版本已有稳定的 API。
- 当 API 的兼容性发生变化时，X 必须递增，Y 和 Z 同时设置为 0；当新增功能（不影响 API 的兼容性）或者 API 被标记为 Deprecated 时，Y 必须递增，同时 Z 设置为 0；当进行漏洞修复时，Z 必须递增。
- 先行版本号（Pre-release）意味着该版本不稳定，可能存在兼容性问题，其格式为 $X.Y.Z.$[a-c][正整数]，如 1.0.0.a1、1.0.0.b99、1.0.0.c1000。
- 开发版本号常用于 CI-CD，格式为 $X.Y.Z.$dev[正整数]，如 1.0.1.dev4。
- 版本号的排序规则为依次比较主版本号、次版本号和修订号的数值，如 1.0.0<1.0.1<1.1.1<2.0.0；对于先行版本号和开发版本号，如 1.0.0.a100<1.0.0，2.1.0.dev3<2.1.0；当存在字母时，以 ASCII 的排序来比较，如 1.0.0.a1 < 1.0.0.b1。

注意：版本一经发布，不得修改其内容，有任何修改都必须发布新版本！

5.3 商业软件中常见的修饰词

商业软件中常见的修饰词如下表所示。

描述方式	说明	含义
Snapshot	快照版	尚不稳定、尚处于开发中的版本
Alpha	内部版	严重缺陷基本完成修正并通过复测，但需要完整的功能测试
Beta	测试版	相对Alpha版有很大的改进，消除了严重的错误，但还存在一些缺陷
RC	终测版	Release Candidate（最终测试），即将作为正式版发布
Demo	演示版	只集成了正式版部分功能，无法升级
SP	SP1	是Service Pack的意思，表示升级包，相信大家在windows中都见过
Release	稳定版	功能相对稳定，可以对外发行，但有时间限制
Trial	试用版	试用版，仅对部分用户发行
Full Version	完整版	即正式版，已发布
Unregistered	未注册	有功能或时间限制的版本
Standard	标准版	能满足正常使用的功能的版本
Lite	精简版	只含有正式版的核心功能
Enhance	增强版	正式版，功能优化的版本
Ultimate	旗舰版	标配版本的升级，体验更好
Professiona	专业版	针对要求更高、专业性更强的使用群体发行的版本
Free	自由版	自由免费使用的版本
Upgrade	升级版	有功能增强或修复了已知缺陷
Retail	零售版	单独发售
Cardware	共享版	公用许可证（iOS签证）
LTS	维护版	该版本需要长期维护

5.4 软件版本号使用限定

为了方便理解，版本限定的语法简述为 [范围描述]<版本号描述>，范围描述可选，必须配和版本描述确定范围，无法独立存在。

- <：小于某一版本号。
- <=：小于或等于某一版本号。
- >：大于某一版本号。
- >=：大于或等于某一版本号。
- =：等于某一版本号，没有意义和直接写该版本号一样。
- ~：基于版本号描述的最新补丁版本。
- ^：基于版本号描述的最新兼容版本。
- -：某个范围，应该出现在两个版本描述中间，实际上语法应为 <版本描述>-<版本描述>，写在此处为了统一。

严格来讲，对~和^的表述需要结合具体的包管理工具和版本号规则来确定，但是一般使用应记住如下原则：

- ^ 是确保版本兼容性时默认对次版本号的限定约束。
- ~ 是确保版本兼容性时默认对补丁号的约束。

5.5 Spring 版本命名规则

Spring 版本命名规则如下表所示。

描述方式	说明	含义
Snapshot	快照版	尚不稳定、尚处于开发中的版本
Release	稳定版	功能相对稳定，可以对外发行，但有时间限制
GA	正式版	代表广泛可用的稳定版（General Availability）
M	里程碑版	具有一些全新的功能或具有里程碑意义的版本（M是Milestone的意思）
RC	终测版	Release Candidate（最终测试），即将作为正式版发布

第 6 章
Spring 源码下载及构建技巧

6.1 Spring 5 源码下载

首先你的 JDK 需要升级到 1.8 以上版本。从 Spring 3.0 开始，Spring 源码采用 GitHub 托管，不再提供官网下载链接。这里不做过多赘述，大家可自行去 GitHub 网站下载，我使用的版本下载链接为 https://github.com/spring-projects/spring-framework/archive/v5.0.2.RELEASE.zip，下载完成后，解压源码包会看到如下图所示的文件目录。

名称	修改日期	类型	大小
.gradle	2017/12/27 14:26	文件夹	
.settings	2017/12/27 14:39	文件夹	
bin	2017/12/27 15:03	文件夹	
build	2017/12/27 14:26	文件夹	
buildSrc	2017/12/27 14:39	文件夹	
gradle	2017/11/27 18:52	文件夹	
spring-aop	2017/12/27 15:05	文件夹	
spring-aspects	2017/12/27 15:08	文件夹	
spring-beans	2017/12/27 15:06	文件夹	
spring-context	2017/12/27 15:06	文件夹	
spring-context-indexer	2017/12/27 15:06	文件夹	
spring-context-support	2017/12/27 15:06	文件夹	
spring-core	2017/12/27 15:17	文件夹	
spring-expression	2017/12/27 15:06	文件夹	
spring-framework-bom	2017/12/27 14:39	文件夹	
spring-instrument	2017/12/27 15:07	文件夹	
spring-jcl	2017/12/27 14:41	文件夹	
spring-jdbc	2017/12/27 15:07	文件夹	
spring-jms	2017/12/27 15:07	文件夹	
spring-messaging	2017/12/27 15:08	文件夹	
spring-orm	2017/12/27 15:08	文件夹	
spring-oxm	2017/12/27 15:08	文件夹	
spring-test	2017/12/27 15:08	文件夹	
spring-tx	2017/12/27 15:08	文件夹	
spring-web	2017/12/27 15:09	文件夹	
spring-webflux	2017/12/27 15:09	文件夹	
spring-webmvc	2017/12/27 15:09	文件夹	
spring-websocket	2017/12/27 15:09	文件夹	
src	2017/11/27 18:52	文件夹	
.editorconfig	2017/11/27 18:52	EDITORCONFIG ...	1 KB
.gitignore	2017/11/27 18:52	文本文档	1 KB
.mailmap	2017/11/27 18:52	MAILMAP 文件	2 KB
.project	2017/12/27 14:39	PROJECT 文件	1 KB
build.gradle	2017/11/27 18:52	GRADLE 文件	11 KB
CODE_OF_CONDUCT.adoc	2017/11/27 18:52	ADOC 文件	3 KB
CONTRIBUTING.md	2017/11/27 18:52	Markdown File	6 KB
gradle.properties	2017/11/27 18:52	PROPERTIES 文件	1 KB
gradlew	2017/11/27 18:52	文件	6 KB
gradlew.bat	2017/11/27 18:52	Windows 批处理...	3 KB
import-into-eclipse.bat	2017/11/27 18:52	Windows 批处理...	5 KB
import-into-eclipse.sh	2017/11/27 18:52	Shell Script	4 KB
import-into-idea.md	2017/11/27 18:52	Markdown File	2 KB
README.md	2017/11/27 18:52	Markdown File	3 KB
settings.gradle	2017/11/27 18:52	GRADLE 文件	1 KB

6.2　基于 Gradle 的源码构建技巧

由于从 Spirng 5 开始都采用 Gradle 编译，所以构建源码前需要先安装 Gradle。Gradle 下载地址为 https://gradle.org/releases，我使用的是 Spring 5 官方推荐的版本 Gradle 4.0，下载链接为 https://gradle.org/next-steps/?version=4.0&format=bin，下载完成后按以下步骤操作，这里以 Windows 操作系统为例。

第一步，配置环境变量，如下图所示。

第二步，添加环境变量"%GRADLE_HOME%\bin"，如下图所示。

第三步，检测环境，输入 gradle -v 命令，得到如下图所示结果。

```
------------------------------------------------------------
Gradle 4.0
------------------------------------------------------------

Build time:2017-06-14 15:11:08 UTC
Revision: 316546a5fcb4e2dfe1d6aaÜb73a4e09e8cecb5a5

Groovy:2. 4. 11
Ant:Apache Ant THD version 1. 9.6 compiled on June 29 2015
JVM:1.8.0_ 131 (Oracle Corporation 25.131-b11)
OS:Windows 10 10. 0 amd64
```

第四步，编译源码，将 cmd 切到 spring-framework-5.0.2.RELEASE 目录，运行 gradlew.bat，

如下图所示。

```
Starting a Gradle Daemon (subeequent builds will be faster)

> Task :help

Welcome to Gradle 4.3.1.

To run a build, run gradlew <task> ...

To see a list of available tasks, run gradlew tasks

To see a list of command-line options, run gradlew –help

To see more detail about a task, run gradlew help --task <task>

DUILD SUCCESSPUL in 8s
1 actionable task: 1 executed
```

第五步，转换为 Eclipse 项目，执行 import-into-eclipse.bat 命令，构建前请确保网络状态良好，按任意键继续：

```
-----------------------------------------------------------------
Spring Framework - Eclipse/STS project impart guide

This script will guide you through the process of importing the Spring
Framerork projects into Eclipse or the Spring Tool Suite STS) It is
recommended that you have a recent version of Eclipse or STS. As a bare
minimum you will need Eclipse with full Java 8 support, the AspectJ
Development Tools (AJDT), and the Groovy Compiler.

If you need to download and install Eclipse or STS, please do that now
by visiting one of the following sites:

- Eclipse dommloads: http://dowload.eclipse.org/eclipse/dowloads
- STS dowmloads: http://spring.io/tools/sts/all
- STS nightly builds:http://dist.springsource.com/snapshot/STS/nightly-distributions.html
- ADJT: http://www.eclipse.org/ajdt/downloads/
- Groovy Eclipse: https://github.com/groovy/groovy-eclipse/wiki

Otheryise. prgss enter and we'11 begin.
请按任意键继续. . .

-----------------------------------------------------------------
STEP 1: Generate subproject Eclipse metadata
```

• 203 •

```
The first step will be to generate Eclipse project metadata for each
of the spring-* subprojects. This happens via the built-in
"Gradle wrapper" script (./gradlew in this directory). If this is your
first time using the Gradle wrapper, this step may take a fewr minutes
while a Gradle distribution is downloaded for you.
```

第六步,等待构建成功(若中途出现错误,大部分情况是由于网络中断造成的,一般重试一下都能解决问题),构建成功后会出现如下信息:

```
BUILD SUOCCESSFUL in 6s
6 actionable tasks: 6 executed

------------------------------------------------------------------
STEP 4: Import root project into Eclipse/STS

Follow the project inport steps listed in step 2 above to inport the
root project.

Press. enter, when complete, and move on to the final step.
请按任意键继续. . .

------------------------------------------------------------------
STEP 5: Enable Git support for all projects

- In the Eclipse/STS Package Explorer, select all spring* projects.
- Right-click to cpen the context menu and select Team > Share Project...
- In the Share Project dialog that appears, select Git and press Next
- Check "Use or create repository in parent folder of project"
- Click Finish

When complete, you'll have Git support enabled for all projects.

You're ready to code! Goodbye!
```

到此为止,已经可以将项目导入 Eclipse 了。我们推荐使用的 IDEA 也比较智能,可以直接兼容 Eclipse 项目。接下来继续看下面的步骤。

第七步,导入 IDEA。打开 IntelliJ IDEA,单击"Import Project",在弹出的界面中选择 spring-framework-5.0.2.RELEASE 文件夹,单击"OK"按钮,再单击"Finish"按钮,如下三图所示。

第 6 章 Spring 源码下载及构建技巧

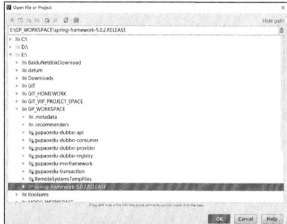

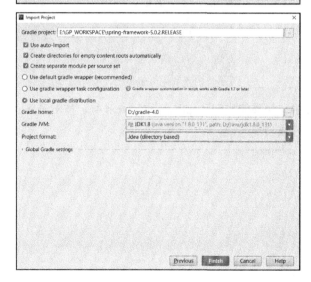

第八步，等待构建完成，在网络良好的情况下大约需要 10 分钟便可自动构建完成，你会看到如下图所示界面。

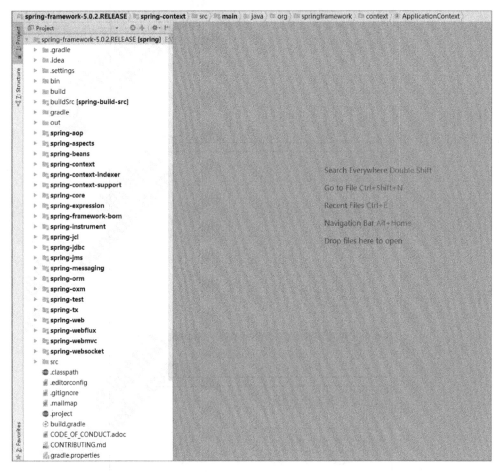

第九步，在 IDEA 中，如果能找到 ApplicationContext 类，按 Ctrl+Shift+Alt+U 键，出现如下图所示类图界面说明构建成功。

第 6 章　Spring 源码下载及构建技巧

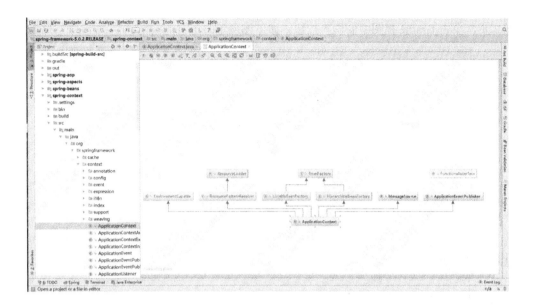

6.3　Gradle 构建过程中的坑

如果项目环境一直无法构建成功，类图无法自动生成，那么你一定是踩到了一个坑。

第一步，单击 View→Tool Windows→Gradle 命令，如下图所示。

第二步，单击 Gradle 视图中的刷新按钮，如下图所示。

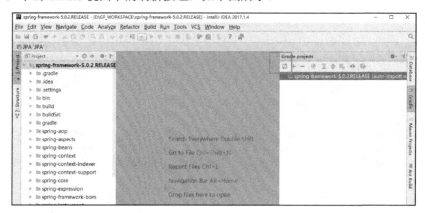

这时会出现如下图所示的错误。

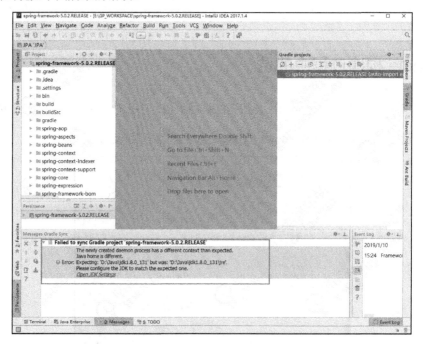

第三步，出现错误显然跟 Gradle 没有任何关系，解决办法：

（1）关闭 IDEA，打开任务管理器，结束跟 Java 有关的所有进程。

（2）找到 JAVA_HOME→jre→lib 目录，将 tools.jar 重命名为 tools.jar.bak。

（3）重启 IDEA，再次单击刷新按钮，等待构建完成。

第 3 篇
Spring 核心原理

第 7 章　用 300 行代码手写提炼 Spring 核心原理
第 8 章　一步一步手绘 Spring IoC 运行时序图
第 9 章　一步一步手绘 Spring DI 运行时序图
第 10 章　一步一步手绘 Spring AOP 运行时序图
第 11 章　一步一步手绘 Spring MVC 运行时序图

第 7 章
用 300 行代码手写提炼 Spring 核心原理

7.1 自定义配置

7.1.1 配置 application.properties 文件

为了解析方便，我们用 application.properties 来代替 application.xml 文件，具体配置内容如下：

```
scanPackage=com.gupaoedu.demo
```

7.1.2 配置 web.xml 文件

大家都知道，所有依赖于 Web 容器的项目都是从读取 web.xml 文件开始的。我们先配置好 web.xml 中的内容：

```
<?xml version="1.0" encoding="UTF-8"?>
<web-app xmlns:xsi="http://www.w3.org/2001/XMLSchema-instance"
    xmlns="http://java.sun.com/xml/ns/j2ee" xmlns:javaee="http://java.sun.com/xml/ns/javaee"
    xmlns:web="http://java.sun.com/xml/ns/javaee/web-app_2_5.xsd"
```

```xml
xsi:schemaLocation="http://java.sun.com/xml/ns/j2ee
http://java.sun.com/xml/ns/j2ee/web-app_2_4.xsd"
    version="2.4">
    <display-name>Gupao Web Application</display-name>
    <servlet>
        <servlet-name>gpmvc</servlet-name>
        <servlet-class>com.gupaoedu.mvcframework.v1.servlet.GPDispatcherServlet</servlet-class>
        <init-param>
            <param-name>contextConfigLocation</param-name>
            <param-value>application.properties</param-value>
        </init-param>
        <load-on-startup>1</load-on-startup>
    </servlet>
    <servlet-mapping>
        <servlet-name>gpmvc</servlet-name>
        <url-pattern>/*</url-pattern>
    </servlet-mapping>
</web-app>
```

其中的 GPDispatcherServlet 是模拟 Spring 实现的核心功能类。

7.1.3 自定义注解

@GPService 注解如下：

```java
package com.gupaoedu.mvcframework.annotation;
import java.lang.annotation.*;
@Target({ElementType.TYPE})
@Retention(RetentionPolicy.RUNTIME)
@Documented
public @interface GPService {
    String value() default "";
}
```

@GPAutowired 注解如下：

```java
package com.gupaoedu.mvcframework.annotation;
import java.lang.annotation.*;
@Target({ElementType.FIELD})
@Retention(RetentionPolicy.RUNTIME)
@Documented
public @interface GPAutowired {
    String value() default "";
}
```

@GPController 注解如下：

```java
package com.gupaoedu.mvcframework.annotation;
```

```
import java.lang.annotation.*;
@Target({ElementType.TYPE})
@Retention(RetentionPolicy.RUNTIME)
@Documented
public @interface GPController {
    String value() default "";
}
```

@GPRequestMapping 注解如下：

```
package com.gupaoedu.mvcframework.annotation;
import java.lang.annotation.*;
@Target({ElementType.TYPE,ElementType.METHOD})
@Retention(RetentionPolicy.RUNTIME)
@Documented
public @interface GPRequestMapping {
    String value() default "";
}
```

@GPRequestParam 注解如下：

```
package com.gupaoedu.mvcframework.annotation;
import java.lang.annotation.*;
@Target({ElementType.PARAMETER})
@Retention(RetentionPolicy.RUNTIME)
@Documented
public @interface GPRequestParam {
    String value() default "";
}
```

7.1.4 配置注解

配置业务实现类 DemoService：

```
package com.gupaoedu.demo.service.impl;
import com.gupaoedu.demo.service.IDemoService;
import com.gupaoedu.mvcframework.annotation.GPService;
/**
 * 核心业务逻辑
 */
@GPService
public class DemoService implements IDemoService{
    public String get(String name) {
        return "My name is " + name;
    }
}
```

配置请求入口类 DemoAction：

```
package com.gupaoedu.demo.mvc.action;
import java.io.IOException;
import javax.servlet.http.HttpServletRequest;
import javax.servlet.http.HttpServletResponse;
import com.gupaoedu.demo.service.IDemoService;
import com.gupaoedu.mvcframework.annotation.GPAutowired;
import com.gupaoedu.mvcframework.annotation.GPController;
import com.gupaoedu.mvcframework.annotation.GPRequestMapping;
import com.gupaoedu.mvcframework.annotation.GPRequestParam;
@GPController
@GPRequestMapping("/demo")
public class DemoAction {
    @GPAutowired private IDemoService demoService;
    @GPRequestMapping("/query")
    public void query(HttpServletRequest req, HttpServletResponse resp,
                @GPRequestParam("name") String name){
        String result = demoService.get(name);
        try {
            resp.getWriter().write(result);
        } catch (IOException e) {
            e.printStackTrace();
        }
    }
    @GPRequestMapping("/add")
    public void add(HttpServletRequest req, HttpServletResponse resp,
            @GPRequestParam("a") Integer a, @GPRequestParam("b") Integer b){
        try {
            resp.getWriter().write(a + "+" + b + "=" + (a + b));
        } catch (IOException e) {
            e.printStackTrace();
        }
    }
    @GPRequestMapping("/remove")
    public void remove(HttpServletRequest req,HttpServletResponse resp,
            @GPRequestParam("id") Integer id){
    }
}
```

至此，配置全部完成。

7.2 容器初始化

7.2.1 实现 1.0 版本

所有的核心逻辑全部写在 init() 方法中，代码如下：

```java
package com.gupaoedu.mvcframework.v1.servlet;
import com.gupaoedu.mvcframework.annotation.GPAutowired;
import com.gupaoedu.mvcframework.annotation.GPController;
import com.gupaoedu.mvcframework.annotation.GPRequestMapping;
import com.gupaoedu.mvcframework.annotation.GPService;
import javax.servlet.ServletConfig;
import javax.servlet.ServletException;
import javax.servlet.http.HttpServlet;
import javax.servlet.http.HttpServletRequest;
import javax.servlet.http.HttpServletResponse;
import java.io.File;
import java.io.IOException;
import java.io.InputStream;
import java.lang.reflect.Field;
import java.lang.reflect.Method;
import java.net.URL;
import java.util.*;

public class GPDispatcherServlet extends HttpServlet {
    private Map<String,Object> mapping = new HashMap<String, Object>();
    @Override
    protected void doGet(HttpServletRequest req, HttpServletResponse resp) throws ServletException, IOException {this.doPost(req,resp);}
    @Override
    protected void doPost(HttpServletRequest req, HttpServletResponse resp) throws ServletException, IOException {
        try {
            doDispatch(req,resp);
        } catch (Exception e) {
            resp.getWriter().write("500 Exception " + Arrays.toString(e.getStackTrace()));
        }
    }
    private void doDispatch(HttpServletRequest req, HttpServletResponse resp) throws Exception {
        String url = req.getRequestURI();
        String contextPath = req.getContextPath();
        url = url.replace(contextPath, "").replaceAll("/+", "/");
        if(!this.mapping.containsKey(url)){resp.getWriter().write("404 Not Found!!");return;}
        Method method = (Method) this.mapping.get(url);
        Map<String,String[]> params = req.getParameterMap();
        method.invoke(this.mapping.get(method.getDeclaringClass().getName()),new Object[]{req,resp,params.get("name")[0]});
    }
    @Override
    public void init(ServletConfig config) throws ServletException {
        InputStream is = null;
        try{
```

```java
            Properties configContext = new Properties();
            is = this.getClass().getClassLoader().getResourceAsStream(config.getInitParameter
("contextConfigLocation"));
            configContext.load(is);
            String scanPackage = configContext.getProperty("scanPackage");
            doScanner(scanPackage);
            for (String className : mapping.keySet()) {
                if(!className.contains(".")){continue;}
                Class<?> clazz = Class.forName(className);
                if(clazz.isAnnotationPresent(GPController.class)){
                    mapping.put(className,clazz.newInstance());
                    String baseUrl = "";
                    if (clazz.isAnnotationPresent(GPRequestMapping.class)) {
                        GPRequestMapping requestMapping = clazz.getAnnotation
(GPRequestMapping.class);
                        baseUrl = requestMapping.value();
                    }
                    Method[] methods = clazz.getMethods();
                    for (Method method : methods) {
                        if(!method.isAnnotationPresent(GPRequestMapping.class)){ continue; }
                        GPRequestMapping requestMapping = method.getAnnotation
(GPRequestMapping.class);
                        String url = (baseUrl + "/" + requestMapping.value()).replaceAll("/+", "/");
                        mapping.put(url, method);
                        System.out.println("Mapped " + url + "," + method);
                    }
                }else if(clazz.isAnnotationPresent(GPService.class)){
                    GPService service = clazz.getAnnotation(GPService.class);
                    String beanName = service.value();
                    if("".equals(beanName)){beanName = clazz.getName();}
                    Object instance = clazz.newInstance();
                    mapping.put(beanName,instance);
                    for (Class<?> i : clazz.getInterfaces()) {
                        mapping.put(i.getName(),instance);
                    }
                }else {continue;}
            }
            for (Object object : mapping.values()) {
                if(object == null){continue;}
                Class clazz = object.getClass();
                if(clazz.isAnnotationPresent(GPController.class)){
                    Field [] fields = clazz.getDeclaredFields();
                    for (Field field : fields) {
                        if(!field.isAnnotationPresent(GPAutowired.class)){continue; }
                        GPAutowired autowired = field.getAnnotation(GPAutowired.class);
                        String beanName = autowired.value();
                        if("".equals(beanName)){beanName = field.getType().getName();}
```

```java
                    field.setAccessible(true);
                    try {
                        field.set(mapping.get(clazz.getName()),mapping.get(beanName));
                    } catch (IllegalAccessException e) {
                        e.printStackTrace();
                    }
                }
            }
        }
    } catch (Exception e) {
    }finally {
        if(is != null){
            try {is.close();} catch (IOException e) {
                e.printStackTrace();
            }
        }
    }
    System.out.print("GP MVC Framework is init");
}
private void doScanner(String scanPackage) {
    URL url = this.getClass().getClassLoader().getResource("/" + scanPackage.replaceAll("\\.","/"));
    File classDir = new File(url.getFile());
    for (File file : classDir.listFiles()) {
        if(file.isDirectory()){ doScanner(scanPackage + "." + file.getName());}else {
            if(!file.getName().endsWith(".class")){continue;}
            String clazzName = (scanPackage + "." + file.getName().replace(".class",""));
            mapping.put(clazzName,null);
        }
    }
}
```

7.2.2 实现 2.0 版本

在 1.0 版本上进行优化，采用常用的设计模式（工厂模式、单例模式、委派模式、策略模式），将 init()方法中的代码进行封装。按照之前的实现思路，先搭基础框架，再"填肉注血"，具体代码如下：

```java
//初始化阶段
@Override
public void init(ServletConfig config) throws ServletException {

    //1. 加载配置文件
    doLoadConfig(config.getInitParameter("contextConfigLocation"));
```

```java
//2. 扫描相关的类
doScanner(contextConfig.getProperty("scanPackage"));

//3. 初始化扫描到的类,并且将它们放入 IoC 容器中
doInstance();

//4. 完成依赖注入
doAutowired();

//5. 初始化 HandlerMapping
initHandlerMapping();

System.out.println("GP Spring framework is init.");
}
```

声明全局成员变量,其中 IoC 容器就是注册时单例的具体案例:

```java
//保存 application.properties 配置文件中的内容
private Properties contextConfig = new Properties();

//保存扫描的所有的类名
private List<String> classNames = new ArrayList<String>();

//传说中的 IoC 容器,我们来揭开它的神秘面纱
//为了简化程序,暂时不考虑 ConcurrentHashMap
//主要还是关注设计思想和原理
private Map<String,Object> ioc = new HashMap<String,Object>();

//保存 url 和 Method 的对应关系
private Map<String,Method> handlerMapping = new HashMap<String,Method>();
```

实现 doLoadConfig()方法:

```java
//加载配置文件
private void doLoadConfig(String contextConfigLocation) {
    //直接通过类路径找到 Spring 主配置文件所在的路径
    //并且将其读取出来放到 Properties 对象中
    //相当于将 scanPackage=com.gupaoedu.demo 保存到了内存中
    InputStream fis = this.getClass().getClassLoader().getResourceAsStream(contextConfigLocation);
    try {
        contextConfig.load(fis);
    } catch (IOException e) {
        e.printStackTrace();
    }finally {
        if(null != fis){
            try {
```

```
            fis.close();
        } catch (IOException e) {
            e.printStackTrace();
        }
    }
  }
}
```

实现 doScanner()方法：

```
//扫描相关的类
private void doScanner(String scanPackage) {
    //scanPackage = com.gupaoedu.demo，存储的是包路径
    //转换为文件路径，实际上就是把.替换为/
    URL url = this.getClass().getClassLoader().getResource("/" + scanPackage.replaceAll
("\\.","/"));
    File classPath = new File(url.getFile());
    for (File file : classPath.listFiles()) {
        if(file.isDirectory()){
            doScanner(scanPackage + "." + file.getName());
        }else{
            if(!file.getName().endsWith(".class")){ continue;}
            String className = (scanPackage + "." + file.getName().replace(".class",""));
            classNames.add(className);
        }
    }
}
```

实现 doInstance()方法，doInstance()方法就是工厂模式的具体实现：

```
private void doInstance() {
    //初始化，为 DI 做准备
    if(classNames.isEmpty()){return;}

    try {
        for (String className : classNames) {
            Class<?> clazz = Class.forName(className);

            //什么样的类才需要初始化呢？
            //加了注解的类才初始化，怎么判断？
            //为了简化代码逻辑，主要体会设计思想，只用@Controller 和@Service 举例，
            //@Component 等就不一一举例了
            if(clazz.isAnnotationPresent(GPController.class)){
                Object instance = clazz.newInstance();
                //Spring 默认类名首字母小写
                String beanName = toLowerFirstCase(clazz.getSimpleName());
                ioc.put(beanName,instance);
            }else if(clazz.isAnnotationPresent(GPService.class)){
```

```java
            //1. 自定义的 beanName
            GPService service = clazz.getAnnotation(GPService.class);
            String beanName = service.value();
            //2. 默认类名首字母小写
            if("".equals(beanName.trim())){
                beanName = toLowerFirstCase(clazz.getSimpleName());
            }

            Object instance = clazz.newInstance();
            ioc.put(beanName,instance);
            //3. 根据类型自动赋值，这是投机取巧的方式
            for (Class<?> i : clazz.getInterfaces()) {
                if(ioc.containsKey(i.getName())){
                    throw new Exception("The "" + i.getName() + "" is exists!!");
                }
                //把接口的类型直接当成 key
                ioc.put(i.getName(),instance);
            }
        }else {
            continue;
        }

        }
    }catch (Exception e){
        e.printStackTrace();
    }

}
```

为了处理方便，自己实现了 toLowerFirstCase() 方法，来实现类名首字母小写，具体代码如下：

```java
//将类名首字母改为小写
private String toLowerFirstCase(String simpleName) {
    char [] chars = simpleName.toCharArray();
    //之所以要做加法，是因为大、小写字母的 ASCII 码相差 32
    //而且大写字母的 ASCII 码要小于小写字母的 ASCII 码
    //在 Java 中，对 char 做算术运算实际上就是对 ASCII 码做算术运算
    chars[0] += 32;
    return String.valueOf(chars);
}
```

实现 doAutowired() 方法：

```java
//自动进行依赖注入
private void doAutowired() {
    if(ioc.isEmpty()){return;}

    for (Map.Entry<String, Object> entry : ioc.entrySet()) {
```

```java
        //获取所有的字段，包括 private、protected、default 类型的
        //正常来说，普通的 OOP 编程只能获得 public 类型的字段
        Field[] fields = entry.getValue().getClass().getDeclaredFields();
        for (Field field : fields) {
            if(!field.isAnnotationPresent(GPAutowired.class)){continue;}
            GPAutowired autowired = field.getAnnotation(GPAutowired.class);

            //如果用户没有自定义 beanName，默认就根据类型注入
            //这个地方省去了对类名首字母小写的情况的判断，这个作为课后作业请"小伙伴们"自己去实现
            String beanName = autowired.value().trim();
            if("".equals(beanName)){
                //获得接口的类型，作为 key，稍后用这个 key 到 IoC 容器中取值
                beanName = field.getType().getName();
            }

            //如果是 public 以外的类型，只要加了@Autowired 注解都要强制赋值
            //反射中叫作暴力访问
            field.setAccessible(true);

            try {
                //用反射机制动态给字段赋值
                field.set(entry.getValue(),ioc.get(beanName));
            } catch (IllegalAccessException e) {
                e.printStackTrace();
            }


        }
    }

}
```

实现 initHandlerMapping()方法，HandlerMapping 就是策略模式的应用案例：

```java
//初始化 url 和 Method 的一对一关系
private void initHandlerMapping() {
    if(ioc.isEmpty()){ return; }

    for (Map.Entry<String, Object> entry : ioc.entrySet()) {
        Class<?> clazz = entry.getValue().getClass();

        if(!clazz.isAnnotationPresent(GPController.class)){continue;}


        //保存写在类上面的@GPRequestMapping("/demo")
```

```java
    String baseUrl = "";
    if(clazz.isAnnotationPresent(GPRequestMapping.class)){
        GPRequestMapping requestMapping = clazz.getAnnotation(GPRequestMapping.class);
        baseUrl = requestMapping.value();
    }

    //默认获取所有的 public 类型的方法
    for (Method method : clazz.getMethods()) {
        if(!method.isAnnotationPresent(GPRequestMapping.class)){continue;}

        GPRequestMapping requestMapping = method.getAnnotation(GPRequestMapping.class);
        //优化
        String url = ("/" + baseUrl + "/" + requestMapping.value())
                .replaceAll("/+","/");
        handlerMapping.put(url,method);
        System.out.println("Mapped :" + url + "," + method);

    }

    }

}
```

到这里初始化的工作完成，接下来实现运行的逻辑，来看 doGet()和 doPost()方法的代码：

```java
@Override
protected void doGet(HttpServletRequest req, HttpServletResponse resp) throws ServletException, IOException {
    this.doPost(req,resp);
}

@Override
protected void doPost(HttpServletRequest req, HttpServletResponse resp) throws ServletException, IOException {

    //运行阶段
    try {
        doDispatch(req,resp);
    } catch (Exception e) {
        e.printStackTrace();
        resp.getWriter().write("500 Exection,Detail : " + Arrays.toString(e.getStackTrace()));
    }
}
```

doPost()方法中用了委派模式，委派模式的具体逻辑在 doDispatch()方法中实现：

```java
private void doDispatch(HttpServletRequest req, HttpServletResponse resp)throws Exception {
    String url = req.getRequestURI();
    String contextPath = req.getContextPath();
    url = url.replaceAll(contextPath,"").replaceAll("/+","/");
    if(!this.handlerMapping.containsKey(url)){
        resp.getWriter().write("404 Not Found!!");
        return;
    }
    Method method = this.handlerMapping.get(url);
    //第一个参数：方法所在的实例
    //第二个参数：调用时所需要的实参

    Map<String,String[]> params = req.getParameterMap();
    //投机取巧的方式
    String beanName = toLowerFirstCase(method.getDeclaringClass().getSimpleName());
    method.invoke(ioc.get(beanName),new Object[]{req,resp,params.get("name")[0]});
    //System.out.println(method);
}
```

在以上代码中，doDispatch()虽然完成了动态委派并进行了反射调用，但对 url 参数的处理还是静态的。要实现 url 参数的动态获取，其实有些复杂。我们可以优化 doDispatch()方法的实现，代码如下：

```java
private void doDispatch(HttpServletRequest req, HttpServletResponse resp)throws Exception {
    String url = req.getRequestURI();
    String contextPath = req.getContextPath();
    url = url.replaceAll(contextPath,"").replaceAll("/+","/");
    if(!this.handlerMapping.containsKey(url)){
        resp.getWriter().write("404 Not Found!!");
        return;
    }

    Method method = this.handlerMapping.get(url);
    //第一个参数：方法所在的实例
    //第二个参数：调用时所需要的实参
    Map<String,String[]> params = req.getParameterMap();
    //获取方法的形参列表
    Class<?> [] parameterTypes = method.getParameterTypes();
    //保存请求的 url 参数列表
    Map<String,String[]> parameterMap = req.getParameterMap();
    //保存赋值参数的位置
    Object [] paramValues = new Object[parameterTypes.length];
    //根据参数位置动态赋值
    for (int i = 0; i < parameterTypes.length; i ++){
        Class parameterType = parameterTypes[i];
        if(parameterType == HttpServletRequest.class){
```

```java
            paramValues[i] = req;
            continue;
        }else if(parameterType == HttpServletResponse.class){
            paramValues[i] = resp;
            continue;
        }else if(parameterType == String.class){

            //提取方法中加了注解的参数
            Annotation[] [] pa = method.getParameterAnnotations();
            for (int j = 0; j < pa.length ; j ++) {
                for(Annotation a : pa[i]){
                    if(a instanceof GPRequestParam){
                        String paramName = ((GPRequestParam) a).value();
                        if(!"".equals(paramName.trim())){
                            String value = Arrays.toString(parameterMap.get(paramName))
                                    .replaceAll("\\[|\\]","")
                                    .replaceAll("\\s",",");
                            paramValues[i] = value;
                        }
                    }
                }
            }
        }
    }
    //投机取巧的方式
    //通过反射获得 Method 所在的 Class,获得 Class 之后还要获得 Class 的名称
    //再调用 toLowerFirstCase 获得 beanName
    String beanName = toLowerFirstCase(method.getDeclaringClass().getSimpleName());
    method.invoke(ioc.get(beanName),new Object[]{req,resp,params.get("name")[0]});
}
```

7.2.3 实现 3.0 版本

在 2.0 版本中,基本功能已经实现,但代码的优雅程度还不太高。譬如 HandlerMapping 还不能像 Spring MVC 一样支持正则,url 参数还不支持强制类型转换,在反射调用前还需要重新获取 beanName,在 3.0 版本中我们继续优化。

首先,改造 HandlerMapping,在真实的 Spring 源码中,HandlerMapping 其实是一个 List 而非 Map。List 中的元素是自定义类型的。现在我们来仿真写一段代码,先定义一个内部类 Handler:

```java
/**
 * Handler 记录 Controller 中的 RequestMapping 和 Method 的对应关系
 * @author Tom
```

```java
 * 内部类
 */
private class Handler{
    protected Object controller;    //保存方法对应的实例
    protected Method method;        //保存映射的方法
    protected Pattern pattern;
    protected Map<String,Integer> paramIndexMapping;    //参数顺序
    /**
     * 构造一个 Handler 的基本参数
     * @param controller
     * @param method
     */
    protected Handler(Pattern pattern,Object controller,Method method){
        this.controller = controller;
        this.method = method;
        this.pattern = pattern;
        paramIndexMapping = new HashMap<String,Integer>();
        putParamIndexMapping(method);
    }
    private void putParamIndexMapping(Method method){
        //提取方法中加了注解的参数
        Annotation [] [] pa = method.getParameterAnnotations();
        for (int i = 0; i < pa.length ; i ++) {
            for(Annotation a : pa[i]){
                if(a instanceof GPRequestParam){
                    String paramName = ((GPRequestParam) a).value();
                    if(!"".equals(paramName.trim())){
                        paramIndexMapping.put(paramName, i);
                    }
                }
            }
        }
        //提取方法中的 request 和 response 参数
        Class<?> [] paramsTypes = method.getParameterTypes();
        for (int i = 0; i < paramsTypes.length ; i ++) {
            Class<?> type = paramsTypes[i];
            if(type == HttpServletRequest.class ||
                type == HttpServletResponse.class){
                paramIndexMapping.put(type.getName(),i);
            }
        }
    }
}
```

然后，优化 HandlerMapping 的结构，代码如下：

```java
//保存所有的 url 和 Method 的映射关系
private List<Handler> handlerMapping = new ArrayList<Handler>();
```

修改 initHandlerMapping()方法：

```java
private void initHandlerMapping(){
    if(ioc.isEmpty()){ return; }
    for (Entry<String, Object> entry : ioc.entrySet()) {
        Class<?> clazz = entry.getValue().getClass();
        if(!clazz.isAnnotationPresent(GPController.class)){ continue; }
        String url = "";
        //获取 Controller 的 url 配置
        if(clazz.isAnnotationPresent(GPRequestMapping.class)){
            GPRequestMapping requestMapping = clazz.getAnnotation(GPRequestMapping.class);
            url = requestMapping.value();
        }
        //获取 Method 的 url 配置
        Method [] methods = clazz.getMethods();
        for (Method method : methods) {
            //没有加 RequestMapping 注解的直接忽略
            if(!method.isAnnotationPresent(GPRequestMapping.class)){ continue; }
            //映射 url
            GPRequestMapping requestMapping = method.getAnnotation(GPRequestMapping.class);
            String regex = ("/" + url + requestMapping.value()).replaceAll("/+", "/");
            Pattern pattern = Pattern.compile(regex);
            handlerMapping.add(new Handler(pattern,entry.getValue(),method));
            System.out.println("mapping " + regex + "," + method);
        }
    }
}
```

修改 doDispatch()方法：

```java
/**
 * 匹配 URL
 * @param req
 * @param resp
 * @return
 * @throws Exception
 */
private void doDispatch(HttpServletRequest req, HttpServletResponse resp) throws Exception {

        Handler handler = getHandler(req);
        if(handler == null){
//          if(!this.handlerMapping.containsKey(url)){
            resp.getWriter().write("404 Not Found!!!");
            return;
        }
```

```java
        //获得方法的形参列表
        Class<?> [] paramTypes = handler.getParamTypes();

        Object [] paramValues = new Object[paramTypes.length];

        Map<String,String[]> params = req.getParameterMap();
        for (Map.Entry<String, String[]> parm : params.entrySet()) {
            String value = Arrays.toString(parm.getValue()).replaceAll("\\[|\\]","")
                    .replaceAll("\\s",",");

            if(!handler.paramIndexMapping.containsKey(parm.getKey())){continue;}

            int index = handler.paramIndexMapping.get(parm.getKey());
            paramValues[index] = convert(paramTypes[index],value);
        }

        if(handler.paramIndexMapping.containsKey(HttpServletRequest.class.getName())) {
            int reqIndex = handler.paramIndexMapping.get(HttpServletRequest.class.getName());
            paramValues[reqIndex] = req;
        }

        if(handler.paramIndexMapping.containsKey(HttpServletResponse.class.getName())) {
            int respIndex = handler.paramIndexMapping.get(HttpServletResponse.class.getName());
            paramValues[respIndex] = resp;
        }

        Object returnValue = handler.method.invoke(handler.controller,paramValues);
        if(returnValue == null || returnValue instanceof Void){ return; }
        resp.getWriter().write(returnValue.toString());
}

private Handler getHandler(HttpServletRequest req) throws Exception{
    if(handlerMapping.isEmpty()){ return null; }
    String url = req.getRequestURI();
    String contextPath = req.getContextPath();
    url = url.replace(contextPath, "").replaceAll("/+", "/");
    for (Handler handler : handlerMapping) {
        try{
            Matcher matcher = handler.pattern.matcher(url);
            //如果没有匹配上，继续匹配下一个
            if(!matcher.matches()){ continue; }
            return handler;
        }catch(Exception e){
            throw e;
        }
    }
}
```

```
    return null;
}

//url 传过来的参数都是 String 类型的，由于 HTTP 基于字符串协议
//只需要把 String 转换为任意类型
private Object convert(Class<?> type,String value){
    if(Integer.class == type){
        return Integer.valueOf(value);
    }
    //如果还有 double 或者其他类型的参数，继续加 if
    //这时候，我们应该想到策略模式了
    //在这里暂时不实现，希望"小伙伴们"自己实现
    return value;
}
```

在以上代码中增加了两个方法：一个是 getHandler()方法，主要负责处理 url 的正则匹配；另一个是 convert()方法，主要负责 url 参数的强制类型转换。

至此，手写 Mini 版 Spring MVC 框架就全部完成了。

7.3 运行效果演示

在浏览器中输入 localhost:8080/demo/query.json?name=Tom，就会得到下面的结果：

当然，真正的 Spring 要复杂得多，本章主要通过手写的形式了解 Spring 的基本设计思路，以及设计模式的应用。

第 8 章
一步一步手绘 Spring IoC 运行时序图

8.1 Spring 核心之 IoC 容器初体验

8.1.1 再谈 IoC 与 DI

IoC（Inversion of Control，控制反转）就是把原来代码里需要实现的对象创建、依赖，反转给容器来帮忙实现。我们需要创建一个容器，同时需要一种描述来让容器知道要创建的对象与对象的关系。这个描述最具体的表现就是我们所看到的配置文件。

DI（Dependency Injection，依赖注入）就是指对象被动接受依赖类而不自己主动去找，换句话说，就是指对象不是从容器中查找它依赖的类，而是在容器实例化对象时主动将它依赖的类注入给它。

我们先从自己设计的视角来考虑。

（1）对象与对象的关系怎么表示？

可以用 XML、properties 等语义化配置文件表示。

（2）描述对象关系的文件存放在哪里？

可能是 classpath、filesystem 或者 URL 网络资源、servletContext 等。

（3）不同的配置文件对对象的描述不一样，如标准的、自定义声明式的，如何统一？

在内部需要有一个统一的关于对象的定义，所有外部的描述都必须转化成统一的描述定义。

（4）如何对不同的配置文件进行解析？

需要对不同的配置文件语法采用不同的解析器。

8.1.2 Spring 核心容器类图

1. BeanFactory

Spring 中 Bean 的创建是典型的工厂模式，这一系列的 Bean 工厂，即 IoC 容器，为开发者管理对象之间的依赖关系提供了很多便利和基础服务，在 Spring 中有许多 IoC 容器的实现供用户选择，其相互关系如下图所示。

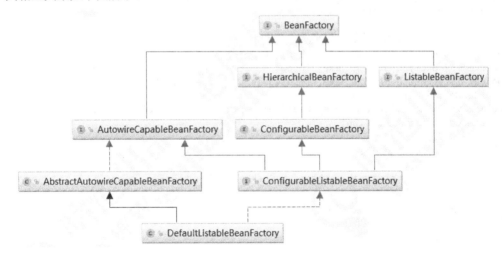

其中，BeanFactory 作为最顶层的一个接口类，定义了 IoC 容器的基本功能规范，BeanFactory 有三个重要的子类：ListableBeanFactory、HierarchicalBeanFactory 和 AutowireCapableBeanFactory。但是从类图中我们可以发现最终的默认实现类是 DefaultListableBeanFactory，它实现了所有的接口。那么为何要定义这么多层次的接口呢？查阅这些接口的源码和说明发现，每个接口都有它的使用场合，主要是为了区分在 Spring 内部操作过程中对象的传递和转化，对对象的数据访问所做的限制。例如，ListableBeanFactory 接口表示这些 Bean 可列表化，而 HierarchicalBeanFactory 表示这些

Bean 是有继承关系的,也就是每个 Bean 可能有父 Bean。AutowireCapableBeanFactory 接口定义 Bean 的自动装配规则。这三个接口共同定义了 Bean 的集合、Bean 之间的关系及 Bean 行为。最基本的 IoC 容器接口是 BeanFactory,来看一下它的源码:

```java
public interface BeanFactory {

  //对 FactoryBean 的转义定义,因为如果使用 Bean 的名字检索 FactoryBean 得到的对象是工厂生成的对象
  //如果需要得到工厂本身,需要转义
  String FACTORY_BEAN_PREFIX = "&";

  //根据 Bean 的名字,获取在 IoC 容器中得到的 Bean 实例
  Object getBean(String name) throws BeansException;

  //根据 Bean 的名字和 Class 类型来得到 Bean 实例,增加了类型安全验证机制
  <T> T getBean(String name, @Nullable Class<T> requiredType) throws BeansException;

  Object getBean(String name, Object... args) throws BeansException;
  <T> T getBean(Class<T> requiredType) throws BeansException;
  <T> T getBean(Class<T> requiredType, Object... args) throws BeansException;

  //提供对 Bean 的检索,看看在 IoC 容器中是否有这个名字的 Bean
  boolean containsBean(String name);

  //根据 Bean 的名字得到 Bean 实例,同时判断这个 Bean 是不是单例
  boolean isSingleton(String name) throws NoSuchBeanDefinitionException;
  boolean isPrototype(String name) throws NoSuchBeanDefinitionException;
  boolean isTypeMatch(String name, ResolvableType typeToMatch) throws NoSuchBeanDefinitionException;
  boolean isTypeMatch(String name, @Nullable Class<?> typeToMatch) throws NoSuchBeanDefinitionException;

  //得到 Bean 实例的 Class 类型
  @Nullable
  Class<?> getType(String name) throws NoSuchBeanDefinitionException;

  //得到 Bean 的别名,如果根据别名检索,那么其原名也会被检索出来
  String[] getAliases(String name);

}
```

在 BeanFactory 里只对 IoC 容器的基本行为做了定义,根本不关心你的 Bean 是如何定义及怎样加载的。正如我们只关心能从工厂里得到什么产品,不关心工厂是怎么生产这些产品的。

要知道工厂是如何产生对象的,我们需要看具体的 IoC 容器实现,Spring 提供了许多 IoC 容器实现,比如 GenericApplicationContext、ClasspathXmlApplicationContext 等。

ApplicationContext 是 Spring 提供的一个高级的 IoC 容器,它除了能够提供 IoC 容器的基本功

能，还为用户提供了以下附加服务。

（1）支持信息源，可以实现国际化（实现 MessageSource 接口）。

（2）访问资源（实现 ResourcePatternResolver 接口，后面章节会讲到）。

（3）支持应用事件（实现 ApplicationEventPublisher 接口）。

2. BeanDefinition

Spring IoC 容器管理我们定义的各种 Bean 对象及其相互关系，Bean 对象在 Spring 实现中是以 BeanDefinition 来描述的，其继承体系如下图所示。

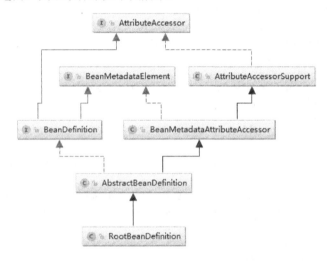

3. BeanDefinitionReader

Bean 的解析过程非常复杂，功能被分得很细，因为这里需要被扩展的地方很多，必须保证足够的灵活性，以应对可能的变化。Bean 的解析主要就是对 Spring 配置文件的解析。这个解析过程主要通过 BeanDefinitionReader 来完成，看看 Spring 中 BeanDefinitionReader 的类结构图，如下图所示。

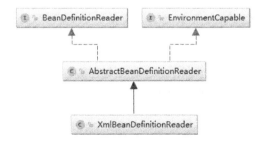

通过前面的分析，我们对 Spring 框架体系有了一个基本的宏观了解，希望"小伙伴们"好好理解，最好在脑海中形成画面，为以后的学习打下良好的基础。

8.1.3　Web IoC 容器初体验

我们还是从大家最熟悉的 DispatcherServlet 开始，最先想到的应该是 DispatcherServlet 的 init() 方法。我们在 DispatherServlet 中并没有找到 init()方法，经过探索，在其父类 HttpServletBean 中找到了，代码如下：

```java
@Override
public final void init() throws ServletException {
    if (logger.isDebugEnabled()) {
        logger.debug("Initializing servlet '" + getServletName() + "'");
    }

    PropertyValues pvs = new ServletConfigPropertyValues(getServletConfig(),
    this.requiredProperties);
    if (!pvs.isEmpty()) {
        try {
            //定位资源
            BeanWrapper bw = PropertyAccessorFactory.forBeanPropertyAccess(this);
            //加载配置信息
            ResourceLoader resourceLoader = new ServletContextResourceLoader(getServletContext());
            bw.registerCustomEditor(Resource.class, new ResourceEditor(resourceLoader,
getEnvironment()));
            initBeanWrapper(bw);
            bw.setPropertyValues(pvs, true);
        }
        catch (BeansException ex) {
            if (logger.isErrorEnabled()) {
                logger.error("Failed to set bean properties on servlet '" + getServletName() + "'", ex);
            }
            throw ex;
        }
    }

    initServletBean();

    if (logger.isDebugEnabled()) {
        logger.debug("Servlet '" + getServletName() + "' configured successfully");
    }
}
```

在 init()方法中，真正完成初始化容器动作的代码其实在 initServletBean()方法中，我们继续跟进：

```java
@Override
protected final void initServletBean() throws ServletException {
  getServletContext().log("Initializing Spring FrameworkServlet '" + getServletName() + "'");
  if (this.logger.isInfoEnabled()) {
    this.logger.info("FrameworkServlet '" + getServletName() + "': initialization started");
  }
  long startTime = System.currentTimeMillis();

  try {
    this.webApplicationContext = initWebApplicationContext();
    initFrameworkServlet();
  }
  catch (ServletException ex) {
    this.logger.error("Context initialization failed", ex);
    throw ex;
  }
  catch (RuntimeException ex) {
    this.logger.error("Context initialization failed", ex);
    throw ex;
  }

  if (this.logger.isInfoEnabled()) {
    long elapsedTime = System.currentTimeMillis() - startTime;
    this.logger.info("FrameworkServlet '" + getServletName() + "': initialization completed
        in " + elapsedTime + " ms");
  }
}
```

在上面的代码中终于看到了似曾相识的代码 initWebApplicationContext()，继续跟进：

```java
protected WebApplicationContext initWebApplicationContext() {

  //先从 ServletContext 中获得父容器 WebApplicationContext
  WebApplicationContext rootContext =
      WebApplicationContextUtils.getWebApplicationContext(getServletContext());
  //声明子容器
  WebApplicationContext wac = null;

  //建立父、子容器之间的关联关系
  if (this.webApplicationContext != null) {
    wac = this.webApplicationContext;
```

```java
        if (wac instanceof ConfigurableWebApplicationContext) {
            ConfigurableWebApplicationContext cwac = (ConfigurableWebApplicationContext) wac;
            if (!cwac.isActive()) {
                if (cwac.getParent() == null) {
                    cwac.setParent(rootContext);
                }
                configureAndRefreshWebApplicationContext(cwac);
            }
        }
    }
    //先去 ServletContext 中查找 Web 容器的引用是否存在,并创建好默认的空 IoC 容器
    if (wac == null) {
        wac = findWebApplicationContext();
    }
    //给上一步创建好的 IoC 容器赋值
    if (wac == null) {
        wac = createWebApplicationContext(rootContext);
    }
    //触发 onRefresh()方法
    if (!this.refreshEventReceived) {
        onRefresh(wac);
    }

    if (this.publishContext) {
        String attrName = getServletContextAttributeName();
        getServletContext().setAttribute(attrName, wac);
        if (this.logger.isDebugEnabled()) {
            this.logger.debug("Published WebApplicationContext of servlet '" + getServletName() +
                    "' as ServletContext attribute with name [" + attrName + "]");
        }
    }

    return wac;
}

@Nullable
protected WebApplicationContext findWebApplicationContext() {
    String attrName = getContextAttribute();
    if (attrName == null) {
        return null;
    }
    WebApplicationContext wac =
            WebApplicationContextUtils.getWebApplicationContext(getServletContext(),
```

```java
      attrName);
   if (wac == null) {
      throw new IllegalStateException("No WebApplicationContext found: initializer not registered?");
   }
   return wac;
}

protected WebApplicationContext createWebApplicationContext(@Nullable ApplicationContext parent) {
   Class<?> contextClass = getContextClass();
   if (this.logger.isDebugEnabled()) {
      this.logger.debug("Servlet with name '" + getServletName() +
         "' will try to create custom WebApplicationContext context of class '" +
         contextClass.getName() + "'" + ", using parent context [" + parent + "]");
   }
   if (!ConfigurableWebApplicationContext.class.isAssignableFrom(contextClass)) {
      throw new ApplicationContextException(
         "Fatal initialization error in servlet with name '" + getServletName() +
         "': custom WebApplicationContext class [" + contextClass.getName() +
         "] is not of type ConfigurableWebApplicationContext");
   }
   ConfigurableWebApplicationContext wac =
      (ConfigurableWebApplicationContext) BeanUtils.instantiateClass(contextClass);

   wac.setEnvironment(getEnvironment());
   wac.setParent(parent);
   String configLocation = getContextConfigLocation();
   if (configLocation != null) {
      wac.setConfigLocation(configLocation);
   }
   configureAndRefreshWebApplicationContext(wac);

   return wac;
}
protected void configureAndRefreshWebApplicationContext(ConfigurableWebApplicationContext wac) {
   if (ObjectUtils.identityToString(wac).equals(wac.getId())) {
      if (this.contextId != null) {
         wac.setId(this.contextId);
      }
      else {
         wac.setId(ConfigurableWebApplicationContext.APPLICATION_CONTEXT_ID_PREFIX +
```

```
                ObjectUtils.getDisplayString(getServletContext().getContextPath()) + '/' +
getServletName());
    }
  }

  wac.setServletContext(getServletContext());
  wac.setServletConfig(getServletConfig());
  wac.setNamespace(getNamespace());
  wac.addApplicationListener(new SourceFilteringListener(wac, new ContextRefreshListener()));

  ConfigurableEnvironment env = wac.getEnvironment();
  if (env instanceof ConfigurableWebEnvironment) {
    ((ConfigurableWebEnvironment) env).initPropertySources(getServletContext(),
 getServletConfig());
  }

  postProcessWebApplicationContext(wac);
  applyInitializers(wac);
  wac.refresh();
}
```

从上面的代码可以看出，在 configAndRefreshWebApplicationContext()方法中调用了 refresh()方法，这是真正启动 IoC 容器的入口，后面会详细介绍。IoC 容器初始化以后，调用了 DispatcherServlet 的 onRefresh()方法，在 onRefresh()方法中又直接调用 initStrategies()方法初始化 Spring MVC 的九大组件：

```
@Override
protected void onRefresh(ApplicationContext context) {
  initStrategies(context);
}

//初始化策略
protected void initStrategies(ApplicationContext context) {
  //多文件上传的组件
  initMultipartResolver(context);
  //初始化本地语言环境
  initLocaleResolver(context);
  //初始化模板处理器
  initThemeResolver(context);
  //初始化 handlerMapping
  initHandlerMappings(context);
  //初始化参数适配器
  initHandlerAdapters(context);
  //初始化异常拦截器
  initHandlerExceptionResolvers(context);
```

```
    //初始化视图预处理器
    initRequestToViewNameTranslator(context);
    //初始化视图转换器
    initViewResolvers(context);
    //初始化 Flashmap 管理器
    initFlashMapManager(context);
}
```

8.2 基于 XML 的 IoC 容器的初始化

IoC 容器的初始化包括 BeanDefinition 的 Resource 定位、加载和注册三个基本的过程。我们以 ApplicationContext 为例讲解，ApplicationContext 系列容器也是我们最熟悉的，因为 Web 项目中使用的 XmlWebApplicationContext 就属于这个系列，还有 ClasspathXmlApplicationContext 等，其继承体系如下图所示。

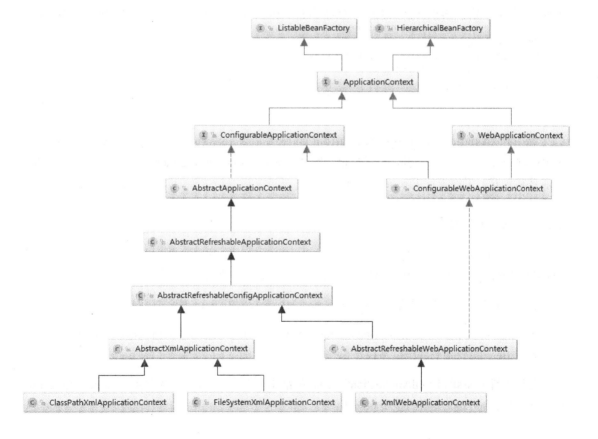

ApplicationContext 允许上下文嵌套，通过保持父上下文可以维持一个上下文体系。对于 Bean 的查找可以在这个上下文体系中进行，首先检查当前上下文，其次检查父上下文，逐级向上，这样可以为不同的 Spring 应用提供一个共享的 Bean 定义环境。

8.2.1 寻找入口

还有一个我们用得比较多的 ClassPathXmlApplicationContext，通过 main() 方法启动：

```
ApplicationContext app = new ClassPathXmlApplicationContext("application.xml");
```

先看其构造函数的调用：

```java
public ClassPathXmlApplicationContext(String configLocation) throws BeansException {
    this(new String[]{configLocation}, true, (ApplicationContext)null);
}
```

实际调用的构造函数为：

```java
public ClassPathXmlApplicationContext(String[] configLocations, boolean refresh, @Nullable ApplicationContext parent) throws BeansException {
    super(parent);
    this.setConfigLocations(configLocations);
    if(refresh) {
        this.refresh();
    }
}
```

还有 AnnotationConfigApplicationContext、FileSystemXmlApplicationContext、XmlWebApplicationContext 等，都继承自父容器 AbstractApplicationContext，主要用到了装饰者模式和策略模式，最终都调用 refresh() 方法。

8.2.2 获得配置路径

通过分析 ClassPathXmlApplicationContext 的源代码可以知道，在创建 ClassPathXmlApplicationContext 容器时，构造方法做了以下两项重要工作。

首先，调用父容器的构造方法 super(parent) 为容器设置好 Bean 资源加载器。

然后，调用父类 AbstractRefreshableConfigApplicationContext 的 setConfigLocations(configLocations) 方法设置 Bean 配置信息的定位路径。

通过追踪 ClassPathXmlApplicationContext 的继承体系，发现其父类的父类 AbstractApplicationContext 中初始化 IoC 容器的主要源码如下：

```java
public abstract class AbstractApplicationContext extends DefaultResourceLoader
    implements ConfigurableApplicationContext {
//静态初始化块,在整个容器创建过程中只执行一次
static {
    //为了避免应用程序在Weblogic 8.1关闭时出现类加载异常问题,加载IoC容器关闭事件类
    ContextClosedEvent.class.getName();
}
public AbstractApplicationContext() {
    this.resourcePatternResolver = getResourcePatternResolver();
}
public AbstractApplicationContext(@Nullable ApplicationContext parent) {
    this();
    setParent(parent);
}
//获取一个Spring Source的加载器用于读入Spring Bean配置信息
protected ResourcePatternResolver getResourcePatternResolver() {
    //AbstractApplicationContext继承DefaultResourceLoader,因此也是一个资源加载器
    //Spring资源加载器的getResource(String location)方法用于载入资源
    return new PathMatchingResourcePatternResolver(this);
}
...
}
```

在 AbstractApplicationContext 的默认构造方法中调用 PathMatchingResourcePatternResolver 的构造方法创建 Spring 资源加载器:

```java
public PathMatchingResourcePatternResolver(ResourceLoader resourceLoader) {
    Assert.notNull(resourceLoader, "ResourceLoader must not be null");
    //设置Spring的资源加载器
    this.resourceLoader = resourceLoader;
}
```

在设置容器的资源加载器之后,接下来 ClassPathXmlApplicationContext 执行 setConfigLocations() 方法,通过调用其父类 AbstractRefreshableConfigApplicationContext 的方法进行 Bean 配置信息的定位,该方法的源码如下:

```java
//处理单个资源文件路径为一个字符串的情况
public void setConfigLocation(String location) {
    //String CONFIG_LOCATION_DELIMITERS = ",; /t/n";
    //即多个资源文件路径之间用",; \t\n"分隔,解析成数组形式
    setConfigLocations(StringUtils.tokenizeToStringArray(location,
CONFIG_LOCATION_DELIMITERS));
}
//解析Bean定义资源文件的路径,处理多个资源文件字符串数组
public void setConfigLocations(@Nullable String... locations) {
    if (locations != null) {
```

```
    Assert.noNullElements(locations, "Config locations must not be null");
    this.configLocations = new String[locations.length];
    for (int i = 0; i < locations.length; i++) {
        // resolvePath 为同一个类中将字符串解析为路径的方法
        this.configLocations[i] = resolvePath(locations[i]).trim();
    }
}
else {
    this.configLocations = null;
}
```

从这两个方法的源码可以看出，我们既可以使用一个字符串来配置多个 Spring Bean 信息，也可以使用字符串数组来配置。

多个资源文件路径之间可以是用 ",; \t\n" 等分隔。

`ClassPathResource res =new ClassPathResource(new String[]{"a.xml","b.xml"});`

至此，Spring IoC 容器在初始化时将配置的 Bean 信息定位为 Spring 封装的 Resource。

8.2.3 开始启动

Spring IoC 容器对 Bean 配置资源的载入是从 refresh()方法开始的。refresh()方法是一个模板方法，规定了 IoC 容器的启动流程，有些逻辑要交给其子类实现。它对 Bean 配置资源进行载入，ClassPathXmlApplicationContext 通过调用其父类 AbstractApplicationContext 的 refresh()方法启动整个 IoC 容器对 Bean 定义的载入过程。现在我们来详细看看 refresh()方法的代码：

```
@Override
public void refresh() throws BeansException, IllegalStateException {
    synchronized (this.startupShutdownMonitor) {
        //1. 调用容器准备刷新的方法，获取容器的当前时间，同时给容器设置同步标识
        prepareRefresh();

        //2. 告诉子类启动 refreshBeanFactory()方法,Bean 定义资源文件的载入从子类的 refreshBeanFactory()方法启动
        ConfigurableListableBeanFactory beanFactory = obtainFreshBeanFactory();

        //3. 为 BeanFactory 配置容器特性，例如类加载器、事件处理器等
        prepareBeanFactory(beanFactory);

        try {
            //4. 为容器的某些子类指定特殊的 Post 事件处理器
            postProcessBeanFactory(beanFactory);
```

```
    //5. 调用所有注册的 BeanFactoryPostProcessor 的 Bean
    invokeBeanFactoryPostProcessors(beanFactory);

    //6. 为 BeanFactory 注册 Post 事件处理器
    //BeanPostProcessor 是 Bean 后置处理器,用于监听容器触发的事件
    registerBeanPostProcessors(beanFactory);

    //7. 初始化信息源,和国际化相关
    initMessageSource();

    //8. 初始化容器事件传播器
    initApplicationEventMulticaster();

    //9. 调用子类的某些特殊 Bean 的初始化方法
    onRefresh();

    //10. 为事件传播器注册事件监听器
    registerListeners();

    //11. 初始化所有剩余的单例 Bean
    finishBeanFactoryInitialization(beanFactory);

    //12. 初始化容器的生命周期事件处理器,并发布容器的生命周期事件
    finishRefresh();
}

catch (BeansException ex) {
    if (logger.isWarnEnabled()) {
        logger.warn("Exception encountered during context initialization - " +
                "cancelling refresh attempt: " + ex);
    }

    //13. 销毁已创建的 Bean
    destroyBeans();

    //14. 取消刷新操作,重置容器的同步标识
    cancelRefresh(ex);

    throw ex;
}
finally {
    //15. 重设公共缓存
    resetCommonCaches();
}
}
}
```

refresh()方法主要为 IoC 容器 Bean 的生命周期管理提供条件，Spring IoC 容器载入 Bean 配置信息从其子类容器的 refreshBeanFactory() 方法启动，所以整个 refresh() 方法中"ConfigurableListableBeanFactory beanFactory = obtainFreshBeanFactory();"以后的代码都是在注册容器的信息源和生命周期事件，我们前面说的载入就通过这句代码启动。

refresh()方法的主要作用是：在创建 IoC 容器前，如果已经有容器存在，需要把已有的容器销毁和关闭，以保证在 refresh()方法之后使用的是新创建的 IoC 容器。它类似于对 IoC 容器的重启，在新创建的容器中对容器进行初始化，对 Bean 配置资源进行载入。

8.2.4 创建容器

obtainFreshBeanFactory()方法调用子类容器的 refreshBeanFactory()方法，启动容器载入 Bean 配置信息的过程，代码如下：

```
protected ConfigurableListableBeanFactory obtainFreshBeanFactory() {
    //这里使用了委派模式，父类定义了抽象的 refreshBeanFactory()方法
    //具体实现调用子类容器的 refreshBeanFactory()方法
    refreshBeanFactory();
    ConfigurableListableBeanFactory beanFactory = getBeanFactory();
    if (logger.isDebugEnabled()) {
        logger.debug("Bean factory for " + getDisplayName() + ": " + beanFactory);
    }
    return beanFactory;
}
```

AbstractApplicationContext 类中只抽象定义了 refreshBeanFactory()方法，容器真正调用的是其子类 AbstractRefreshableApplicationContext 实现的 refreshBeanFactory()方法，方法的源码如下：

```
protected final void refreshBeanFactory() throws BeansException {
    //如果已经有容器，销毁容器中的 Bean，关闭容器
    if (hasBeanFactory()) {
        destroyBeans();
        closeBeanFactory();
    }
    try {
        //创建 IoC 容器
        DefaultListableBeanFactory beanFactory = createBeanFactory();
        beanFactory.setSerializationId(getId());
        //对 IoC 容器进行定制化，如设置启动参数、开启注解的自动装配等
        customizeBeanFactory(beanFactory);
        //调用载入 Bean 定义的方法，这里又使用了一个委派模式
        //在当前类中只定义了抽象的 loadBeanDefinitions()方法，调用子类容器实现
        loadBeanDefinitions(beanFactory);
```

```java
      synchronized (this.beanFactoryMonitor) {
         this.beanFactory = beanFactory;
      }
   }
   catch (IOException ex) {
      throw new ApplicationContextException("I/O error parsing bean definition source for "
+ getDisplayName(), ex);
   }
}
```

在这个方法中，先判断 beanFactory 是否存在，如果存在则先销毁 Bean 并关闭 beanFactory，接着创建 DefaultListableBeanFactory，并调用 loadBeanDefinitions()方法装载 Bean 定义。

8.2.5 载入配置路径

在 AbstractRefreshableApplicationContext 中只定义了抽象父类的 loadBeanDefinitions()方法，容器真正调用的是其子类 AbstractXmlApplicationContext 对该方法的实现，AbstractXmlApplicationContext 的主要源码如下。loadBeanDefinitions()方法同样是抽象方法，是由其子类实现的，即在 AbstractXmlApplicationContext 中。

```java
public abstract class AbstractXmlApplicationContext extends
AbstractRefreshableConfigApplicationContext {
   ...
   //实现父类抽象的载入 Bean 定义方法
   @Override
   protected void loadBeanDefinitions(DefaultListableBeanFactory beanFactory) throws
BeansException, IOException {
      //创建 XmlBeanDefinitionReader，即创建 Bean 读取器，
      //并通过回调设置到容器中，容器使用该读取器读取 Bean 配置资源
      XmlBeanDefinitionReader beanDefinitionReader = new XmlBeanDefinitionReader(beanFactory);

      //为 Bean 读取器设置 Spring 资源加载器
      //AbstractXmlApplicationContext 的祖先父类 AbstractApplicationContext 继承 DefaultResourceLoader
      //因此容器本身也是一个资源加载器
      beanDefinitionReader.setEnvironment(this.getEnvironment());
      beanDefinitionReader.setResourceLoader(this);
      //为 Bean 读取器设置 SAX xml 解析器
      beanDefinitionReader.setEntityResolver(new ResourceEntityResolver(this));

      //当 Bean 读取器读取 Bean 定义的 xml 资源文件时，启用 xml 的校验机制
      initBeanDefinitionReader(beanDefinitionReader);
      //Bean 读取器真正实现加载的方法
      loadBeanDefinitions(beanDefinitionReader);
```

```
        }

        protected void initBeanDefinitionReader(XmlBeanDefinitionReader reader) {
            reader.setValidating(this.validating);
        }

        //xml Bean 读取器加载 Bean 配置资源
        protected void loadBeanDefinitions(XmlBeanDefinitionReader reader) throws BeansException,
IOException {
            //获取 Bean 配置资源的定位
            Resource[] configResources = getConfigResources();
            if (configResources != null) {
                //xml Bean 读取器调用其父类 AbstractBeanDefinitionReader 读取定位的 Bean 配置资源
                reader.loadBeanDefinitions(configResources);
            }
            //如果子类中获取的 Bean 配置资源定位为空,
            //则获取 ClassPathXmlApplicationContext 构造方法中 setConfigLocations 方法设置的资源
            String[] configLocations = getConfigLocations();
            if (configLocations != null) {
                //xml Bean 读取器调用其父类 AbstractBeanDefinitionReader 读取定位的 Bean 配置资源
                reader.loadBeanDefinitions(configLocations);
            }
        }

        //这里又使用了一个委派模式,调用子类的获取 Bean 配置资源定位的方法
        //该方法在 ClassPathXmlApplicationContext 中实现
        //我们举例分析源码的 ClassPathXmlApplicationContext 没有使用该方法
        @Nullable
        protected Resource[] getConfigResources() {
            return null;
        }
    }
```

以 xml Bean 读取器的一种策略 XmlBeanDefinitionReader 为例,XmlBeanDefinitionReader 调用其父类 AbstractBeanDefinitionReader 的 reader.loadBeanDefinitions()方法读取 Bean 配置资源。

由于我们使用 ClassPathXmlApplicationContext 作为例子,getConfigResources()方法的返回值为 null,因此程序执行 reader.loadBeanDefinitions(configLocations)分支。

8.2.6　分配路径处理策略

在 XmlBeanDefinitionReader 的抽象父类 AbstractBeanDefinitionReader 中定义了载入过程。

AbstractBeanDefinitionReader 的 loadBeanDefinitions()方法源码如下：

```java
//重载方法，调用下面的 loadBeanDefinitions(String, Set<Resource>)方法
@Override
public int loadBeanDefinitions(String location) throws BeanDefinitionStoreException {
    return loadBeanDefinitions(location, null);
}

public int loadBeanDefinitions(String location, @Nullable Set<Resource> actualResources)
throws BeanDefinitionStoreException {
    //获取在 IoC 容器初始化过程中设置的资源加载器
    ResourceLoader resourceLoader = getResourceLoader();
    if (resourceLoader == null) {
        throw new BeanDefinitionStoreException(
            "Cannot import bean definitions from location [" + location + "]: no ResourceLoader available");
    }

    if (resourceLoader instanceof ResourcePatternResolver) {
        try {
            //将指定位置的 Bean 配置信息解析为 Spring IoC 容器封装的资源
            //加载多个指定位置的 Bean 配置信息
            Resource[] resources = ((ResourcePatternResolver) resourceLoader).getResources(location);
            //委派调用其子类 XmlBeanDefinitionReader 的方法，实现加载功能
            int loadCount = loadBeanDefinitions(resources);
            if (actualResources != null) {
                for (Resource resource : resources) {
                    actualResources.add(resource);
                }
            }
            if (logger.isDebugEnabled()) {
                logger.debug("Loaded " + loadCount + " bean definitions from location pattern [" + location + "]");
            }
            return loadCount;
        }
        catch (IOException ex) {
            throw new BeanDefinitionStoreException(
                "Could not resolve bean definition resource pattern [" + location + "]", ex);
        }
    }
```

```
    else {
        //将指定位置的 Bean 配置信息解析为 Spring IoC 容器封装的资源
        //加载单个指定位置的 Bean 配置信息
        Resource resource = resourceLoader.getResource(location);
        //委派调用其子类 XmlBeanDefinitionReader 的方法，实现加载功能
        int loadCount = loadBeanDefinitions(resource);
        if (actualResources != null) {
            actualResources.add(resource);
        }
        if (logger.isDebugEnabled()) {
            logger.debug("Loaded " + loadCount + " bean definitions from location [" + location
+ "]");
        }
        return loadCount;
    }
}

//重载方法，调用 loadBeanDefinitions(String)
@Override
public int loadBeanDefinitions(String... locations) throws BeanDefinitionStoreException {
    Assert.notNull(locations, "Location array must not be null");
    int counter = 0;
    for (String location : locations) {
        counter += loadBeanDefinitions(location);
    }
    return counter;
}
```

AbstractRefreshableConfigApplicationContext 的 loadBeanDefinitions(Resource...resources)方法实际上调用 AbstractBeanDefinitionReader 的 loadBeanDefinitions()方法。

从 AbstractBeanDefinitionReader 的 loadBeanDefinitions()方法的源码分析可以看出，该方法就做了两件事：首先，调用资源加载器的获取资源方法 resourceLoader.getResource(location)，获取要加载的资源；其次，真正执行加载功能，由其子类 XmlBeanDefinitionReader 的 loadBeanDefinitions()方法完成。

在 loadBeanDefinitions()方法中调用了 AbstractApplicationContext 的 getResources()方法，getResources()方法其实在 ResourcePatternResolver 中定义，此时我们有必要来看一下 ResourcePatternResolver 的类图，如下图所示。

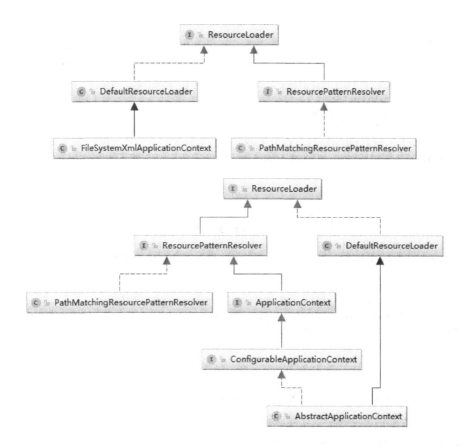

从图中可以看到 ResourceLoader 与 ApplicationContext 的继承关系，实际上调用 DefaultResourceLoader 中的 getSource() 方法定位 Resource，因为 ClassPathXmlApplicationContext 本身就是 DefaultResourceLoader 的实现类，所以此时又回到了 ClassPathXmlApplicationContext 中。

8.2.7　解析配置文件路径

XmlBeanDefinitionReader 通过调用 ClassPathXmlApplicationContext 的父类 DefaultResourceLoader 的 getResource() 方法获取要加载的资源，其源码如下：

```
//获取 Resource 的具体实现方法
@Override
public Resource getResource(String location) {
    Assert.notNull(location, "Location must not be null");

    for (ProtocolResolver protocolResolver : this.protocolResolvers) {
        Resource resource = protocolResolver.resolve(location, this);
```

```
      if (resource != null) {
         return resource;
      }
   }
   //如果是类路径的方式,使用 ClassPathResource 来得到 Bean 文件的资源对象
   if (location.startsWith("/")) {
      return getResourceByPath(location);
   }
   else if (location.startsWith(CLASSPATH_URL_PREFIX)) {
      return new ClassPathResource(location.substring(CLASSPATH_URL_PREFIX.length()),
getClassLoader());
   }
   else {
      try {
         //如果是 URL 方式,使用 UrlResource 作为 Bean 文件的资源对象
         URL url = new URL(location);
         return (ResourceUtils.isFileURL(url) ? new FileUrlResource(url) : new UrlResource(url));
      }
      catch (MalformedURLException ex) {
         //如果既不是 classpath 标识又不是 URL 标识的 Resource 定位,则调用
         //容器本身的 getResourceByPath()方法获取 Resource
         return getResourceByPath(location);
      }
   }
}
```

DefaultResourceLoader 提供了 getResourceByPath()方法的实现,就是为了处理既不是 ClassPath 标识又不是 URL 标识的 Resource 定位的情况。

```
protected Resource getResourceByPath(String path) {
   return new ClassPathContextResource(path, getClassLoader());
}
```

在 ClassPathResource 中完成了对整个路径的解析。这样,就可以从类路径上对 IoC 配置文件进行加载,当然我们可以按照这个逻辑从任何地方加载。在 Spring 中提供了各种资源抽象,比如 ClassPathResource、UrlResource、FileSystemResource 等供我们使用。上面是定位 Resource 的一个过程,而这只是加载过程的一部分,例如 FileSystemXmlApplication 容器就重写了 getResourceByPath() 方法:

```
@Override
protected Resource getResourceByPath(String path) {
   if (path.startsWith("/")) {
      path = path.substring(1);
   }
   //这里使用文件系统资源对象来定义 Bean 文件
   return new FileSystemResource(path);
}
```

它通过子类的覆盖巧妙地将类路径变成了文件路径。

8.2.8 开始读取配置内容

继续回到 XmlBeanDefinitionReader 的 loadBeanDefinitions()方法，看之后的载入过程：

```java
//XmlBeanDefinitionReader 加载资源的入口方法
@Override
public int loadBeanDefinitions(Resource resource) throws BeanDefinitionStoreException {
    //将读入的 XML 资源进行特殊编码处理
    return loadBeanDefinitions(new EncodedResource(resource));
}

//这里载入 XML 形式 Bean 配置信息方法
public int loadBeanDefinitions(EncodedResource encodedResource) throws
BeanDefinitionStoreException {
    ...
    try {
        //将资源文件转为 InputStream 的 I/O 流
        InputStream inputStream = encodedResource.getResource().getInputStream();
        try {
            //从 InputStream 中得到 XML 的解析源
            InputSource inputSource = new InputSource(inputStream);
            if (encodedResource.getEncoding() != null) {
                inputSource.setEncoding(encodedResource.getEncoding());
            }
            //这里是具体的读取过程
            return doLoadBeanDefinitions(inputSource, encodedResource.getResource());
        }
        finally {
            //关闭从 Resource 中得到的 I/O 流
            inputStream.close();
        }
    }
    ...
}
//从特定 XML 文件中实际载入 Bean 配置资源的方法
protected int doLoadBeanDefinitions(InputSource inputSource, Resource resource)
        throws BeanDefinitionStoreException {
    try {
        //将 XML 文件转换为 DOM 对象，解析过程由 documentLoader()方法实现
        Document doc = doLoadDocument(inputSource, resource);
        //这里启动对 Bean 定义解析的详细过程，该解析过程会用到 Spring 的 Bean 配置规则
        return registerBeanDefinitions(doc, resource);
    }
    ...
}
```

通过源码分析可知，载入 Bean 配置信息的最后一步是将 Bean 配置信息转换为文档对象，该过程由 documentLoader()方法实现。

8.2.9 准备文档对象

DocumentLoader 将 Bean 配置信息转换成文档对象的源码如下：

```java
//使用标准的 JAXP 将载入的 Bean 配置信息转换成文档对象
@Override
public Document loadDocument(InputSource inputSource, EntityResolver entityResolver,
    ErrorHandler errorHandler, int validationMode, boolean namespaceAware) throws Exception {

    //创建文件解析器工厂
    DocumentBuilderFactory factory = createDocumentBuilderFactory(validationMode, namespaceAware);
    if (logger.isDebugEnabled()) {
        logger.debug("Using JAXP provider [" + factory.getClass().getName() + "]");
    }
    //创建文档解析器
    DocumentBuilder builder = createDocumentBuilder(factory, entityResolver, errorHandler);
    //解析 Spring 的 Bean 配置信息
    return builder.parse(inputSource);
}

protected DocumentBuilderFactory createDocumentBuilderFactory(int validationMode, boolean namespaceAware)
        throws ParserConfigurationException {

    //创建文档解析工厂
    DocumentBuilderFactory factory = DocumentBuilderFactory.newInstance();
    factory.setNamespaceAware(namespaceAware);

    //设置解析 XML 的校验
    if (validationMode != XmlValidationModeDetector.VALIDATION_NONE) {
        factory.setValidating(true);
        if (validationMode == XmlValidationModeDetector.VALIDATION_XSD) {
            factory.setNamespaceAware(true);
            try {
                factory.setAttribute(SCHEMA_LANGUAGE_ATTRIBUTE, XSD_SCHEMA_LANGUAGE);
            }
            catch (IllegalArgumentException ex) {
                ParserConfigurationException pcex = new ParserConfigurationException(
                    "Unable to validate using XSD: Your JAXP provider [" + factory +
                    "] does not support XML Schema. Are you running on Java 1.4 with Apache Crimson? "
                    + "Upgrade to Apache Xerces (or Java 1.5) for full XSD support.");
                pcex.initCause(ex);
```

```
            throw pcex;
        }
    }
}
    return factory;
}
```

上面的解析过程是调用 Java EE 的 JAXP 标准进行处理的。至此 Spring IoC 容器根据定位的 Bean 配置信息将其读入并转换成为文档对象完成。接下来我们要继续分析 Spring IoC 容器将载入的 Bean 配置信息转换为文档对象之后，是如何将其解析为 Spring IoC 管理的 Bean 对象并将其注册到容器中的。

8.2.10 分配解析策略

XmlBeanDefinitionReader 类中的 doLoadBeanDefinition()方法实际上从特定 XML 文件中载入 Bean 配置信息的方法，该方法在载入 Bean 配置信息之后将其转换为文档对象。接下来调用 registerBeanDefinitions()方法启动 Spring IoC 容器对 Bean 定义的解析过程，registerBeanDefinitions() 方法源码如下：

```
//按照 Spring 的 Bean 语义要求将 Bean 配置信息解析并转换为容器内部数据结构
public int registerBeanDefinitions(Document doc, Resource resource) throws
BeanDefinitionStoreException {
    //得到 BeanDefinitionDocumentReader 来对 XML 格式的 BeanDefinition 进行解析
    BeanDefinitionDocumentReader documentReader = createBeanDefinitionDocumentReader();
    //获得容器中注册的 Bean 数量
    int countBefore = getRegistry().getBeanDefinitionCount();
    //解析过程的入口，这里使用了委派模式，BeanDefinitionDocumentReader 只是一个接口
    //具体的解析过程由实现类 DefaultBeanDefinitionDocumentReader 完成
    documentReader.registerBeanDefinitions(doc, createReaderContext(resource));
    //统计解析的 Bean 数量
    return getRegistry().getBeanDefinitionCount() - countBefore;
}
```

Bean 配置信息的载入和解析过程如下。

首先，通过调用 XML 解析器将 Bean 配置信息转换为文档对象，但是这些文档对象并没有按照 Spring 的 Bean 规则进行解析。这一步是载入的过程。

其次，在完成通用的 XML 解析之后，按照 Spring Bean 的定义规则对文档对象进行解析，其解析过程在接口 BeanDefinitionDocumentReader 的实现类 DefaultBeanDefinitionDocumentReader 中实现。

8.2.11 将配置载入内存

BeanDefinitionDocumentReader 接口通过 registerBeanDefinitions() 方法调用其实现类 DefaultBeanDefinitionDocumentReader 对文档对象进行解析，代码如下：

```java
//根据 Spring DTD 对 Bean 的定义规则解析 Bean 定义的文档对象
@Override
public void registerBeanDefinitions(Document doc, XmlReaderContext readerContext) {
    //获得 XML 描述符
    this.readerContext = readerContext;
    logger.debug("Loading bean definitions");
    //获得 Document 的根元素
    Element root = doc.getDocumentElement();
    doRegisterBeanDefinitions(root);
}
...
protected void doRegisterBeanDefinitions(Element root) {

    //具体的解析过程由 BeanDefinitionParserDelegate 实现
    //BeanDefinitionParserDelegate 中定义了 Spring Bean 定义 XML 文件的各种元素
    BeanDefinitionParserDelegate parent = this.delegate;
    this.delegate = createDelegate(getReaderContext(), root, parent);

    if (this.delegate.isDefaultNamespace(root)) {
        String profileSpec = root.getAttribute(PROFILE_ATTRIBUTE);
        if (StringUtils.hasText(profileSpec)) {
            String[] specifiedProfiles = StringUtils.tokenizeToStringArray(
                    profileSpec, BeanDefinitionParserDelegate.MULTI_VALUE_ATTRIBUTE_DELIMITERS);
            if (!getReaderContext().getEnvironment().acceptsProfiles(specifiedProfiles)) {
                if (logger.isInfoEnabled()) {
                    logger.info("Skipped XML bean definition file due to specified profiles [" +
                            profileSpec + "] not matching: " + getReaderContext().getResource());
                }
                return;
            }
        }
    }

    //在解析 Bean 定义之前，进行自定义解析，增强解析过程的可扩展性
    preProcessXml(root);
    //从文档的根元素开始进行 Bean 定义的文档对象的解析
    parseBeanDefinitions(root, this.delegate);
    //在解析 Bean 定义之后，进行自定义解析，增加解析过程的可扩展性
    postProcessXml(root);

    this.delegate = parent;
```

```java
}
//创建 BeanDefinitionParserDelegate，用于完成真正的解析过程
protected BeanDefinitionParserDelegate createDelegate(
        XmlReaderContext readerContext, Element root, @Nullable BeanDefinitionParserDelegate
parentDelegate) {

    BeanDefinitionParserDelegate delegate = new BeanDefinitionParserDelegate(readerContext);
    //BeanDefinitionParserDelegate 初始化 Document 根元素
    delegate.initDefaults(root, parentDelegate);
    return delegate;
}
//使用 Spring 的 Bean 规则从文档的根元素开始 Bean 定义的文档对象的解析
protected void parseBeanDefinitions(Element root, BeanDefinitionParserDelegate delegate) {
    //Bean 定义的文档对象使用了 Spring 默认的 XML 命名空间
    if (delegate.isDefaultNamespace(root)) {
        //获取 Bean 定义的文档对象根元素的所有子节点
        NodeList nl = root.getChildNodes();
        for (int i = 0; i < nl.getLength(); i++) {
            Node node = nl.item(i);
            //获得的文档节点是 XML 元素节点
            if (node instanceof Element) {
                Element ele = (Element) node;
                //Bean 定义的文档的元素节点使用的是 Spring 默认的 XML 命名空间
                if (delegate.isDefaultNamespace(ele)) {
                    //使用 Spring 的 Bean 规则解析元素节点
                    parseDefaultElement(ele, delegate);
                }
                else {
                    //如果没有使用 Spring 默认的 XML 命名空间，则使用用户自定义的解析规则解析元素节点
                    delegate.parseCustomElement(ele);
                }
            }
        }
    }
    else {
        //文档的根节点没有使用 Spring 默认的命名空间
        //使用自定义的解析规则解析文档的根节点
        delegate.parseCustomElement(root);
    }
}
//使用 Spring 的 Bean 规则解析文档元素节点
private void parseDefaultElement(Element ele, BeanDefinitionParserDelegate delegate) {
    //如果节点是<import>导入元素，进行导入解析
    if (delegate.nodeNameEquals(ele, IMPORT_ELEMENT)) {
        importBeanDefinitionResource(ele);
    }
    //如果节点是<alias>别名元素，进行别名解析
```

```java
   else if (delegate.nodeNameEquals(ele, ALIAS_ELEMENT)) {
      processAliasRegistration(ele);
   }
   //如果节点既不是导入元素,也不是别名元素,而是普通的<bean>元素,则按照 Spring 的 Bean 规则解析元素
   else if (delegate.nodeNameEquals(ele, BEAN_ELEMENT)) {
      processBeanDefinition(ele, delegate);
   }
   else if (delegate.nodeNameEquals(ele, NESTED_BEANS_ELEMENT)) {
      doRegisterBeanDefinitions(ele);
   }
}
//解析<import>导入元素,从给定的导入路径加载 Bean 资源到 Spring IoC 容器中
protected void importBeanDefinitionResource(Element ele) {
   //获取给定的导入元素的 location 属性
   String location = ele.getAttribute(RESOURCE_ATTRIBUTE);
   //如果导入元素的 location 属性值为空,则没有导入任何资源,直接返回
   if (!StringUtils.hasText(location)) {
      getReaderContext().error("Resource location must not be empty", ele);
      return;
   }

   //使用系统变量值解析 location 属性值
   location = getReaderContext().getEnvironment().resolveRequiredPlaceholders(location);

   Set<Resource> actualResources = new LinkedHashSet<>(4);

   //标识给定的导入元素的 location 属性值是否是绝对路径
   boolean absoluteLocation = false;
   try {
      absoluteLocation = ResourcePatternUtils.isUrl(location) || ResourceUtils.toURI(location).isAbsolute();
   }
   catch (URISyntaxException ex) {
      //给定的导入元素的 location 属性值不是绝对路径
   }

   //给定的导入元素的 location 属性值是绝对路径
   if (absoluteLocation) {
      try {
         //使用资源读入器加载给定路径的 Bean 资源
         int importCount = getReaderContext().getReader().loadBeanDefinitions(location, actualResources);
         if (logger.isDebugEnabled()) {
            logger.debug("Imported " + importCount + " bean definitions from URL location [" + location + "]");
         }
      }
```

```java
        catch (BeanDefinitionStoreException ex) {
            getReaderContext().error(
                    "Failed to import bean definitions from URL location [" + location + "]", ele, ex);
        }
    }
    else {
        //给定的导入元素的 location 属性值是相对路径
        try {
            int importCount;
            //将给定导入元素的 location 封装为相对路径资源
            Resource relativeResource = getReaderContext().getResource().createRelative(location);
            //封装的相对路径资源存在
            if (relativeResource.exists()) {
                //使用资源读入器加载 Bean 资源
                importCount = getReaderContext().getReader().loadBeanDefinitions(relativeResource);
                actualResources.add(relativeResource);
            }
            //封装的相对路径资源不存在
            else {
                //获取 Spring IoC 容器资源读入器的基本路径
                String baseLocation = getReaderContext().getResource().getURL().toString();
                //根据 Spring IoC 容器资源读入器的基本路径加载给定导入路径的资源
                importCount = getReaderContext().getReader().loadBeanDefinitions(
                        StringUtils.applyRelativePath(baseLocation, location), actualResources);
            }
            if (logger.isDebugEnabled()) {
                logger.debug("Imported " + importCount + " bean definitions from relative location [" + location + "]");
            }
        }
        catch (IOException ex) {
            getReaderContext().error("Failed to resolve current resource location", ele, ex);
        }
        catch (BeanDefinitionStoreException ex) {
            getReaderContext().error("Failed to import bean definitions from relative location [" + location + "]", ele, ex);
        }
    }
    Resource[] actResArray = actualResources.toArray(new Resource[actualResources.size()]);
    //在解析完<import>元素之后，发送容器导入其他资源处理完成事件
    getReaderContext().fireImportProcessed(location, actResArray, extractSource(ele));
}
//解析<alias>别名元素，为 Bean 向 Spring IoC 容器注册别名
protected void processAliasRegistration(Element ele) {
    //获取<alias>别名元素中 name 的属性值
    String name = ele.getAttribute(NAME_ATTRIBUTE);
    //获取<alias>别名元素中 alias 的属性值
```

```java
            String alias = ele.getAttribute(ALIAS_ATTRIBUTE);
            boolean valid = true;
            //<alias>别名元素的 name 属性值为空
            if (!StringUtils.hasText(name)) {
                getReaderContext().error("Name must not be empty", ele);
                valid = false;
            }
            //<alias>别名元素的 alias 属性值为空
            if (!StringUtils.hasText(alias)) {
                getReaderContext().error("Alias must not be empty", ele);
                valid = false;
            }
            if (valid) {
                try {
                    //向容器的资源读入器注册别名
                    getReaderContext().getRegistry().registerAlias(name, alias);
                }
                catch (Exception ex) {
                    getReaderContext().error("Failed to register alias '" + alias +
                        "' for bean with name '" + name + "'", ele, ex);
                }
                //在解析完<alias>元素之后，发送容器别名处理完成事件
                getReaderContext().fireAliasRegistered(name, alias, extractSource(ele));
            }
        }
    }
    //解析 Bean 资源文档对象的普通元素
    protected void processBeanDefinition(Element ele, BeanDefinitionParserDelegate delegate) {
        BeanDefinitionHolder bdHolder = delegate.parseBeanDefinitionElement(ele);
        //BeanDefinitionHolder 是对 BeanDefinition 的封装，即 Bean 定义的封装类
        //对文档对象中<bean>元素的解析由 BeanDefinitionParserDelegate 实现
        //BeanDefinitionHolder bdHolder = delegate.parseBeanDefinitionElement(ele);
        if (bdHolder != null) {
            bdHolder = delegate.decorateBeanDefinitionIfRequired(ele, bdHolder);
            try {
                //向 Spring IoC 容器注册解析得到的 Bean 定义，这是 Bean 定义向 IoC 容器注册的入口
                BeanDefinitionReaderUtils.registerBeanDefinition(bdHolder, getReaderContext().getRegistry());
            }
            catch (BeanDefinitionStoreException ex) {
                getReaderContext().error("Failed to register bean definition with name '" +
                    bdHolder.getBeanName() + "'", ele, ex);
            }
            //在完成向 Spring IoC 容器注册解析得到的 Bean 定义之后，发送注册事件
            getReaderContext().fireComponentRegistered(new BeanComponentDefinition(bdHolder));
        }
    }
}
```

通过上述 Spring IoC 容器对载入的 Bean 定义的文档解析可以看出,在 Spring 配置文件中可以使用<import>元素来导入 IoC 容器所需要的其他资源,Spring IoC 容器在解析时首先将指定的资源加载到容器中。使用<alias>别名时,Spring IoC 容器首先将别名元素所定义的别名注册到容器中。

对于既不是<import>元素又不是<alias>元素的元素,即 Spring 配置文件中普通的<bean>元素,由 BeanDefinitionParserDelegate 类的 parseBeanDefinitionElement()方法实现解析。这个解析过程非常复杂,我们在 Mini 版本中用 properties 文件代替了。

8.2.12 载入<bean>元素

Bean 配置信息中的<import>和<alias>元素解析在 DefaultBeanDefinitionDocumentReader 中已经完成,Bean 配置信息中使用最多的<bean>元素交由 BeanDefinitionParserDelegate 来解析,其实现源码如下:

```java
//解析<bean>元素的入口
@Nullable
public BeanDefinitionHolder parseBeanDefinitionElement(Element ele) {
    return parseBeanDefinitionElement(ele, null);
}

//解析 Bean 配置信息中的<bean>元素,这个方法中主要处理<bean>元素的 id、name 和别名属性
@Nullable
public BeanDefinitionHolder parseBeanDefinitionElement(Element ele, @Nullable BeanDefinition containingBean) {
    //获取<bean>元素中的 id 属性值
    String id = ele.getAttribute(ID_ATTRIBUTE);
    //获取<bean>元素中的 name 属性值
    String nameAttr = ele.getAttribute(NAME_ATTRIBUTE);
    //获取<bean>元素中的 alias 属性值
    List<String> aliases = new ArrayList<>();

    //将<bean>元素中的所有 name 属性值存放到别名中
    if (StringUtils.hasLength(nameAttr)) {
        String[] nameArr = StringUtils.tokenizeToStringArray(nameAttr, MULTI_VALUE_ATTRIBUTE_DELIMITERS);
        aliases.addAll(Arrays.asList(nameArr));
    }

    String beanName = id;
    //如果<bean>元素中没有配置 id 属性,将别名中的第一个值赋给 beanName
    if (!StringUtils.hasText(beanName) && !aliases.isEmpty()) {
        beanName = aliases.remove(0);
        if (logger.isDebugEnabled()) {
```

```
            logger.debug("No XML 'id' specified - using '" + beanName +
                "' as bean name and " + aliases + " as aliases");
        }
    }

    //检查<bean>元素所配置的 id 或者 name 的唯一性
    //containingBean 标识<bean>元素中是否包含子<bean>元素
    if (containingBean == null) {
        //检查<bean>元素所配置的 id、name 或者别名是否重复
        checkNameUniqueness(beanName, aliases, ele);
    }

    //详细对<bean>元素中配置的 Bean 定义进行解析
    AbstractBeanDefinition beanDefinition = parseBeanDefinitionElement(ele, beanName,
containingBean);
    if (beanDefinition != null) {
        if (!StringUtils.hasText(beanName)) {
            try {
                if (containingBean != null) {
                    //如果<bean>元素中没有配置 id、别名或者 name，且没有包含子元素
                    //<bean>元素，则为解析的 Bean 生成一个唯一 beanName 并注册
                    beanName = BeanDefinitionReaderUtils.generateBeanName(
                        beanDefinition, this.readerContext.getRegistry(), true);
                }
                else {
                    //如果<bean>元素中没有配置 id、别名或者 name，且包含了子元素
                    //<bean>元素，则将解析的 Bean 使用别名向 IoC 容器注册
                    beanName = this.readerContext.generateBeanName(beanDefinition);
                    //为解析的 Bean 使用别名注册时，为了向后兼容
                    //Spring1.2/2.0，给别名添加类名后缀
                    String beanClassName = beanDefinition.getBeanClassName();
                    if (beanClassName != null &&
                        beanName.startsWith(beanClassName) && beanName.length() >
beanClassName.length() &&
                        !this.readerContext.getRegistry().isBeanNameInUse(beanClassName)) {
                        aliases.add(beanClassName);
                    }
                }
                if (logger.isDebugEnabled()) {
                    logger.debug("Neither XML 'id' nor 'name' specified - " +
                        "using generated bean name [" + beanName + "]");
                }
            }
            catch (Exception ex) {
                error(ex.getMessage(), ele);
                return null;
            }
```

```java
        }
        String[] aliasesArray = StringUtils.toStringArray(aliases);
        return new BeanDefinitionHolder(beanDefinition, beanName, aliasesArray);
    }
    //当解析出错时,返回 null
    return null;
}

protected void checkNameUniqueness(String beanName, List<String> aliases, Element beanElement)
{
    String foundName = null;

    if (StringUtils.hasText(beanName) && this.usedNames.contains(beanName)) {
        foundName = beanName;
    }
    if (foundName == null) {
        foundName = CollectionUtils.findFirstMatch(this.usedNames, aliases);
    }
    if (foundName != null) {
        error("Bean name '" + foundName + "' is already used in this <beans> element", beanElement);
    }

    this.usedNames.add(beanName);
    this.usedNames.addAll(aliases);
}

//详细对<bean>元素中配置的 Bean 定义的其他属性进行解析
//由于上面的方法中已经对 Bean 的 id、name 和别名等属性进行了处理
//该方法中主要处理除这三个以外的其他属性
@Nullable
public AbstractBeanDefinition parseBeanDefinitionElement(
        Element ele, String beanName, @Nullable BeanDefinition containingBean) {
    //记录解析的<bean>元素
    this.parseState.push(new BeanEntry(beanName));

    //这里只读取<bean>元素中配置的 class 名字,然后载入 BeanDefinition 中
    //只是记录配置的 class 名字,不做实例化,对象的实例化在依赖注入时完成
    String className = null;

    //如果<bean>元素中配置了 parent 属性,则获取 parent 属性的值
    if (ele.hasAttribute(CLASS_ATTRIBUTE)) {
        className = ele.getAttribute(CLASS_ATTRIBUTE).trim();
    }
    String parent = null;
    if (ele.hasAttribute(PARENT_ATTRIBUTE)) {
        parent = ele.getAttribute(PARENT_ATTRIBUTE);
    }
```

```java
try {
    //根据<bean>元素配置的class名称和parent属性值创建BeanDefinition
    //为载入Bean定义信息做准备
    AbstractBeanDefinition bd = createBeanDefinition(className, parent);

    //对当前的<bean>元素中配置的一些属性进行解析和设置,如配置的单态(singleton)属性等
    parseBeanDefinitionAttributes(ele, beanName, containingBean, bd);
    //为<bean>元素解析的Bean设置描述信息
    bd.setDescription(DomUtils.getChildElementValueByTagName(ele, DESCRIPTION_ELEMENT));

    //对<bean>元素的meta(元信息)属性进行解析
    parseMetaElements(ele, bd);
    //对<bean>元素的lookup-Method属性进行解析
    parseLookupOverrideSubElements(ele, bd.getMethodOverrides());
    //对<bean>元素的replaced-Method属性进行解析
    parseReplacedMethodSubElements(ele, bd.getMethodOverrides());

    //解析<bean>元素的构造方法设置
    parseConstructorArgElements(ele, bd);
    //解析<bean>元素的<property>设置
    parsePropertyElements(ele, bd);
    //解析<bean>元素的qualifier属性
    parseQualifierElements(ele, bd);

    //为当前解析的Bean设置所需的资源和依赖对象
    bd.setResource(this.readerContext.getResource());
    bd.setSource(extractSource(ele));

    return bd;
}
catch (ClassNotFoundException ex) {
    error("Bean class [" + className + "] not found", ele, ex);
}
catch (NoClassDefFoundError err) {
    error("Class that bean class [" + className + "] depends on not found", ele, err);
}
catch (Throwable ex) {
    error("Unexpected failure during bean definition parsing", ele, ex);
}
finally {
    this.parseState.pop();
}

//当解析<bean>元素出错时,返回null
return null;
}
```

使用过 Spring、对 Spring 配置文件比较熟悉的人，通过上述源码分析就会明白，在 Spring 配置文件中<bean>元素配置的属性就是通过上面的方法解析和设置到 Bean 中的。

注意：在解析<bean>元素的过程中没有创建和实例化 Bean 对象，只是创建了 Bean 对象的定义类 BeanDefinition，将<bean>元素中的配置信息设置到 BeanDefinition 中作为记录，当依赖注入时才使用这些记录信息创建和实例化具体的 Bean 对象。

上面的方法中对一些配置（如 meta、qualifier 等）的解析，我们在 Spring 中使用得不多，在使用 Spring 的<bean>元素时，配置最多的是<property>子元素，因此下面继续分析源码，了解在解析时 Bean 的属性在解析时是如何设置的。

8.2.13 载入<property>元素

BeanDefinitionParserDelegate 在解析<bean>时调用 parsePropertyElements()方法解析<bean>元素中的<property>元素，源码如下：

```java
//解析<bean>元素中的<property>元素
public void parsePropertyElements(Element beanEle, BeanDefinition bd) {
    //获取<bean>元素中所有的子元素
    NodeList nl = beanEle.getChildNodes();
    for (int i = 0; i < nl.getLength(); i++) {
        Node node = nl.item(i);
        //如果子元素是<property>子元素，则调用解析<property>子元素的方法解析
        if (isCandidateElement(node) && nodeNameEquals(node, PROPERTY_ELEMENT)) {
            parsePropertyElement((Element) node, bd);
        }
    }
}

//解析<property>元素
public void parsePropertyElement(Element ele, BeanDefinition bd) {
    //获取<property>元素的名字
    String propertyName = ele.getAttribute(NAME_ATTRIBUTE);
    if (!StringUtils.hasLength(propertyName)) {
        error("Tag 'property' must have a 'name' attribute", ele);
        return;
    }
    this.parseState.push(new PropertyEntry(propertyName));
    try {
        //如果一个 Bean 中已经有同名的<property>元素存在，则不进行解析，直接返回
        //即如果在同一个 Bean 中配置同名的<property>元素，则只有第一个起作用
        if (bd.getPropertyValues().contains(propertyName)) {
```

```java
            error("Multiple 'property' definitions for property '" + propertyName + "'", ele);
        return;
    }
    //解析获取<property>元素的值
    Object val = parsePropertyValue(ele, bd, propertyName);
    //根据<property>元素的名字和值创建实例
    PropertyValue pv = new PropertyValue(propertyName, val);
    //解析<property>元素中的属性
    parseMetaElements(ele, pv);
    pv.setSource(extractSource(ele));
    bd.getPropertyValues().addPropertyValue(pv);
    }
    finally {
        this.parseState.pop();
    }
}


//解析获取<property>元素的值
@Nullable
public Object parsePropertyValue(Element ele, BeanDefinition bd, @Nullable String propertyName) {
    String elementName = (propertyName != null) ?
            "<property> element for property '" + propertyName + "'" :
            "<constructor-arg> element";
    //获取<property>的所有子元素，只能是 ref、value、list、etc 中的一种类型
    NodeList nl = ele.getChildNodes();
    Element subElement = null;
    for (int i = 0; i < nl.getLength(); i++) {
        Node node = nl.item(i);
        //子元素不是 description 和 meta 属性
        if (node instanceof Element && !nodeNameEquals(node, DESCRIPTION_ELEMENT) &&
                !nodeNameEquals(node, META_ELEMENT)) {
            if (subElement != null) {
                error(elementName + " must not contain more than one sub-element", ele);
            }
            else {
                //当前<property>元素包含子元素
                subElement = (Element) node;
            }
        }
    }

    //判断属性值是 ref 还是 value，不允许既是 ref 又是 value
    boolean hasRefAttribute = ele.hasAttribute(REF_ATTRIBUTE);
    boolean hasValueAttribute = ele.hasAttribute(VALUE_ATTRIBUTE);
    if ((hasRefAttribute && hasValueAttribute) ||
            ((hasRefAttribute || hasValueAttribute) && subElement != null)) {
```

```java
        error(elementName +
            " is only allowed to contain either 'ref' attribute OR 'value' attribute OR sub-element", ele);
    }

    //如果属性值是 ref，创建一个 ref 的数据对象 RuntimeBeanReference，这个对象封装了 ref
    if (hasRefAttribute) {
        String refName = ele.getAttribute(REF_ATTRIBUTE);
        if (!StringUtils.hasText(refName)) {
            error(elementName + " contains empty 'ref' attribute", ele);
        }
        //一个指向运行时所依赖对象的引用
        RuntimeBeanReference ref = new RuntimeBeanReference(refName);
        //设置这个 ref 的数据对象被当前对象所引用
        ref.setSource(extractSource(ele));
        return ref;
    }
    //如果属性值是 value，创建一个 value 的数据对象 TypedStringValue，这个对象封装了 value
    else if (hasValueAttribute) {
        //一个持有 String 类型值的对象
        TypedStringValue valueHolder = new TypedStringValue(ele.getAttribute(VALUE_ATTRIBUTE));
        //设置这个 value 的数据对象被当前对象所引用
        valueHolder.setSource(extractSource(ele));
        return valueHolder;
    }
    //如果当前<property>元素还有子元素
    else if (subElement != null) {
        //解析<property>的子元素
        return parsePropertySubElement(subElement, bd);
    }
    else {
        //属性值既不是 ref 也不是 value，解析出错，返回 null
        error(elementName + " must specify a ref or value", ele);
        return null;
    }
}
```

通过上述源码分析，我们了解了 Spring 配置文件中<bean>元素中<property>子元素的相关配置是如何处理的，注意以下三点：

（1）ref 被封装为指向依赖对象的一个引用。

（2）value 被封装成一个字符串类型的对象。

（3）ref 和 value 都通过 "解析的数据类型属性值.setSource(extractSource(ele));" 方法将属性值（或引用）与所引用的属性关联起来。

最后<property>元素的子元素通过 parsePropertySubElement ()方法解析,下面我们继续分析该方法的源码,了解其解析过程。

8.2.14 载入<property>子元素

BeanDefinitionParserDelegate 类中的 parsePropertySubElement()方法用于对<property>元素中的子元素进行解析,源码如下:

```
//解析<property>元素中的 ref、value 或者集合等子元素
@Nullable
public Object parsePropertySubElement(Element ele, @Nullable BeanDefinition bd, @Nullable
String defaultValueType) {
    //如果<property>元素没有使用 Spring 默认的命名空间,则使用用户自定义的规则解析内嵌元素
    if (!isDefaultNamespace(ele)) {
        return parseNestedCustomElement(ele, bd);
    }
    //如果子元素是 Bean,则使用解析<bean>元素的方法解析
    else if (nodeNameEquals(ele, BEAN_ELEMENT)) {
        BeanDefinitionHolder nestedBd = parseBeanDefinitionElement(ele, bd);
        if (nestedBd != null) {
            nestedBd = decorateBeanDefinitionIfRequired(ele, nestedBd, bd);
        }
        return nestedBd;
    }
    //如果子元素是 ref,ref 只能有 3 个属性:bean、local、parent
    else if (nodeNameEquals(ele, REF_ELEMENT)) {
        //可以不在同一个 Spring 配置文件中,具体请参考 Spring 对 ref 的配置规则
        String refName = ele.getAttribute(BEAN_REF_ATTRIBUTE);
        boolean toParent = false;
        if (!StringUtils.hasLength(refName)) {
            //获取<property>元素中的 parent 属性值,引用父容器中的 Bean
            refName = ele.getAttribute(PARENT_REF_ATTRIBUTE);
            toParent = true;
            if (!StringUtils.hasLength(refName)) {
                error("'bean' or 'parent' is required for <ref> element", ele);
                return null;
            }
        }
        if (!StringUtils.hasText(refName)) {
            error("<ref> element contains empty target attribute", ele);
            return null;
        }
        //创建 ref 类型数据,指向被引用的对象
        RuntimeBeanReference ref = new RuntimeBeanReference(refName, toParent);
        //设置引用类型值被当前子元素所引用
```

```java
        ref.setSource(extractSource(ele));
        return ref;
    }
    //如果子元素是<idref>，使用解析<ref>元素的方法解析
    else if (nodeNameEquals(ele, IDREF_ELEMENT)) {
        return parseIdRefElement(ele);
    }
    //如果子元素是<value>，使用解析<value>元素的方法解析
    else if (nodeNameEquals(ele, VALUE_ELEMENT)) {
        return parseValueElement(ele, defaultValueType);
    }
    //如果子元素是null，为<property>元素设置一个封装null值的字符串数据
    else if (nodeNameEquals(ele, NULL_ELEMENT)) {
        TypedStringValue nullHolder = new TypedStringValue(null);
        nullHolder.setSource(extractSource(ele));
        return nullHolder;
    }
    //如果子元素是<array>，使用解析<array>集合子元素的方法解析
    else if (nodeNameEquals(ele, ARRAY_ELEMENT)) {
        return parseArrayElement(ele, bd);
    }
    //如果子元素是<list>，使用解析<list>集合子元素的方法解析
    else if (nodeNameEquals(ele, LIST_ELEMENT)) {
        return parseListElement(ele, bd);
    }
    //如果子元素是<set>，使用解析<set>集合子元素的方法解析
    else if (nodeNameEquals(ele, SET_ELEMENT)) {
        return parseSetElement(ele, bd);
    }
    //如果子元素是<map>，使用解析<map>集合子元素的方法解析
    else if (nodeNameEquals(ele, MAP_ELEMENT)) {
        return parseMapElement(ele, bd);
    }
    //如果子元素是<props>，使用解析<props>集合子元素的方法解析
    else if (nodeNameEquals(ele, PROPS_ELEMENT)) {
        return parsePropsElement(ele);
    }
    //既不是 ref 又不是 value，也不是集合，则说明子元素配置错误，返回 null
    else {
        error("Unknown property sub-element: [" + ele.getNodeName() + "]", ele);
        return null;
    }
}
```

通过上述源码分析可知，在 Spring 配置文件中，对<property>元素中配置的<array>、<list>、<set>、<map>、<props>等各种集合子元素都通过上述方法解析，生成对应的数据对象，比如

ManagedList、ManagedArray、ManagedSet 等。这些 Managed 类是 Spring 对象 BeanDefinition 的数据封装，对集合数据类型的具体解析由各自的解析方法实现，解析方法的命名非常规范，一目了然。

8.2.15　载入<list>子元素

BeanDefinitionParserDelegate 类中的 parseListElement()方法用于解析<property>元素中的<list>集合子元素，源码如下：

```
//解析<list>集合子元素
public List<Object> parseListElement(Element collectionEle, @Nullable BeanDefinition bd) {
    //获取<list>元素中的 value-type 属性，即获取集合元素的数据类型
    String defaultElementType = collectionEle.getAttribute(VALUE_TYPE_ATTRIBUTE);
    //获取<list>集合子元素中的所有子节点
    NodeList nl = collectionEle.getChildNodes();
    //Spring 将 List 封装为 ManagedList
    ManagedList<Object> target = new ManagedList<>(nl.getLength());
    target.setSource(extractSource(collectionEle));
    //设置集合目标数据类型
    target.setElementTypeName(defaultElementType);
    target.setMergeEnabled(parseMergeAttribute(collectionEle));
    //具体的<list>子元素解析
    parseCollectionElements(nl, target, bd, defaultElementType);
    return target;
}

//具体解析<list>集合子元素，<array>、<list>和<set>都使用以下方法解析
protected void parseCollectionElements(
        NodeList elementNodes, Collection<Object> target, @Nullable BeanDefinition bd, String defaultElementType) {
    //遍历集合的所有节点
    for (int i = 0; i < elementNodes.getLength(); i++) {
        Node node = elementNodes.item(i);
        //节点不是 description 节点
        if (node instanceof Element && !nodeNameEquals(node, DESCRIPTION_ELEMENT)) {
            //将解析的元素加入集合，递归调用下一个子元素
            target.add(parsePropertySubElement((Element) node, bd, defaultElementType));
        }
    }
}
```

经过对 Spring Bean 配置信息转换文档对象中的元素的层层解析，Spring IoC 现在已经将 XML 形式定义的 Bean 配置信息转换为 Spring IoC 所识别的数据结构——BeanDefinition。它是 Bean 配置信息中配置的 POJO 对象在 Spring IoC 容器中的映射，我们可以通过 AbstractBeanDefinition 入

口,看到如何对 Spring IoC 容器进行索引、查询和其他操作。

通过 Spring IoC 容器对 Bean 配置信息的解析,Spring IoC 容器大致完成了管理 Bean 对象的准备工作,即初始化过程。但是最重要的依赖注入还没有发生,在 Spring IoC 容器中 BeanDefinition 存储的还只是一些静态信息,接下来需要向容器注册 Bean 定义信息,才能真正完成 Spring IoC 容器的初始化。

8.2.16 分配注册策略

我们继续跟踪程序的执行顺序,接下来在 DefaultBeanDefinitionDocumentReader 对 Bean 定义转换的文档对象解析的流程中,在 parseDefaultElement()方法中完成对文档对象的解析后得到封装 BeanDefinition 的 BeanDefinitionHold 对象,然后调用 BeanDefinitionReaderUtils 的 registerBeanDefinition()方法向 Spring IoC 容器注册解析的 Bean。BeanDefinitionReaderUtils 的注册源码如下:

```java
//将解析的 BeanDefinitionHold 注册到 Spring IoC 容器中
public static void registerBeanDefinition(
    BeanDefinitionHolder definitionHolder, BeanDefinitionRegistry registry)
    throws BeanDefinitionStoreException {
  //获取解析的 BeanDefinition 的名称
  String beanName = definitionHolder.getBeanName();
  //向 Spring IoC 容器注册 BeanDefinition
  registry.registerBeanDefinition(beanName, definitionHolder.getBeanDefinition());
  //如果解析的 BeanDefinition 有别名,向 Spring IoC 容器注册别名
  String[] aliases = definitionHolder.getAliases();
  if (aliases != null) {
    for (String alias : aliases) {
      registry.registerAlias(beanName, alias);
    }
  }
}
```

当调用 BeanDefinitionReaderUtils 向 Spring IoC 容器注册解析的 BeanDefinition 时,真正完成注册功能的是 DefaultListableBeanFactory。

8.2.17 向容器注册

DefaultListableBeanFactory 中使用一个 HashMap 的集合对象存放 Spring IoC 容器中注册解析的 BeanDefinition,下面看向 Spring IoC 容器注册的主要源码,类图如下图所示。

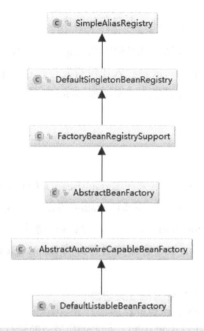

```
//存储注册信息的 BeanDefinition
private final Map<String, BeanDefinition> beanDefinitionMap = new ConcurrentHashMap<>(256);

//向 Spring IoC 容器注册解析的 BeanDefinition
@Override
public void registerBeanDefinition(String beanName, BeanDefinition beanDefinition)
        throws BeanDefinitionStoreException {

    Assert.hasText(beanName, "Bean name must not be empty");
    Assert.notNull(beanDefinition, "BeanDefinition must not be null");

    //校验解析的 BeanDefinition
    if (beanDefinition instanceof AbstractBeanDefinition) {
        try {
            ((AbstractBeanDefinition) beanDefinition).validate();
        }
        catch (BeanDefinitionValidationException ex) {
            throw new BeanDefinitionStoreException(beanDefinition.getResourceDescription(), beanName,
                    "Validation of bean definition failed", ex);
        }
    }

    BeanDefinition oldBeanDefinition;

    oldBeanDefinition = this.beanDefinitionMap.get(beanName);
```

```java
if (oldBeanDefinition != null) {
    if (!isAllowBeanDefinitionOverriding()) {
        throw new BeanDefinitionStoreException(beanDefinition.getResourceDescription(),
beanName, "Cannot register bean definition [" + beanDefinition + "] for bean '" + beanName +
"': There is already [" + oldBeanDefinition + "] bound.");
    }
    else if (oldBeanDefinition.getRole() < beanDefinition.getRole()) {
        if (this.logger.isWarnEnabled()) {
            this.logger.warn("Overriding user-defined bean definition for bean '" + beanName +
                "' with a framework-generated bean definition: replacing [" +
                oldBeanDefinition + "] with [" + beanDefinition + "]");
        }
    }
    else if (!beanDefinition.equals(oldBeanDefinition)) {
        if (this.logger.isInfoEnabled()) {
            this.logger.info("Overriding bean definition for bean '" + beanName +
                "' with a different definition: replacing [" + oldBeanDefinition +
                "] with [" + beanDefinition + "]");
        }
    }
    else {
        if (this.logger.isDebugEnabled()) {
            this.logger.debug("Overriding bean definition for bean '" + beanName +
                "' with an equivalent definition: replacing [" + oldBeanDefinition +
                "] with [" + beanDefinition + "]");
        }
    }
    this.beanDefinitionMap.put(beanName, beanDefinition);
}
else {
    if (hasBeanCreationStarted()) {
        //注册的过程中需要线程同步，以保证数据的一致性
        synchronized (this.beanDefinitionMap) {
            this.beanDefinitionMap.put(beanName, beanDefinition);
            List<String> updatedDefinitions = new ArrayList<>(this.beanDefinitionNames.size() + 1);
            updatedDefinitions.addAll(this.beanDefinitionNames);
            updatedDefinitions.add(beanName);
            this.beanDefinitionNames = updatedDefinitions;
            if (this.manualSingletonNames.contains(beanName)) {
                Set<String> updatedSingletons = new LinkedHashSet<>(this.manualSingletonNames);
                updatedSingletons.remove(beanName);
                this.manualSingletonNames = updatedSingletons;
            }
        }
    }
    else {
```

```
        this.beanDefinitionMap.put(beanName, beanDefinition);
        this.beanDefinitionNames.add(beanName);
        this.manualSingletonNames.remove(beanName);
    }
    this.frozenBeanDefinitionNames = null;
}

//检查是否已经注册过同名的 BeanDefinition
if (oldBeanDefinition != null || containsSingleton(beanName)) {
    //重置所有已经注册过的 BeanDefinition 的缓存
    resetBeanDefinition(beanName);
}
}
```

至此，Bean 配置信息中配置的 Bean 被解析后已经注册到 Spring IoC 容器中，被容器管理起来，真正完成了 Spring IoC 容器初始化的全部工作。现在 Spring IoC 容器中已经建立了所有 Bean 的配置信息，Bean 定义信息已经可以使用，并且可以被检索。Spring IoC 容器的作用就是对这些注册的 Bean 定义信息进行处理和维护。注册的 Bean 定义信息是 Spring IoC 容器控制反转的基础，正是有了这些信息，容器才可以进行依赖注入。

8.3 基于注解的 IoC 初始化

8.3.1 注解的前世今生

在 Spring 2.0 以后的版本中，引入了基于注解（Annotation）方式的配置，注解（Annotation）是 JDK 1.5 引入的一个新特性，用于简化 Bean 的配置，可以取代 XML 配置文件。开发人员对注解的态度也是萝卜青菜各有所爱，个人认为注解可以大大简化配置，提高开发速度，但也给后期维护增加了难度。目前来说，XML 方式相对成熟，便于统一管理。随着 Spring Boot 的兴起，基于注解的开发甚至实现了零配置。但作为个人的习惯，我还是倾向于 XML 配置文件和注解相互配合使用。Spring IoC 容器对于类级别的注解和类内部的注解处理策略如下。

（1）**类级别的注解**：如@Component、@Repository、@Controller、@Service，以及 Java EE 6 的@ManagedBean 和@Named，都是添加在类上的类级别注解，Spring IoC 容器根据注解的过滤规则扫描读取注解 Bean 定义类，并将其注册到 Spring IoC 容器中。

（2）**类内部的注解**：如@Autowire、@Value、@Resource，以及 EJB 和 WebService 相关的注解等，都是添加在类内部的字段或者方法上的类内部注解，Spring IoC 容器通过 Bean 后置注解处理器解析 Bean 内部的注解。

下面将分析 Spring 处理注解相关的源码。

8.3.2 定位 Bean 扫描路径

在 Spring 中管理注解的 Bean 定义的容器有两个：AnnotationConfigApplicationContext 和 AnnotationConfigWebApplicationContex。这两个是专门处理 Spring 注解方式配置的容器，直接依赖于将注解作为容器配置信息来源的 IoC 容器。AnnotationConfigWebApplicationContext 是 AnnotationConfigApplicationContext 的 Web 版本，两者的用法及对注解的处理方式几乎没有差别。我们以 AnnotationConfigApplicationContext 为例分析源码：

```java
public class AnnotationConfigApplicationContext extends GenericApplicationContext implements
AnnotationConfigRegistry {

    //保存一个读取注解的 Bean 定义读取器，并将其设置到容器中
    private final AnnotatedBeanDefinitionReader reader;

    //保存一个扫描指定类路径中注解 Bean 定义的扫描器，并将其设置到容器中
    private final ClassPathBeanDefinitionScanner scanner;

    //默认构造函数，初始化一个空容器，容器不包含任何 Bean 信息，需要稍后通过调用其 register()
    //方法注册配置类，并调用 refresh()方法刷新容器，触发容器对注解 Bean 的载入、解析和注册
    public AnnotationConfigApplicationContext() {
        this.reader = new AnnotatedBeanDefinitionReader(this);
        this.scanner = new ClassPathBeanDefinitionScanner(this);
    }

    public AnnotationConfigApplicationContext(DefaultListableBeanFactory beanFactory) {
        super(beanFactory);
        this.reader = new AnnotatedBeanDefinitionReader(this);
        this.scanner = new ClassPathBeanDefinitionScanner(this);
    }

    //最常用的构造函数，通过将涉及的配置类传递给该构造函数，实现将相应配置类中的 Bean 自动注册到容器中
    public AnnotationConfigApplicationContext(Class<?>... annotatedClasses) {
        this();
        register(annotatedClasses);
        refresh();
    }

    //该构造函数会自动扫描已给定的包及其子包下的所有类，并自动识别所有的 Spring Bean，将其注册到容器中
    public AnnotationConfigApplicationContext(String... basePackages) {
        this();
        scan(basePackages);
        refresh();
```

```java
    }

    @Override
    public void setEnvironment(ConfigurableEnvironment environment) {
        super.setEnvironment(environment);
        this.reader.setEnvironment(environment);
        this.scanner.setEnvironment(environment);
    }

    //为容器的注解 Bean 读取器和注解 Bean 扫描器设置 Bean 名称产生器
    public void setBeanNameGenerator(BeanNameGenerator beanNameGenerator) {
        this.reader.setBeanNameGenerator(beanNameGenerator);
        this.scanner.setBeanNameGenerator(beanNameGenerator);
        getBeanFactory().registerSingleton(
                AnnotationConfigUtils.CONFIGURATION_BEAN_NAME_GENERATOR, beanNameGenerator);
    }

    //为容器的注解 Bean 读取器和注解 Bean 扫描器设置作用范围元信息解析器
    public void setScopeMetadataResolver(ScopeMetadataResolver scopeMetadataResolver) {
        this.reader.setScopeMetadataResolver(scopeMetadataResolver);
        this.scanner.setScopeMetadataResolver(scopeMetadataResolver);
    }

    //为容器注册一个要被处理的注解 Bean，新注册的 Bean，必须手动调用容器的
    //refresh()方法刷新容器，触发容器对新注册的 Bean 的处理
    public void register(Class<?>... annotatedClasses) {
        Assert.notEmpty(annotatedClasses, "At least one annotated class must be specified");
        this.reader.register(annotatedClasses);
    }

    //扫描指定包路径及其子包下的注解类，为了使新添加的类被处理，必须手动调用 refresh()方法刷新容器
    public void scan(String... basePackages) {
        Assert.notEmpty(basePackages, "At least one base package must be specified");
        this.scanner.scan(basePackages);
    }

    ...
}
```

通过上面的源码可以看到，Spring 对注解的处理分为以下两种方式。

（1）直接将注解 Bean 注册到容器中：可以在初始化容器时注册；也可以在容器创建之后手动调用注册方法向容器注册，然后通过手动刷新容器使容器对注册的注解 Bean 进行处理。

（2）通过扫描指定的包及其子包下的所有类处理：在初始化注解容器时指定要自动扫描的路

径，如果容器创建以后向给定路径动态添加了注解 Bean，则需要手动调用容器扫描的方法手动刷新容器，使容器对所注册的注解 Bean 进行处理。

接下来，将会详细分析两种处理方式的实现过程。

8.3.3 读取注解的元数据

1. AnnotationConfigApplicationContext 通过调用注解 Bean 定义读取器注册注解 Bean

AnnotatedBeanDefinitionReader 的 register()方法向容器注册指定的注解 Bean，注解 Bean 定义读取器向容器注册注解 Bean 的源码如下：

```java
//注册多个注解 Bean 定义类
public void register(Class<?>... annotatedClasses) {
    for (Class<?> annotatedClass : annotatedClasses) {
        registerBean(annotatedClass);
    }
}

//注册一个注解 Bean 定义类
public void registerBean(Class<?> annotatedClass) {
    doRegisterBean(annotatedClass, null, null, null);
}

public <T> void registerBean(Class<T> annotatedClass, @Nullable Supplier<T> instanceSupplier) {
    doRegisterBean(annotatedClass, instanceSupplier, null, null);
}

public <T> void registerBean(Class<T> annotatedClass, String name, @Nullable Supplier<T> instanceSupplier) {
    doRegisterBean(annotatedClass, instanceSupplier, name, null);
}

//Bean 定义读取器注册注解 Bean 定义的入口方法
@SuppressWarnings("unchecked")
public void registerBean(Class<?> annotatedClass, Class<? extends Annotation>... qualifiers) {
    doRegisterBean(annotatedClass, null, null, qualifiers);
}

//Bean 定义读取器向容器注册注解 Bean 定义类
@SuppressWarnings("unchecked")
public void registerBean(Class<?> annotatedClass, String name, Class<? extends Annotation>... qualifiers) {
    doRegisterBean(annotatedClass, null, name, qualifiers);
}
```

```java
//Bean 定义读取器向容器注册注解 Bean 定义类
<T> void doRegisterBean(Class<T> annotatedClass, @Nullable Supplier<T> instanceSupplier,
@Nullable String name,
    @Nullable Class<? extends Annotation>[] qualifiers, BeanDefinitionCustomizer...
definitionCustomizers) {

    //根据指定的注解 Bean 定义类，创建 Spring 容器中对注解 Bean 的封装的数据结构
    AnnotatedGenericBeanDefinition abd = new AnnotatedGenericBeanDefinition(annotatedClass);
    if (this.conditionEvaluator.shouldSkip(abd.getMetadata())) {
        return;
    }

    abd.setInstanceSupplier(instanceSupplier);
    //解析注解 Bean 定义的作用域，若@Scope("prototype")，则 Bean 为原型类型
    //若@Scope("singleton")，则 Bean 为单态类型
    ScopeMetadata scopeMetadata = this.scopeMetadataResolver.resolveScopeMetadata(abd);
    //为注解 Bean 定义设置作用域
    abd.setScope(scopeMetadata.getScopeName());
    //为注解 Bean 定义生成 Bean 名称
    String beanName = (name != null ? name : this.beanNameGenerator.generateBeanName(abd,
this.registry));

    //处理注解 Bean 定义中的通用注解
    AnnotationConfigUtils.processCommonDefinitionAnnotations(abd);
    //如果在向容器注册注解 Bean 定义时，使用了额外的限定符注解，则解析限定符注解
    //主要配置 autowiring 自动依赖注入装配的限定条件，即@Qualifier 注解
    //Spring 自动依赖注入默认按类型装配，如果使用@Qualifier 则按名称装配
    if (qualifiers != null) {
        for (Class<? extends Annotation> qualifier : qualifiers) {
            //如果配置了@Primary 注解，设置该 Bean 为 autowiring 自动依赖注入装配时的首选
            if (Primary.class == qualifier) {
                abd.setPrimary(true);
            }
            //如果配置了@Lazy 注解，则设置该 Bean 为非延迟初始化，如果没有配置，则该 Bean 为预实例化
            else if (Lazy.class == qualifier) {
                abd.setLazyInit(true);
            }
            //如果使用了除@Primary 和@Lazy 以外的其他注解，则为该 Bean 添加一
            //个 autowiring 自动依赖注入装配限定符，该 Bean 在进 autowiring
            //自动依赖注入装配时，根据名称装配限定符指定的 Bean
            else {
                abd.addQualifier(new AutowireCandidateQualifier(qualifier));
            }
        }
    }
    for (BeanDefinitionCustomizer customizer : definitionCustomizers) {
```

```
    customizer.customize(abd);
}
//创建一个指定 Bean 名称的 Bean 定义对象,封装注解 Bean 定义类数据
BeanDefinitionHolder definitionHolder = new BeanDefinitionHolder(abd, beanName);
//根据注解 Bean 定义类中配置的作用域,创建相应的代理对象
definitionHolder        =        AnnotationConfigUtils.applyScopedProxyMode(scopeMetadata,
definitionHolder, this.registry);
//向 IoC 容器注册注解 Bean 类定义对象
BeanDefinitionReaderUtils.registerBeanDefinition(definitionHolder, this.registry);
}
```

从上面的源码可以看出,注册注解 Bean 定义类的基本步骤如下:

(1)使用注解元数据解析器解析注解 Bean 中关于作用域的配置。

(2)使用 AnnotationConfigUtils 的 processCommonDefinitionAnnotations()方法处理注解 Bean 定义类中通用的注解。

(3)使用 AnnotationConfigUtils 的 applyScopedProxyMode()方法创建作用域的代理对象。

(4)通过 BeanDefinitionReaderUtils 向容器注册 Bean。

下面继续分析这 4 步的具体实现过程。

2. AnnotationScopeMetadataResolver 解析作用域元信息

AnnotationScopeMetadataResolver 通过 resolveScopeMetadata()方法解析注解 Bean 定义类的作用域元信息,即判断注册的 Bean 是原生类型(prototype)还是单态(singleton)类型,其源码如下:

```
//解析注解 Bean 定义类中的作用域元信息
@Override
public ScopeMetadata resolveScopeMetadata(BeanDefinition definition) {
    ScopeMetadata metadata = new ScopeMetadata();
    if (definition instanceof AnnotatedBeanDefinition) {
        AnnotatedBeanDefinition annDef = (AnnotatedBeanDefinition) definition;
        //从注解 Bean 定义类的属性中查找属性为 Scope 的值,即@Scope 注解的值
        //annDef.getMetadata()方法将 Bean 中所有的注解和注解的值存放在一个 map 集合中
        AnnotationAttributes attributes = AnnotationConfigUtils.attributesFor(
                annDef.getMetadata(), this.scopeAnnotationType);
        //将获取的@Scope 注解的值设置到要返回的对象中
        if (attributes != null) {
            metadata.setScopeName(attributes.getString("value"));
            //获取@Scope 注解中的 proxyMode 属性值,在创建代理对象时会用到
            ScopedProxyMode proxyMode = attributes.getEnum("proxyMode");
            //如果@Scope 的 proxyMode 属性为 DEFAULT 或者 NO
            if (proxyMode == ScopedProxyMode.DEFAULT) {
```

```
        //设置 proxyMode 为 NO
        proxyMode = this.defaultProxyMode;
    }
    //为返回的元数据设置 proxyMode
    metadata.setScopedProxyMode(proxyMode);
  }
}
//返回解析的作用域元信息对象
return metadata;
}
```

上述代码中的 annDef.getMetadata()方法用于获取对象中指定类型的注解的值。

3. AnnotationConfigUtils 处理注解 Bean 定义类中的通用注解

AnnotationConfigUtils 的 processCommonDefinitionAnnotations()方法在向容器注册 Bean 之前，首先对注解 Bean 定义类中的通用注解进行处理，源码如下：

```
//处理 Bean 定义中的通用注解
static void processCommonDefinitionAnnotations(AnnotatedBeanDefinition abd,
 AnnotatedTypeMetadata metadata) {
  AnnotationAttributes lazy = attributesFor(metadata, Lazy.class);
  //如果 Bean 定义中有@Lazy 注解，则将该 Bean 预实例化属性设置为@lazy 注解的值
  if (lazy != null) {
    abd.setLazyInit(lazy.getBoolean("value"));
  }

  else if (abd.getMetadata() != metadata) {
    lazy = attributesFor(abd.getMetadata(), Lazy.class);
    if (lazy != null) {
      abd.setLazyInit(lazy.getBoolean("value"));
    }
  }
  //如果 Bean 定义中有@Primary 注解，则将该 Bean 设置为 autowiring 自动依赖注入装配的首选对象
  if (metadata.isAnnotated(Primary.class.getName())) {
    abd.setPrimary(true);
  }
  //如果 Bean 定义中有@DependsOn 注解，则为该 Bean 设置所依赖的 Bean 名称，
  //容器将确保在实例化该 Bean 之前首先实例化所依赖的 Bean
  AnnotationAttributes dependsOn = attributesFor(metadata, DependsOn.class);
  if (dependsOn != null) {
    abd.setDependsOn(dependsOn.getStringArray("value"));
  }

  if (abd instanceof AbstractBeanDefinition) {
    AbstractBeanDefinition absBd = (AbstractBeanDefinition) abd;
    AnnotationAttributes role = attributesFor(metadata, Role.class);
```

```
    if (role != null) {
        absBd.setRole(role.getNumber("value").intValue());
    }
    AnnotationAttributes description = attributesFor(metadata, Description.class);
    if (description != null) {
        absBd.setDescription(description.getString("value"));
    }
}
```

4. AnnotationConfigUtils 根据注解 Bean 定义类中配置的作用域为其应用相应的代理策略

AnnotationConfigUtils 的 applyScopedProxyMode()方法根据注解 Bean 定义类中配置的作用域 @Scope 注解的值，为 Bean 定义应用相应的代理模式，主要在 Spring 面向切面编程（AOP）中使用，源码如下：

```
//根据作用域为 Bean 定义应用的代理模式
static BeanDefinitionHolder applyScopedProxyMode(
    ScopeMetadata metadata, BeanDefinitionHolder definition, BeanDefinitionRegistry
registry) {
    //获取注解 Bean 定义类中@Scope 注解的 proxyMode 属性值
    ScopedProxyMode scopedProxyMode = metadata.getScopedProxyMode();
    //如果配置的@Scope 注解的 proxyMode 属性值为 NO，则不应用代理模式
    if (scopedProxyMode.equals(ScopedProxyMode.NO)) {
        return definition;
    }
    //获取配置的@Scope 注解的 proxyMode 属性值，如果为 TARGET_CLASS
    //则返回 true，如果为 INTERFACES 则返回 false
    boolean proxyTargetClass = scopedProxyMode.equals(ScopedProxyMode.TARGET_CLASS);
    //为注册的 Bean 创建相应模式的代理对象
    return ScopedProxyCreator.createScopedProxy(definition, registry, proxyTargetClass);
}
```

这里如何为 Bean 引用创建相应模式的代理，我们不再做深入的分析。

5. BeanDefinitionReaderUtils 向容器注册 Bean

BeanDefinitionReaderUtils 主要校验 BeanDefinition 信息，然后将 Bean 添加到容器中一个管理 BeanDefinition 的 HashMap 中。

8.3.4 扫描指定包并解析为 BeanDefinition

当创建注解处理容器时，如果传入的初始参数是注解 Bean 定义类所在的包，注解容器将扫描给定的包及其子包，将扫描到的注解 Bean 定义载入并进行注册。

1. ClassPathBeanDefinitionScanner 扫描给定的包及其子包

AnnotationConfigApplicationContext 通过调用类路径 Bean 定义扫描器 ClassPathBeanDefinition-Scanner 扫描给定包及其子包下的所有类，主要源码如下：

```java
public class ClassPathBeanDefinitionScanner extends ClassPathScanningCandidateComponentProvider {

    //创建一个类路径 Bean 定义扫描器
    public ClassPathBeanDefinitionScanner(BeanDefinitionRegistry registry) {
        this(registry, true);
    }

    //为容器创建一个类路径 Bean 定义扫描器，并指定是否使用默认的扫描过滤规则
    //即 Spring 默认扫描配置@Component、@Repository、@Service、@Controller
    //注解的 Bean，同时也支持 Java EE 6 的@ManagedBean 和 JSR-330 的@Named 注解
    public ClassPathBeanDefinitionScanner(BeanDefinitionRegistry registry, boolean useDefaultFilters) {
        this(registry, useDefaultFilters, getOrCreateEnvironment(registry));
    }

    public ClassPathBeanDefinitionScanner(BeanDefinitionRegistry registry, boolean
            useDefaultFilters, Environment environment) {

        this(registry, useDefaultFilters, environment,
                (registry instanceof ResourceLoader ? (ResourceLoader) registry : null));
    }

    public ClassPathBeanDefinitionScanner(BeanDefinitionRegistry registry, boolean
            useDefaultFilters, Environment environment, @Nullable ResourceLoader resourceLoader) {

        Assert.notNull(registry, "BeanDefinitionRegistry must not be null");
        //为容器设置加载 Bean 定义的注册器
        this.registry = registry;

        if (useDefaultFilters) {
            registerDefaultFilters();
        }
        setEnvironment(environment);
        //为容器设置资源加载器
        setResourceLoader(resourceLoader);
    }

    //调用类路径 Bean 定义扫描器入口方法
    public int scan(String... basePackages) {
        //获取容器中已经注册的 Bean 的个数
        int beanCountAtScanStart = this.registry.getBeanDefinitionCount();
```

```java
    //启动扫描器扫描给定包
    doScan(basePackages);

    //注册注解配置(Annotation config)处理器
    if (this.includeAnnotationConfig) {
        AnnotationConfigUtils.registerAnnotationConfigProcessors(this.registry);
    }

    //返回注册的 Bean 的个数
    return (this.registry.getBeanDefinitionCount() - beanCountAtScanStart);
}

//类路径 Bean 定义扫描器扫描给定包及其子包
protected Set<BeanDefinitionHolder> doScan(String... basePackages) {
    Assert.notEmpty(basePackages, "At least one base package must be specified");
    //创建一个集合，存放扫描到 Bean 定义的封装类
    Set<BeanDefinitionHolder> beanDefinitions = new LinkedHashSet<>();
    //遍历扫描所有给定的包
    for (String basePackage : basePackages) {
        //调用父类 ClassPathScanningCandidateComponentProvider 的方法
        //扫描给定类路径，获取符合条件的 Bean 定义
        Set<BeanDefinition> candidates = findCandidateComponents(basePackage);
        //遍历扫描到的 Bean
        for (BeanDefinition candidate : candidates) {
            //获取 Bean 定义类中@Scope 注解的值，即获取 Bean 的作用域
            ScopeMetadata scopeMetadata = this.scopeMetadataResolver.resolveScopeMetadata(candidate);
            //为 Bean 设置注解配置的作用域
            candidate.setScope(scopeMetadata.getScopeName());
            //为 Bean 生成名称
            String beanName = this.beanNameGenerator.generateBeanName(candidate, this.registry);
            //如果扫描到的 Bean 不是 Spring 的注解 Bean，则为 Bean 设置默认值
            //设置 Bean 的自动依赖注入装配属性等
            if (candidate instanceof AbstractBeanDefinition) {
                postProcessBeanDefinition((AbstractBeanDefinition) candidate, beanName);
            }
            //如果扫描到的 Bean 是 Spring 的注解 Bean，则处理其通用的注解
            if (candidate instanceof AnnotatedBeanDefinition) {
                //处理注解 Bean 中通用的注解，在分析注解 Bean 定义类读取器时已经分析过
                AnnotationConfigUtils.processCommonDefinitionAnnotations((AnnotatedBeanDefinition) candidate);
            }
            //根据 Bean 名称检查指定的 Bean 是否需要在容器中注册，或者是否在容器中存在冲突
            if (checkCandidate(beanName, candidate)) {
                BeanDefinitionHolder definitionHolder = new BeanDefinitionHolder(candidate, beanName);
                //根据注解中配置的作用域，为 Bean 应用相应的代理模式
```

```
                definitionHolder =
                    AnnotationConfigUtils.applyScopedProxyMode(scopeMetadata,
definitionHolder, this.registry);
                beanDefinitions.add(definitionHolder);
                //向容器注册扫描到的 Bean
                registerBeanDefinition(definitionHolder, this.registry);
            }
        }
    }
    return beanDefinitions;
}
...
}
```

类路径 Bean 定义扫描器 ClassPathBeanDefinitionScanner 主要通过 findCandidateComponents() 方法调用其父类 ClassPathScanningCandidateComponentProvider 来扫描获取给定包及其子包的类。

2. ClassPathScanningCandidateComponentProvider 扫描给定包及其子包的类

ClassPathScanningCandidateComponentProvider 类的 findCandidateComponents()方法具体实现扫描给定类路径包的功能，主要源码如下：

```
public class ClassPathScanningCandidateComponentProvider implements EnvironmentCapable, ResourceLoaderAware {

    //保存过滤规则包含的注解，即 Spring 默认的@Component、@Repository、@Service、
    //@Controller 注解的 Bean，以及 Java EE 6 的@ManagedBean 和 JSR-330 的@Named 注解
    private final List<TypeFilter> includeFilters = new LinkedList<>();

    //保存过滤规则要排除的注解
    private final List<TypeFilter> excludeFilters = new LinkedList<>();

    //构造方法，该方法在子类 ClassPathBeanDefinitionScanner 的构造方法中被调用
    public ClassPathScanningCandidateComponentProvider(boolean useDefaultFilters) {
        this(useDefaultFilters, new StandardEnvironment());
    }

    public ClassPathScanningCandidateComponentProvider(boolean useDefaultFilters, Environment environment) {
        //如果使用 Spring 默认的过滤规则，则向容器注册过滤规则
        if (useDefaultFilters) {
            registerDefaultFilters();
        }
        setEnvironment(environment);
        setResourceLoader(null);
```

```java
}
//向容器注册过滤规则
@SuppressWarnings("unchecked")
protected void registerDefaultFilters() {
    //向要包含的过滤规则中添加@Component 注解类，注意 Spring 中@Repository、
    //@Service 和@Controller 都是 Component，因为这些注解都添加了@Component 注解
    this.includeFilters.add(new AnnotationTypeFilter(Component.class));
    //获取当前类的类加载器
    ClassLoader cl = ClassPathScanningCandidateComponentProvider.class.getClassLoader();
    try {
        //向要包含的过滤规则添加 Java EE 6 的@ManagedBean 注解
        this.includeFilters.add(new AnnotationTypeFilter(
                ((Class<? extends Annotation>) ClassUtils.forName("javax.annotation.ManagedBean", cl)), false));
        logger.debug("JSR-250 'javax.annotation.ManagedBean' found and supported for component scanning");
    }
    catch (ClassNotFoundException ex) {
        // JSR-250 1.1 API (as included in Java EE 6) not available - simply skip.
    }
    try {
        //向要包含的过滤规则添加@Named 注解
        this.includeFilters.add(new AnnotationTypeFilter(
                ((Class<? extends Annotation>) ClassUtils.forName("javax.inject.Named", cl)), false));
        logger.debug("JSR-330 'javax.inject.Named' annotation found and supported for component scanning");
    }
    catch (ClassNotFoundException ex) {
        // JSR-330 API not available - simply skip.
    }
}


//扫描给定类路径的包
public Set<BeanDefinition> findCandidateComponents(String basePackage) {
    if (this.componentsIndex != null && indexSupportsIncludeFilters()) {
        return addCandidateComponentsFromIndex(this.componentsIndex, basePackage);
    }
    else {
        return scanCandidateComponents(basePackage);
    }
}

private Set<BeanDefinition> addCandidateComponentsFromIndex(CandidateComponentsIndex index, String basePackage) {
```

```java
//创建存储扫描到的类的集合
Set<BeanDefinition> candidates = new LinkedHashSet<>();
try {
    Set<String> types = new HashSet<>();
    for (TypeFilter filter : this.includeFilters) {
        String stereotype = extractStereotype(filter);
        if (stereotype == null) {
            throw new IllegalArgumentException("Failed to extract stereotype from " + filter);
        }
        types.addAll(index.getCandidateTypes(basePackage, stereotype));
    }
    boolean traceEnabled = logger.isTraceEnabled();
    boolean debugEnabled = logger.isDebugEnabled();
    for (String type : types) {
        //为指定资源获取元数据读取器,元数据读取器通过汇编(ASM)读取资源的元信息
        MetadataReader metadataReader = getMetadataReaderFactory().getMetadataReader(type);
        //如果扫描到的类符合容器配置的过滤规则
        if (isCandidateComponent(metadataReader)) {
            //通过汇编(ASM)读取资源字节码中的 Bean 定义的元信息
            AnnotatedGenericBeanDefinition sbd = new AnnotatedGenericBeanDefinition(
                    metadataReader.getAnnotationMetadata());
            if (isCandidateComponent(sbd)) {
                if (debugEnabled) {
                    logger.debug("Using candidate component class from index: " + type);
                }
                candidates.add(sbd);
            }
            else {
                if (debugEnabled) {
                    logger.debug("Ignored because not a concrete top-level class: " + type);
                }
            }
        }
        else {
            if (traceEnabled) {
                logger.trace("Ignored because matching an exclude filter: " + type);
            }
        }
    }
}
catch (IOException ex) {
    throw new BeanDefinitionStoreException("I/O failure during classpath scanning", ex);
}
return candidates;
}
//判断元信息读取器读取的类是否符合容器定义的注解过滤规则
```

```java
protected boolean isCandidateComponent(MetadataReader metadataReader) throws IOException {
    //如果读取的类的注解在排除注解过滤规则中,返回false
    for (TypeFilter tf : this.excludeFilters) {
        if (tf.match(metadataReader, getMetadataReaderFactory())) {
            return false;
        }
    }
    //如果读取的类的注解在包含的注解过滤规则中,则返回ture
    for (TypeFilter tf : this.includeFilters) {
        if (tf.match(metadataReader, getMetadataReaderFactory())) {
            return isConditionMatch(metadataReader);
        }
    }
    //如果读取的类的注解既不在排除规则中,也不在包含规则中,则返回false
    return false;
}
```

8.3.5 注册注解 BeanDefinition

AnnotationConfigWebApplicationContext 是 AnnotationConfigApplicationContext 的 Web 版,它们对于注解 Bean 的注册和扫描是基本相同的,但是 AnnotationConfigWebApplicationContext 对注解 Bean 定义的载入稍有不同。AnnotationConfigWebApplicationContext 注入注解 Bean 定义源码如下:

```java
//载入注解Bean定义资源
@Override
protected void loadBeanDefinitions(DefaultListableBeanFactory beanFactory) {
    //为容器设置注解Bean定义读取器
    AnnotatedBeanDefinitionReader reader = getAnnotatedBeanDefinitionReader(beanFactory);
    //为容器设置类路径Bean定义扫描器
    ClassPathBeanDefinitionScanner scanner = getClassPathBeanDefinitionScanner(beanFactory);

    //获取容器的Bean名称生成器
    BeanNameGenerator beanNameGenerator = getBeanNameGenerator();
    //为注解Bean定义读取器和类路径扫描器设置Bean名称生成器
    if (beanNameGenerator != null) {
        reader.setBeanNameGenerator(beanNameGenerator);
        scanner.setBeanNameGenerator(beanNameGenerator);
        beanFactory.registerSingleton(AnnotationConfigUtils.CONFIGURATION_BEAN_NAME_GENERATOR,
beanNameGenerator);
    }
```

```java
//获取容器的作用域元信息解析器
ScopeMetadataResolver scopeMetadataResolver = getScopeMetadataResolver();
//为注解 Bean 定义读取器和类路径扫描器设置作用域元信息解析器
if (scopeMetadataResolver != null) {
    reader.setScopeMetadataResolver(scopeMetadataResolver);
    scanner.setScopeMetadataResolver(scopeMetadataResolver);
}

if (!this.annotatedClasses.isEmpty()) {
    if (logger.isInfoEnabled()) {
        logger.info("Registering annotated classes: [" +
            StringUtils.collectionToCommaDelimitedString(this.annotatedClasses) + "]");
    }
    reader.register(this.annotatedClasses.toArray(new
Class<?>[this.annotatedClasses.size()]));
}

if (!this.basePackages.isEmpty()) {
    if (logger.isInfoEnabled()) {
        logger.info("Scanning base packages: [" +
            StringUtils.collectionToCommaDelimitedString(this.basePackages) + "]");
    }
    scanner.scan(this.basePackages.toArray(new String[this.basePackages.size()]));
}

//获取容器定义的 Bean 定义资源路径
String[] configLocations = getConfigLocations();
//如果定位的 Bean 定义资源路径不为空
if (configLocations != null) {
    for (String configLocation : configLocations) {
        try {
            //使用当前容器的类加载器加载定位路径的字节码文件
            Class<?> clazz = ClassUtils.forName(configLocation, getClassLoader());
            if (logger.isInfoEnabled()) {
                logger.info("Successfully resolved class for [" + configLocation + "]");
            }
            reader.register(clazz);
        }
        catch (ClassNotFoundException ex) {
            if (logger.isDebugEnabled()) {
                logger.debug("Could not load class for config location [" + configLocation +
                    "] - trying package scan. " + ex);
            }
            //如果容器类加载器加载定义路径的 Bean 定义资源失败
```

```
            //则启用容器类路径扫描器扫描给定路径包及其子包中的类
            int count = scanner.scan(configLocation);
            if (logger.isInfoEnabled()) {
                if (count == 0) {
                    logger.info("No annotated classes found for specified class/package [" +
configLocation + "]");
                }
                else {
                    logger.info("Found " + count + " annotated classes in package [" +
configLocation + "]");
                }
            }
        }
    }
}
```

以上就是解析和注入注解配置资源的全过程分析。

8.4 IoC 容器初始化小结

下面总结一下 IoC 容器初始化的基本步骤：

（1）初始化的入口由容器实现中的 refresh()方法调用来完成。

（2）对 Bean 定义载入 IoC 容器使用的方法是 loadBeanDefinition()。

大致过程如下：通过 ResourceLoader 来完成资源文件的定位，DefaultResourceLoader 是默认的实现，同时上下文本身就给出了 ResourceLoader 的实现，可以通过类路径、文件系统、URL 等方式来定位资源。如果是 XmlBeanFactory 作为 IoC 容器，那么需要为它指定 Bean 定义的资源，也就是说 Bean 定义文件时通过抽象成 Resource 来被 IoC 容器处理，容器通过 BeanDefinitionReader 来完成定义信息的解析和 Bean 信息的注册，往往使用 XmlBeanDefinitionReader 来解析 Bean 的 XML 定义文件——实际的处理过程是委托给 BeanDefinitionParserDelegate 来完成的，从而得到 Bean 的定义信息，这些信息在 Spring 中使用 BeanDefinition 来表示——这个名字可以让我们想到 loadBeanDefinition()、registerBeanDefinition()这些相关方法。它们都是为处理 BeanDefinition 服务的，容器解析得到 BeanDefinition 以后，需要在 IoC 容器中注册，这由 IoC 实现 BeanDefinitionRegistry 接口来实现。注册过程就是在 IoC 容器内部维护的一个 HashMap 来保存得到的 BeanDefinition 的过程。这个 HashMap 是 IoC 容器持有 Bean 信息的场所，以后对 Bean 的操作都是围绕这个 HashMap 来实现的。

之后我们就可以通过 BeanFactory 和 ApplicationContext 来享受 Spring IoC 的服务了。在使用 IoC 容器的时候我们注意到，除了少量黏合代码，绝大多数以正确 IoC 风格编写的应用程序代码完全不用关心如何到达工厂，因为容器将把这些对象与容器管理的其他对象钩在一起了。基本的策略是把工厂放到已知的地方，最好放在对预期使用的上下文有意义的地方，以及代码将实际需要访问工厂的地方。Spring 本身提供了对声明式载入 Web 应用程序用法的应用程序上下文，并将其存储在 ServletContext 的框架实现中。

最后请到本书"下载资源"中查看 Spring IoC 初始化全过程的运行时序图。

第 9 章 一步一步手绘 Spring DI 运行时序图

9.1 Spring 自动装配之依赖注入

9.1.1 依赖注入发生的时间

当 Spring IoC 容器完成了 Bean 定义资源的定位、载入和解析注册，IoC 容器就可以管理 Bean 定义的相关数据了，但是此时 IoC 容器还没有对所管理的 Bean 进行依赖注入（DI），依赖注入在以下两种情况下发生：

（1）用户第一次调用 getBean()方法时，IoC 容器触发依赖注入。

（2）当用户在配置文件中将<bean>元素配置了 lazy-init=false 属性时，即让容器在解析注册 Bean 定义时进行预实例化，触发依赖注入。

BeanFactory 接口定义了 Spring IoC 容器的基本功能规范，是 Spring IoC 容器所应遵守的最低层和最基本的编程规范。BeanFactory 接口中定义了几个 getBean()方法，用于用户向 IoC 容器索取被管理的 Bean 的方法，我们通过分析其子类的具体实现来理解 Spring IoC 容器在用户索取 Bean 时如何完成依赖注入。

在 BeanFactory 中我们可以看到 getBean(String...)方法，但它的具体实现在 AbstractBeanFactory 中，如下图所示。

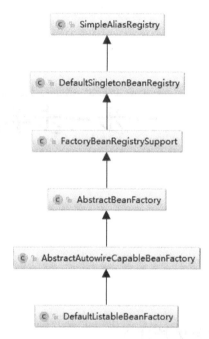

9.1.2 寻找获取 Bean 的入口

AbstractBeanFactory 的 getBean()相关方法的源码如下：

```
//获取 IoC 容器中指定名称的 Bean
@Override
public Object getBean(String name) throws BeansException {
    //doGetBean 才真正向 IoC 容器获取被管理的 Bean
    return doGetBean(name, null, null, false);
}

//获取 IoC 容器中指定名称和类型的 Bean
@Override
public <T> T getBean(String name, @Nullable Class<T> requiredType) throws BeansException {
    //doGetBean 才真正向 IoC 容器获取被管理的 Bean
    return doGetBean(name, requiredType, null, false);
}
```

```java
//获取 IoC 容器中指定名称和参数的 Bean
@Override
public Object getBean(String name, Object... args) throws BeansException {
    //doGetBean 才真正向 IoC 容器获取被管理的 Bean
    return doGetBean(name, null, args, false);
}

//获取 IoC 容器中指定名称、类型和参数的 Bean
public <T> T getBean(String name, @Nullable Class<T> requiredType, @Nullable Object... args)
        throws BeansException {
    //doGetBean 才真正向 IoC 容器获取被管理的 Bean
    return doGetBean(name, requiredType, args, false);
}

@SuppressWarnings("unchecked")
//真正实现向 IoC 容器获取 Bean 的功能,也是触发依赖注入的地方
protected <T> T doGetBean(final String name, @Nullable final Class<T> requiredType,
        @Nullable final Object[] args, boolean typeCheckOnly) throws BeansException {

    //根据指定的名称获取被管理的 Bean 的名称,剥离指定名称中对容器的相关依赖
    //如果指定的是别名,将别名转换为规范的 Bean 名称
    final String beanName = transformedBeanName(name);
    Object bean;

    //先从缓存中读取是否已经有被创建过的单例模式的 Bean
    //对于单例模式的 Bean 整个 IoC 容器中只创建一次,不需要重复创建
    Object sharedInstance = getSingleton(beanName);
    //IoC 容器创建单例模式的 Bean 实例对象
    if (sharedInstance != null && args == null) {
        if (logger.isDebugEnabled()) {
            //如果在容器中已有指定名称的单例模式的 Bean 被创建,直接返回已经创建的 Bean
            if (isSingletonCurrentlyInCreation(beanName)) {
                logger.debug("Returning eagerly cached instance of singleton bean '" + beanName +
                        "' that is not fully initialized yet - a consequence of a circular reference");
            }
            else {
                logger.debug("Returning cached instance of singleton bean '" + beanName + "'");
            }
        }
        //获取给定 Bean 的实例对象,主要完成 FactoryBean 的相关处理
        //注意:BeanFactory 是管理 Bean 的工厂,FactoryBean 是创建对象的工厂 Bean,两者之间有区别
        bean = getObjectForBeanInstance(sharedInstance, name, beanName, null);
    }

    else {
        //缓存中没有正在创建的单例模式的 Bean
```

```java
//缓存中已经有原型模式的 Bean
//但是由于循环引用导致实例化对象失败
if (isPrototypeCurrentlyInCreation(beanName)) {
    throw new BeanCurrentlyInCreationException(beanName);
}

//对 IoC 容器中是否存在指定名称的 BeanDefinition 进行检查,首先检查是否
//能在当前的 BeanFactory 中获取所需要的 Bean,如果不能则委托当前容器
//的父容器去查找,如果还是找不到则沿着容器的继承体系继续向父容器查找
BeanFactory parentBeanFactory = getParentBeanFactory();
//当前容器的父容器存在,且当前容器中不存在指定名称的 Bean
if (parentBeanFactory != null && !containsBeanDefinition(beanName)) {
    //解析指定 Bean 名称的原始名称
    String nameToLookup = originalBeanName(name);
    if (parentBeanFactory instanceof AbstractBeanFactory) {
        return ((AbstractBeanFactory) parentBeanFactory).doGetBean(
            nameToLookup, requiredType, args, typeCheckOnly);
    }
    else if (args != null) {
        //委派父容器根据指定名称和显式的参数查找
        return (T) parentBeanFactory.getBean(nameToLookup, args);
    }
    else {
        //委派父容器根据指定名称和类型查找
        return parentBeanFactory.getBean(nameToLookup, requiredType);
    }
}

//创建的 Bean 是否需要进行类型验证,一般不需要
if (!typeCheckOnly) {
    //向容器标记指定的 Bean 已经被创建
    markBeanAsCreated(beanName);
}

try {
    //根据指定 Bean 名称获取其父级 Bean 定义
    //主要解决 Bean 继承时子类和父类公共属性问题
    final RootBeanDefinition mbd = getMergedLocalBeanDefinition(beanName);
    checkMergedBeanDefinition(mbd, beanName, args);

    //获取当前 Bean 所有依赖 Bean 的名称
    String[] dependsOn = mbd.getDependsOn();
    //如果当前 Bean 有依赖 Bean
    if (dependsOn != null) {
        for (String dep : dependsOn) {
            if (isDependent(beanName, dep)) {
                throw new BeanCreationException(mbd.getResourceDescription(), beanName,
```

```java
                        "Circular depends-on relationship between '" + beanName + "' and '" +
dep + "'");
            }
            //把被依赖 Bean 注册给当前依赖的 Bean
            registerDependentBean(dep, beanName);
            //递归调用 getBean()方法，获取当前 Bean 的依赖 Bean
            getBean(dep);
        }
    }

    //创建单例模式的 Bean 的实例对象
    if (mbd.isSingleton()) {
        //这里使用了一个匿名内部类创建 Bean 实例对象，并且注册给所依赖的对象
        sharedInstance = getSingleton(beanName, () -> {
            try {
                //创建一个指定 Bean 的实例对象，如果有父级继承，则合并子类和父类的定义
                return createBean(beanName, mbd, args);
            }
            catch (BeansException ex) {
                //显式地从容器中单例模式的 Bean 缓存中清除实例对象
                destroySingleton(beanName);
                throw ex;
            }
        });
        //获取给定 Bean 的实例对象
        bean = getObjectForBeanInstance(sharedInstance, name, beanName, mbd);
    }

    //IoC 容器创建原型模式的 Bean 的实例对象
    else if (mbd.isPrototype()) {
        //原型模式（Prototype）每次都会创建一个新的对象
        Object prototypeInstance = null;
        try {
            //回调 beforePrototypeCreation()方法，默认的功能是注册当前创建的原型对象
            beforePrototypeCreation(beanName);
            //创建指定 Bean 的对象实例
            prototypeInstance = createBean(beanName, mbd, args);
        }
        finally {
            //回调 afterPrototypeCreation()方法，默认的功能是告诉 IoC 容器不再创建指定 Bean 的原型对象
            afterPrototypeCreation(beanName);
        }
        //获取指定 Bean 的实例对象
        bean = getObjectForBeanInstance(prototypeInstance, name, beanName, mbd);
    }

    //要创建的 Bean 既不是单例模式的，也不是原型模式的，则根据 Bean 定义资源中
```

```java
            //配置的生命周期范围，选择实例化 Bean 的合适方法，这种方式在 Web 应用程序中
            //比较常用，如 request、session、application 等生命周期
            else {
                String scopeName = mbd.getScope();
                final Scope scope = this.scopes.get(scopeName);
                //如果 Bean 定义资源中没有配置生命周期范围，则 Bean 定义不合法
                if (scope == null) {
                    throw new IllegalStateException("No Scope registered for scope name '" + scopeName
 + "'");
                }
                try {
                    //这里又使用了一个匿名内部类，获取一个指定生命周期范围的实例
                    Object scopedInstance = scope.get(beanName, () -> {
                        beforePrototypeCreation(beanName);
                        try {
                            return createBean(beanName, mbd, args);
                        }
                        finally {
                            afterPrototypeCreation(beanName);
                        }
                    });
                    //获取指定 Bean 的实例对象
                    bean = getObjectForBeanInstance(scopedInstance, name, beanName, mbd);
                }
                catch (IllegalStateException ex) {
                    throw new BeanCreationException(beanName,
                        "Scope '" + scopeName + "' is not active for the current thread; consider "
                        + "defining a scoped proxy for this bean if you intend to refer to it from
                         a singleton", ex);
                }
            }
        }
        catch (BeansException ex) {
            cleanupAfterBeanCreationFailure(beanName);
            throw ex;
        }
    }

    //对创建的 Bean 实例对象进行类型检查
    if (requiredType != null && !requiredType.isInstance(bean)) {
        try {
            T convertedBean = getTypeConverter().convertIfNecessary(bean, requiredType);
            if (convertedBean == null) {
                throw new BeanNotOfRequiredTypeException(name, requiredType, bean.getClass());
            }
            return convertedBean;
        }
```

```
        catch (TypeMismatchException ex) {
            if (logger.isDebugEnabled()) {
                logger.debug("Failed to convert bean '" + name + "' to required type '" +
                        ClassUtils.getQualifiedName(requiredType) + "'", ex);
            }
            throw new BeanNotOfRequiredTypeException(name, requiredType, bean.getClass());
        }
    }
    return (T) bean;
}
```

通过向 IoC 容器获取 Bean 的方法的分析，我们可以看到，在 Spring 中如果 Bean 定义为单例模式（Singleton）的，则容器在创建之前先从缓存中查找，以确保整个容器中只存在一个实例对象。如果 Bean 定义为原型模式（Prototype）的，则容器每次都会创建一个新的实例对象。除此之外，Bean 定义还可以指定其生命周期范围。

上面的源码只定义了根据 Bean 定义的不同模式采取的创建 Bean 实例对象的不同策略，具体的 Bean 实例对象的创建过程由实现了 ObjectFactory 接口的匿名内部类的 createBean()方法完成，ObjectFactory 接口使用委派模式，具体的 Bean 实例创建过程交由其实现类 AbstractAutowireCapableBeanFactory 完成。下面我们继续分析 AbstractAutowireCapableBeanFactory 的 createBean()方法的源码，理解创建 Bean 实例的具体过程。

9.1.3 开始实例化

AbstractAutowireCapableBeanFactory 类实现了 ObjectFactory 接口，创建容器指定的 Bean 实例对象，同时还对创建的 Bean 实例对象进行初始化。创建 Bean 实例对象的方法源码如下：

```
//创建 Bean 实例对象
@Override
protected Object createBean(String beanName, RootBeanDefinition mbd, @Nullable Object[] args)
        throws BeanCreationException {

    if (logger.isDebugEnabled()) {
        logger.debug("Creating instance of bean '" + beanName + "'");
    }
    RootBeanDefinition mbdToUse = mbd;

    //判断需要创建的 Bean 是否可以实例化，即是否可以通过当前的类加载器加载
    Class<?> resolvedClass = resolveBeanClass(mbd, beanName);
    if (resolvedClass != null && !mbd.hasBeanClass() && mbd.getBeanClassName() != null) {
        mbdToUse = new RootBeanDefinition(mbd);
        mbdToUse.setBeanClass(resolvedClass);
    }
```

```java
//校验和准备 Bean 中的方法覆盖
try {
    mbdToUse.prepareMethodOverrides();
}
catch (BeanDefinitionValidationException ex) {
    throw new BeanDefinitionStoreException(mbdToUse.getResourceDescription(),
            beanName, "Validation of Method overrides failed", ex);
}

try {
    //如果 Bean 配置了初始化前和初始化后的处理器,则试图返回一个需要创建 Bean 的代理对象
    Object bean = resolveBeforeInstantiation(beanName, mbdToUse);
    if (bean != null) {
        return bean;
    }
}
catch (Throwable ex) {
    throw new BeanCreationException(mbdToUse.getResourceDescription(), beanName,
            "BeanPostProcessor before instantiation of bean failed", ex);
}

try {
    //创建 Bean 的入口
    Object beanInstance = doCreateBean(beanName, mbdToUse, args);
    if (logger.isDebugEnabled()) {
        logger.debug("Finished creating instance of bean '" + beanName + "'");
    }
    return beanInstance;
}
catch (BeanCreationException ex) {
    throw ex;
}
catch (ImplicitlyAppearedSingletonException ex) {
    throw ex;
}
catch (Throwable ex) {
    throw new BeanCreationException(
            mbdToUse.getResourceDescription(), beanName, "Unexpected exception during bean creation", ex);
}
}

//真正创建 Bean 的方法
protected Object doCreateBean(final String beanName, final RootBeanDefinition mbd, final @Nullable Object[] args)
        throws BeanCreationException {
```

```java
//封装被创建的 Bean 对象
BeanWrapper instanceWrapper = null;
if (mbd.isSingleton()) {
   instanceWrapper = this.factoryBeanInstanceCache.remove(beanName);
}
if (instanceWrapper == null) {
   instanceWrapper = createBeanInstance(beanName, mbd, args);
}
final Object bean = instanceWrapper.getWrappedInstance();
//获取实例化对象的类型
Class<?> beanType = instanceWrapper.getWrappedClass();
if (beanType != NullBean.class) {
   mbd.resolvedTargetType = beanType;
}

//调用 PostProcessor 后置处理器
synchronized (mbd.postProcessingLock) {
   if (!mbd.postProcessed) {
      try {
         applyMergedBeanDefinitionPostProcessors(mbd, beanType, beanName);
      }
      catch (Throwable ex) {
         throw new BeanCreationException(mbd.getResourceDescription(), beanName,
               "Post-processing of merged bean definition failed", ex);
      }
      mbd.postProcessed = true;
   }
}

//向容器中缓存单例模式的 Bean 对象，以防止循环引用
boolean earlySingletonExposure = (mbd.isSingleton() && this.allowCircularReferences &&
      isSingletonCurrentlyInCreation(beanName));
if (earlySingletonExposure) {
   if (logger.isDebugEnabled()) {
      logger.debug("Eagerly caching bean '" + beanName +
            "' to allow for resolving potential circular references");
   }
   //这里是一个匿名内部类，为了防止循环引用，尽早持有对象的引用
   addSingletonFactory(beanName, () -> getEarlyBeanReference(beanName, mbd, bean));
}

//Bean 对象的初始化，依赖注入在此触发
//这个 exposedObject 对象在初始化完成之后返回依赖注入完成后的 Bean
Object exposedObject = bean;
try {
   //将 Bean 实例对象封装，并且将 Bean 定义中配置的属性值赋给实例对象
```

```java
        populateBean(beanName, mbd, instanceWrapper);
        //初始化 Bean 对象
        exposedObject = initializeBean(beanName, exposedObject, mbd);
    }
    catch (Throwable ex) {
        if (ex instanceof BeanCreationException && beanName.equals(((BeanCreationException) ex).getBeanName()))) {
            throw (BeanCreationException) ex;
        }
        else {
            throw new BeanCreationException(
                mbd.getResourceDescription(), beanName, "Initialization of bean failed", ex);
        }
    }

    if (earlySingletonExposure) {
        //获取指定名称的已注册的单例模式的 Bean 对象
        Object earlySingletonReference = getSingleton(beanName, false);
        if (earlySingletonReference != null) {
            //根据名称获取的已注册的 Bean 和正在实例化的 Bean 是同一个
            if (exposedObject == bean) {
                //当前实例化的 Bean 初始化完成
                exposedObject = earlySingletonReference;
            }
            //当前 Bean 依赖其他 Bean，并且当发生循环引用时不允许创建新的实例对象
            else if (!this.allowRawInjectionDespiteWrapping && hasDependentBean(beanName)) {
                String[] dependentBeans = getDependentBeans(beanName);
                Set<String> actualDependentBeans = new LinkedHashSet<>(dependentBeans.length);
                //获取当前 Bean 所依赖的其他 Bean
                for (String dependentBean : dependentBeans) {
                    //对依赖 Bean 进行类型检查
                    if (!removeSingletonIfCreatedForTypeCheckOnly(dependentBean)) {
                        actualDependentBeans.add(dependentBean);
                    }
                }
                if (!actualDependentBeans.isEmpty()) {
                    throw new BeanCurrentlyInCreationException(beanName,
                        "Bean with name '" + beanName + "' has been injected into other beans [" 
                        + StringUtils.collectionToCommaDelimitedString(actualDependentBeans) + 
                        "] in its raw version as part of a circular reference, but has eventually been " + "wrapped. This means that said other beans do not use the final version of the " + "bean. This is often the result of over-eager type matching - consider using " + "'getBeanNamesOfType' with the 'allowEagerInit' flag turned off, for example.");
                }
            }
        }
    }
```

```
   }
   //注册完成依赖注入的 Bean
   try {
      registerDisposableBeanIfNecessary(beanName, bean, mbd);
   }
   catch (BeanDefinitionValidationException ex) {
      throw new BeanCreationException(
            mbd.getResourceDescription(), beanName, "Invalid destruction signature", ex);
   }

   return exposedObject;
}
```

通过源码注释可以看到，具体的依赖注入实现其实就在以下两个方法中：

（1）createBeanInstance()方法，生成 Bean 所包含的 Java 对象实例。

（2）populateBean()方法，对 Bean 属性的依赖注入进行处理。

下面继续分析这两个方法的代码实现。

9.1.4 选择 Bean 实例化策略

在 createBeanInstance()方法中，根据指定的初始化策略，使用简单工厂、工厂方法或者容器的自动装配特性生成 Java 实例对象。创建对象的源码如下：

```
//创建 Bean 的实例对象
protected BeanWrapper createBeanInstance(String beanName, RootBeanDefinition mbd, @Nullable
Object[] args) {
   //确认 Bean 是可实例化的
   Class<?> beanClass = resolveBeanClass(mbd, beanName);

   //使用工厂方法对 Bean 进行实例化
   if (beanClass != null && !Modifier.isPublic(beanClass.getModifiers())
&& !mbd.isNonPublicAccessAllowed()) {
      throw new BeanCreationException(mbd.getResourceDescription(), beanName,
            "Bean class isn't public, and non-public access not allowed: " + beanClass.getName());
   }

   Supplier<?> instanceSupplier = mbd.getInstanceSupplier();
   if (instanceSupplier != null) {
      return obtainFromSupplier(instanceSupplier, beanName);
   }

   if (mbd.getFactoryMethodName() != null) {
```

```java
        //调用工厂方法进行实例化
        return instantiateUsingFactoryMethod(beanName, mbd, args);
    }

    //使用容器的自动装配方法进行实例化
    boolean resolved = false;
    boolean autowireNecessary = false;
    if (args == null) {
        synchronized (mbd.constructorArgumentLock) {
            if (mbd.resolvedConstructorOrFactoryMethod != null) {
                resolved = true;
                autowireNecessary = mbd.constructorArgumentsResolved;
            }
        }
    }
    if (resolved) {
        if (autowireNecessary) {
            //配置了自动装配属性,使用容器的自动装配进行实例化
            //容器的自动装配根据参数类型匹配 Bean 的构造方法
            return autowireConstructor(beanName, mbd, null, null);
        }
        else {
            //使用默认的无参构造方法进行实例化
            return instantiateBean(beanName, mbd);
        }
    }

    //使用 Bean 的构造方法进行实例化
    Constructor<?>[] ctors = determineConstructorsFromBeanPostProcessors(beanClass, beanName);
    if (ctors != null ||
            mbd.getResolvedAutowireMode() == RootBeanDefinition.AUTOWIRE_CONSTRUCTOR ||
            mbd.hasConstructorArgumentValues() || !ObjectUtils.isEmpty(args)) {
        //使用容器的自动装配特性,调用匹配的构造方法进行实例化
        return autowireConstructor(beanName, mbd, ctors, args);
    }

    //使用默认的无参构造方法进行实例化
    return instantiateBean(beanName, mbd);
}

//使用默认的无参构造方法实例化 Bean 对象
protected BeanWrapper instantiateBean(final String beanName, final RootBeanDefinition mbd) {
    try {
        Object beanInstance;
        final BeanFactory parent = this;
        //获取系统的安全管理接口,JDK 标准的安全管理 API
        if (System.getSecurityManager() != null) {
```

```java
        //这里是一个匿名内部类,根据实例化策略创建实例对象
        beanInstance = AccessController.doPrivileged((PrivilegedAction<Object>) () ->
            getInstantiationStrategy().instantiate(mbd, beanName, parent),
            getAccessControlContext());
    }
    else {
        //将实例化的对象封装起来
        beanInstance = getInstantiationStrategy().instantiate(mbd, beanName, parent);
    }
    BeanWrapper bw = new BeanWrapperImpl(beanInstance);
    initBeanWrapper(bw);
    return bw;
}
catch (Throwable ex) {
    throw new BeanCreationException(
        mbd.getResourceDescription(), beanName, "Instantiation of bean failed", ex);
}
}
```

从上面的代码可以看出,对使用工厂方法和自动装配特性的 Bean,调用相应的工厂方法或者参数匹配的构造方法即可完成实例化对象的工作,但是最常使用的默认无参构造方法就需要使用相应的初始化策略(JDK 的反射机制或者 CGLib)来进行初始化,在 getInstantiationStrategy().instantiate() 方法中实现了实例化。

9.1.5 执行 Bean 实例化

在使用默认的无参构造方法创建 Bean 的实例化对象时,getInstantiationStrategy().instantiate() 方法调用了 SimpleInstantiationStrategy 类中的实例化 Bean 的方法,其源码如下:

```java
//使用初始化策略实例化 Bean 对象
@Override
public Object instantiate(RootBeanDefinition bd, @Nullable String beanName, BeanFactory owner) {
    //如果 Bean 定义中没有方法覆盖,就不需要 CGLib 父类的方法
    if (!bd.hasMethodOverrides()) {
        Constructor<?> constructorToUse;
        synchronized (bd.constructorArgumentLock) {
            //获取对象的构造方法或工厂方法
            constructorToUse = (Constructor<?>) bd.resolvedConstructorOrFactoryMethod;
            //如果没有构造方法且没有工厂方法
            if (constructorToUse == null) {
                //使用 JDK 的反射机制,判断要实例化的 Bean 是否是接口
                final Class<?> clazz = bd.getBeanClass();
                if (clazz.isInterface()) {
                    throw new BeanInstantiationException(clazz, "Specified class is an interface");
                }
```

```java
            try {
                if (System.getSecurityManager() != null) {
                    //这里是一个匿名内部类,使用反射机制获取Bean的构造方法
                    constructorToUse = AccessController.doPrivileged(
                            (PrivilegedExceptionAction<Constructor<?>>) () ->
clazz.getDeclaredConstructor());
                }
                else {
                    constructorToUse = clazz.getDeclaredConstructor();
                }
                bd.resolvedConstructorOrFactoryMethod = constructorToUse;
            }
            catch (Throwable ex) {
                throw new BeanInstantiationException(clazz, "No default constructor found", ex);
            }
        }
    }
    //使用BeanUtils进行实例化,通过反射机制调用"构造方法.newInstance(arg)"来进行实例化
    return BeanUtils.instantiateClass(constructorToUse);
    }
    else {
        //使用CGLib来实例化对象
        return instantiateWithMethodInjection(bd, beanName, owner);
    }
}
```

通过上面的代码分析可知,如果 Bean 的方法被覆盖了,则使用 CGLib 进行实例化,否则使用 JDK 的反射机制进行实例化。

instantiateWithMethodInjection()方法调用 SimpleInstantiationStrategy 的子类 CGLibSubclassingInstantiationStrategy 通过 CGLib 来进行初始化,其源码如下:

```java
//使用CGLib进行Bean对象的实例化
public Object instantiate(@Nullable Constructor<?> ctor, @Nullable Object... args) {
    //创建代理子类
    Class<?> subclass = createEnhancedSubclass(this.beanDefinition);
    Object instance;
    if (ctor == null) {
        instance = BeanUtils.instantiateClass(subclass);
    }
    else {
        try {
            Constructor<?> enhancedSubclassConstructor = subclass.getConstructor
(ctor.getParameterTypes());
            instance = enhancedSubclassConstructor.newInstance(args);
        }
        catch (Exception ex) {
```

```java
        throw new BeanInstantiationException(this.beanDefinition.getBeanClass(),
                "Failed to invoke constructor for CGLib enhanced subclass [" +
subclass.getName() + "]", ex);
    }
}

Factory factory = (Factory) instance;
factory.setCallbacks(new Callback[] {NoOp.INSTANCE,
    new LookupOverrideMethodInterceptor(this.beanDefinition, this.owner),
    new ReplaceOverrideMethodInterceptor(this.beanDefinition, this.owner)});
return instance;
}

private Class<?> createEnhancedSubclass(RootBeanDefinition beanDefinition) {
    //CGLib 中的类
    Enhancer enhancer = new Enhancer();
    //将 Bean 本身作为基类
    enhancer.setSuperclass(beanDefinition.getBeanClass());
    enhancer.setNamingPolicy(SpringNamingPolicy.INSTANCE);
    if (this.owner instanceof ConfigurableBeanFactory) {
        ClassLoader cl = ((ConfigurableBeanFactory) this.owner).getBeanClassLoader();
        enhancer.setStrategy(new ClassLoaderAwareGeneratorStrategy(cl));
    }
    enhancer.setCallbackFilter(new MethodOverrideCallbackFilter(beanDefinition));
    enhancer.setCallbackTypes(CALLBACK_TYPES);
    //使用 CGLib 的 createClass()方法生成实例对象
    return enhancer.createClass();
}
}
```

CGLib 是一个常用的字节码生成器的类库，它提供了一系列 API 实现 Java 字节码的生成和转换功能。我们在学习 JDK 的动态代理时学过，JDK 的动态代理只能针对接口，如果一个类没有实现任何接口，要对其进行动态代理只能使用 CGLib。

9.1.6 准备依赖注入

在前面的分析中我们已经了解到 Bean 的依赖注入主要分为两个步骤，首先调用 createBeanInstance()方法生成 Bean 所包含的 Java 对象实例，然后调用 populateBean()方法对 Bean 属性的依赖注入进行处理。

前面已经分析了容器初始化生成 Bean 所包含的 Java 实例对象的过程，下面继续分析生成对象后，Spring IoC 容器是如何将 Bean 的属性依赖关系注入 Bean 实例对象中并设置好的。回到 populateBean()方法，属性依赖注入相关的代码如下：

```java
// 将 Bean 属性设置到生成的实例对象上
protected void populateBean(String beanName, RootBeanDefinition mbd, @Nullable BeanWrapper bw) {
    if (bw == null) {
        if (mbd.hasPropertyValues()) {
            throw new BeanCreationException(
                    mbd.getResourceDescription(), beanName, "Cannot apply property values to null instance");
        }
        else {
            return;
        }
    }

    boolean continueWithPropertyPopulation = true;
    if (!mbd.isSynthetic() && hasInstantiationAwareBeanPostProcessors()) {
        for (BeanPostProcessor bp : getBeanPostProcessors()) {
            if (bp instanceof InstantiationAwareBeanPostProcessor) {
                InstantiationAwareBeanPostProcessor ibp = (InstantiationAwareBeanPostProcessor) bp;
                if (!ibp.postProcessAfterInstantiation(bw.getWrappedInstance(), beanName)) {
                    continueWithPropertyPopulation = false;
                    break;
                }
            }
        }
    }

    if (!continueWithPropertyPopulation) {
        return;
    }
    // 获取容器在解析 Bean 定义资源时为 BeanDefinition 设置的属性值
    PropertyValues pvs = (mbd.hasPropertyValues() ? mbd.getPropertyValues() : null);

    if (mbd.getResolvedAutowireMode() == RootBeanDefinition.AUTOWIRE_BY_NAME ||
            mbd.getResolvedAutowireMode() == RootBeanDefinition.AUTOWIRE_BY_TYPE) {
        MutablePropertyValues newPvs = new MutablePropertyValues(pvs);

        if (mbd.getResolvedAutowireMode() == RootBeanDefinition.AUTOWIRE_BY_NAME) {
            autowireByName(beanName, mbd, bw, newPvs);
        }

        if (mbd.getResolvedAutowireMode() == RootBeanDefinition.AUTOWIRE_BY_TYPE) {
            autowireByType(beanName, mbd, bw, newPvs);
        }

        pvs = newPvs;
    }
```

```java
    boolean hasInstAwareBpps = hasInstantiationAwareBeanPostProcessors();
    boolean needsDepCheck = (mbd.getDependencyCheck() != RootBeanDefinition.DEPENDENCY_CHECK_NONE);

    if (hasInstAwareBpps || needsDepCheck) {
        if (pvs == null) {
            pvs = mbd.getPropertyValues();
        }
        PropertyDescriptor[] filteredPds = filterPropertyDescriptorsForDependencyCheck(bw, mbd.allowCaching);
        if (hasInstAwareBpps) {
            for (BeanPostProcessor bp : getBeanPostProcessors()) {
                if (bp instanceof InstantiationAwareBeanPostProcessor) {
                    InstantiationAwareBeanPostProcessor ibp = (InstantiationAwareBeanPostProcessor) bp;
                    pvs = ibp.postProcessPropertyValues(pvs, filteredPds, bw.getWrappedInstance(), beanName);
                    if (pvs == null) {
                        return;
                    }
                }
            }
        }
        if (needsDepCheck) {
            checkDependencies(beanName, mbd, filteredPds, pvs);
        }
    }

    if (pvs != null) {
        //对属性进行注入
        applyPropertyValues(beanName, mbd, bw, pvs);
    }
}
//解析并注入依赖属性的过程
protected void applyPropertyValues(String beanName, BeanDefinition mbd, BeanWrapper bw, PropertyValues pvs) {
    if (pvs.isEmpty()) {
        return;
    }
    //封装属性值
    MutablePropertyValues mpvs = null;
    List<PropertyValue> original;

    if (System.getSecurityManager() != null) {
        if (bw instanceof BeanWrapperImpl) {
            //设置安全上下文,JDK 安全机制
            ((BeanWrapperImpl) bw).setSecurityContext(getAccessControlContext());
        }
```

```java
        }
        if (pvs instanceof MutablePropertyValues) {
            mpvs = (MutablePropertyValues) pvs;
            //属性值已经转换
            if (mpvs.isConverted()) {
                try {
                    //为实例化对象设置属性值
                    bw.setPropertyValues(mpvs);
                    return;
                }
                catch (BeansException ex) {
                    throw new BeanCreationException(
                        mbd.getResourceDescription(), beanName, "Error setting property values", ex);
                }
            }
            //获取属性值对象的原始类型值
            original = mpvs.getPropertyValueList();
        }
        else {
            original = Arrays.asList(pvs.getPropertyValues());
        }

        //获取用户自定义的类型转换
        TypeConverter converter = getCustomTypeConverter();
        if (converter == null) {
            converter = bw;
        }
        //创建一个Bean定义属性值解析器,将Bean定义中的属性值解析为Bean实例对象的实际值
        BeanDefinitionValueResolver valueResolver = new BeanDefinitionValueResolver(this, beanName,
mbd, converter);

        //为属性的解析值创建一个副本,将副本的数据注入实例对象
        List<PropertyValue> deepCopy = new ArrayList<>(original.size());
        boolean resolveNecessary = false;
        for (PropertyValue pv : original) {
            //属性值不需要转换
            if (pv.isConverted()) {
                deepCopy.add(pv);
            }
            //属性值需要转换
            else {
                String propertyName = pv.getName();
                //原始的属性值,即转换之前的属性值
                Object originalValue = pv.getValue();
                //转换属性值,例如将引用转换为IoC容器中实例化的对象引用
                Object resolvedValue = valueResolver.resolveValueIfNecessary(pv, originalValue);
```

```java
        //转换之后的属性值
        Object convertedValue = resolvedValue;
        //属性值是否可以转换
        boolean convertible = bw.isWritableProperty(propertyName) &&
                !PropertyAccessorUtils.isNestedOrIndexedProperty(propertyName);
        if (convertible) {
            //使用用户自定义的类型转换器转换属性值
            convertedValue = convertForProperty(resolvedValue, propertyName, bw, converter);
        }
        //存储转换后的属性值，避免每次属性注入时的转换工作
        if (resolvedValue == originalValue) {
            if (convertible) {
                //设置属性转换之后的值
                pv.setConvertedValue(convertedValue);
            }
            deepCopy.add(pv);
        }
        //属性是可转换的，且属性原始值是字符串类型的，属性的原始类型值不是
        //动态生成的字符串，属性的原始值不是集合或者数组类型的
        else if (convertible && originalValue instanceof TypedStringValue &&
                !((TypedStringValue) originalValue).isDynamic() &&
                !(convertedValue instanceof Collection || ObjectUtils.isArray(convertedValue))) {
            pv.setConvertedValue(convertedValue);
            //重新封装属性值
            deepCopy.add(pv);
        }
        else {
            resolveNecessary = true;
            deepCopy.add(new PropertyValue(pv, convertedValue));
        }
    }
}
if (mpvs != null && !resolveNecessary) {
    //标记属性值已经转换过
    mpvs.setConverted();
}

//进行属性的依赖注入
try {
    bw.setPropertyValues(new MutablePropertyValues(deepCopy));
}
catch (BeansException ex) {
    throw new BeanCreationException(
            mbd.getResourceDescription(), beanName, "Error setting property values", ex);
}
}
```

从上述代码可以看出，属性的注入过程分以下两种情况：

（1）属性值类型不需要强制转换时，不需要解析属性值，直接进行依赖注入。

（2）属性值类型需要进行强制转换时，如对其他对象的引用等，首先需要解析属性值，然后对解析后的属性值进行依赖注入。

对属性值的解析是在 BeanDefinitionValueResolver 类的 resolveValueIfNecessary() 方法中进行的，对属性值的依赖注入是通过 bw.setPropertyValues() 方法实现的。

9.1.7 解析属性依赖注入规则

容器对属性进行依赖注入时，如果发现属性值需要进行类型转换，例如属性值是容器中另一个 Bean 实例对象的引用，则容器首先需要根据属性值解析出所引用的对象，然后才能将该引用对象注入到目标实例对象的属性上。对属性进行解析由 resolveValueIfNecessary() 方法实现，其源码如下：

```java
//解析属性值，对注入类型进行转换
@Nullable
public Object resolveValueIfNecessary(Object argName, @Nullable Object value) {
    //对引用类型的属性进行解析
    if (value instanceof RuntimeBeanReference) {
        RuntimeBeanReference ref = (RuntimeBeanReference) value;
        //调用引用类型属性的解析方法
        return resolveReference(argName, ref);
    }
    //对引用容器中另一个 Bean 名称的属性进行解析
    else if (value instanceof RuntimeBeanNameReference) {
        String refName = ((RuntimeBeanNameReference) value).getBeanName();
        refName = String.valueOf(doEvaluate(refName));
        //从容器中获取指定名称的 Bean
        if (!this.beanFactory.containsBean(refName)) {
            throw new BeanDefinitionStoreException(
                "Invalid bean name '" + refName + "' in bean reference for " + argName);
        }
        return refName;
    }
    //对 Bean 类型属性的解析，主要是指 Bean 中的内部类
    else if (value instanceof BeanDefinitionHolder) {
        BeanDefinitionHolder bdHolder = (BeanDefinitionHolder) value;
        return resolveInnerBean(argName, bdHolder.getBeanName(), bdHolder.getBeanDefinition());
    }
    else if (value instanceof BeanDefinition) {
        BeanDefinition bd = (BeanDefinition) value;
```

```java
        String innerBeanName = "(inner bean)" + BeanFactoryUtils.GENERATED_BEAN_NAME_SEPARATOR +
                ObjectUtils.getIdentityHexString(bd);
        return resolveInnerBean(argName, innerBeanName, bd);
    }
    //对集合数组类型的属性进行解析
    else if (value instanceof ManagedArray) {
        ManagedArray array = (ManagedArray) value;
        //获取数组的类型
        Class<?> elementType = array.resolvedElementType;
        if (elementType == null) {
            //获取数组元素的类型
            String elementTypeName = array.getElementTypeName();
            if (StringUtils.hasText(elementTypeName)) {
                try {
                    //使用反射机制创建指定类型的对象
                    elementType = ClassUtils.forName(elementTypeName,
this.beanFactory.getBeanClassLoader());
                    array.resolvedElementType = elementType;
                }
                catch (Throwable ex) {
                    throw new BeanCreationException(
                            this.beanDefinition.getResourceDescription(), this.beanName,
                            "Error resolving array type for " + argName, ex);
                }
            }
            //没有获取到数组的类型,也没有获取到数组元素的类型,
            //则直接设置数组的类型为 Object
            else {
                elementType = Object.class;
            }
        }
        //创建指定类型的数组
        return resolveManagedArray(argName, (List<?>) value, elementType);
    }
    //解析 list 类型的属性值
    else if (value instanceof ManagedList) {
        return resolveManagedList(argName, (List<?>) value);
    }
    //解析 set 类型的属性值
    else if (value instanceof ManagedSet) {
        return resolveManagedSet(argName, (Set<?>) value);
    }
    //解析 map 类型的属性值
    else if (value instanceof ManagedMap) {
        return resolveManagedMap(argName, (Map<?, ?>) value);
    }
    //解析 props 类型的属性值,props 其实就是 key 和 value 均为字符串的 map
```

```java
else if (value instanceof ManagedProperties) {
    Properties original = (Properties) value;
    //创建一个副本，作为解析后的返回值
    Properties copy = new Properties();
    original.forEach((propKey, propValue) -> {
        if (propKey instanceof TypedStringValue) {
            propKey = evaluate((TypedStringValue) propKey);
        }
        if (propValue instanceof TypedStringValue) {
            propValue = evaluate((TypedStringValue) propValue);
        }
        if (propKey == null || propValue == null) {
            throw new BeanCreationException(
                this.beanDefinition.getResourceDescription(), this.beanName,
                "Error converting Properties key/value pair for " + argName + ": resolved to null");
        }
        copy.put(propKey, propValue);
    });
    return copy;
}
//解析字符串类型的属性值
else if (value instanceof TypedStringValue) {
    TypedStringValue typedStringValue = (TypedStringValue) value;
    Object valueObject = evaluate(typedStringValue);
    try {
        //获取属性的目标类型
        Class<?> resolvedTargetType = resolveTargetType(typedStringValue);
        if (resolvedTargetType != null) {
            //对目标类型的属性进行解析，递归调用
            return this.typeConverter.convertIfNecessary(valueObject, resolvedTargetType);
        }
        //没有获取到属性的目标对象，则按 Object 类型返回
        else {
            return valueObject;
        }
    }
    catch (Throwable ex) {
        throw new BeanCreationException(
            this.beanDefinition.getResourceDescription(), this.beanName,
            "Error converting typed String value for " + argName, ex);
    }
}
else if (value instanceof NullBean) {
    return null;
}
else {
    return evaluate(value);
```

```java
    }
}

//解析引用类型的属性值
@Nullable
private Object resolveReference(Object argName, RuntimeBeanReference ref) {
    try {
        Object bean;
        //获取引用的 Bean 名称
        String refName = ref.getBeanName();
        refName = String.valueOf(doEvaluate(refName));
        //如果引用的对象在父类容器中，则从父类容器中获取指定的引用对象
        if (ref.isToParent()) {
            if (this.beanFactory.getParentBeanFactory() == null) {
                throw new BeanCreationException(
                        this.beanDefinition.getResourceDescription(), this.beanName,
                        "Can't resolve reference to bean '" + refName +
                        "' in parent factory: no parent factory available");
            }
            bean = this.beanFactory.getParentBeanFactory().getBean(refName);
        }
        //从当前的容器中获取指定的引用 Bean 对象，如果指定的 Bean 没有被实例化，
        //则会递归触发引用 Bean 的初始化和依赖注入
        else {
            bean = this.beanFactory.getBean(refName);
            //当前实例化对象依赖的引用对象
            this.beanFactory.registerDependentBean(refName, this.beanName);
        }
        if (bean instanceof NullBean) {
            bean = null;
        }
        return bean;
    }
    catch (BeansException ex) {
        throw new BeanCreationException(
                this.beanDefinition.getResourceDescription(), this.beanName,
                "Cannot resolve reference to bean '" + ref.getBeanName() + "' while setting " + argName, ex);
    }
}

//解析 array 类型的属性值
private Object resolveManagedArray(Object argName, List<?> ml, Class<?> elementType) {
    //创建一个指定类型的数组，用于存放和返回解析后的数组
    Object resolved = Array.newInstance(elementType, ml.size());
    for (int i = 0; i < ml.size(); i++) {
        //递归解析 array 的每一个元素，并将解析后的值设置到 resolved 数组中，索引为 i
        Array.set(resolved, i,
```

```
            resolveValueIfNecessary(new KeyedArgName(argName, i), ml.get(i)));
    }
    return resolved;
}
```

通过上面的代码分析，我们明白了 Spring 是如何对引用类型、内部类及集合类型的属性进行解析的，解析完成后就可以进行依赖注入了，依赖注入的过程就是将 Bean 对象实例设置到它所依赖的 Bean 对象属性上。真正的依赖注入是通过 bw.setPropertyValues()方法实现的，该方法也使用了委派模式，在 BeanWrapper 接口中至少定义了方法声明，依赖注入的具体实现交由其实现类 BeanWrapperImpl 完成。下面我们就分析 BeanWrapperImpl 类中依赖注入相关的源码。

9.1.8 注入赋值

BeanWrapperImpl 类负责对容器中完成初始化的 Bean 实例对象进行属性的依赖注入，即把 Bean 对象设置到它所依赖的另一个 Bean 的属性上。BeanWrapperImpl 中的注入方法相关源码如下：

```
//实现属性依赖注入功能
protected void setPropertyValue(PropertyTokenHolder tokens, PropertyValue pv) throws
BeansException {
    if (tokens.keys != null) {
        processKeyedProperty(tokens, pv);
    }
    else {
        processLocalProperty(tokens, pv);
    }
}

//实现属性依赖注入功能
@SuppressWarnings("unchecked")
private void processKeyedProperty(PropertyTokenHolder tokens, PropertyValue pv) {
    //调用属性的 getter(readerMethod)方法，获取属性值
    Object propValue = getPropertyHoldingValue(tokens);
    PropertyHandler ph = getLocalPropertyHandler(tokens.actualName);
    if (ph == null) {
        throw new InvalidPropertyException(
            getRootClass(), this.nestedPath + tokens.actualName, "No property handler found");
    }
    Assert.state(tokens.keys != null, "No token keys");
    String lastKey = tokens.keys[tokens.keys.length - 1];

    //注入 array 类型的属性值
    if (propValue.getClass().isArray()) {
        Class<?> requiredType = propValue.getClass().getComponentType();
        int arrayIndex = Integer.parseInt(lastKey);
```

```java
    Object oldValue = null;
    try {
        if (isExtractOldValueForEditor() && arrayIndex < Array.getLength(propValue)) {
            oldValue = Array.get(propValue, arrayIndex);
        }
        Object convertedValue = convertIfNecessary(tokens.canonicalName, oldValue, pv.getValue(),
                requiredType, ph.nested(tokens.keys.length));
        //获取集合类型属性的长度
        int length = Array.getLength(propValue);
        if (arrayIndex >= length && arrayIndex < this.autoGrowCollectionLimit) {
            Class<?> componentType = propValue.getClass().getComponentType();
            Object newArray = Array.newInstance(componentType, arrayIndex + 1);
            System.arraycopy(propValue, 0, newArray, 0, length);
            setPropertyValue(tokens.actualName, newArray);
            //调用属性的 getter(readerMethod)方法，获取属性值
            propValue = getPropertyValue(tokens.actualName);
        }
        //将属性值赋给数组中的元素
        Array.set(propValue, arrayIndex, convertedValue);
    }
    catch (IndexOutOfBoundsException ex) {
        throw new InvalidPropertyException(getRootClass(), this.nestedPath +
                tokens.canonicalName, "Invalid array index in property path '" +
                tokens.canonicalName + "'", ex);
    }
}

//注入 list 类型的属性值
else if (propValue instanceof List) {
    //获取 list 集合的类型
    Class<?> requiredType = ph.getCollectionType(tokens.keys.length);
    List<Object> list = (List<Object>) propValue;
    //获取 list 集合的 size
    int index = Integer.parseInt(lastKey);
    Object oldValue = null;
    if (isExtractOldValueForEditor() && index < list.size()) {
        oldValue = list.get(index);
    }
    //获取解析后 list 的属性值
    Object convertedValue = convertIfNecessary(tokens.canonicalName, oldValue, pv.getValue(),
            requiredType, ph.nested(tokens.keys.length));
    int size = list.size();
    //如果 list 的长度大于属性值的长度，则将多余的元素赋值为 null
    if (index >= size && index < this.autoGrowCollectionLimit) {
        for (int i = size; i < index; i++) {
            try {
                list.add(null);
```

```java
                } catch (NullPointerException ex) {
                    throw new InvalidPropertyException(getRootClass(), this.nestedPath + tokens.canonicalName, "Cannot set element with index " + index + " in List of size " + size + ", accessed using property path '" + tokens.canonicalName +
"': List does not support filling up gaps with null elements");
                }
                list.add(convertedValue);
            }
            else {
                try {
                    //将值添加到list中
                    list.set(index, convertedValue);
                }
                catch (IndexOutOfBoundsException ex) {
                    throw new InvalidPropertyException(getRootClass(), this.nestedPath +
                        tokens.canonicalName, "Invalid list index in property path '" +
                        tokens.canonicalName + "'", ex);
                }
            }
        }
    }
    //注入map类型的属性值
    else if (propValue instanceof Map) {
        //获取map集合key的类型
        Class<?> mapKeyType = ph.getMapKeyType(tokens.keys.length);
        //获取map集合value的类型
        Class<?> mapValueType = ph.getMapValueType(tokens.keys.length);
        Map<Object, Object> map = (Map<Object, Object>) propValue;
        TypeDescriptor typeDescriptor = TypeDescriptor.valueOf(mapKeyType);
        //解析map类型的属性key值
        Object convertedMapKey = convertIfNecessary(null, null, lastKey, mapKeyType, typeDescriptor);
        Object oldValue = null;
        if (isExtractOldValueForEditor()) {
            oldValue = map.get(convertedMapKey);
        }
        //解析map类型的属性value值
        Object convertedMapValue = convertIfNecessary(tokens.canonicalName, oldValue, pv.getValue(),
            mapValueType, ph.nested(tokens.keys.length));
        //将解析后的key和value值赋给map属性
        map.put(convertedMapKey, convertedMapValue);
    }
    else {
        throw new InvalidPropertyException(getRootClass(), this.nestedPath + tokens.canonicalName,
            "Property referenced in indexed property path '" + tokens.canonicalName +
```

```java
            "' is neither an array nor a List nor a Map; returned value was [" + propValue + "]");
    }
}

private Object getPropertyHoldingValue(PropertyTokenHolder tokens) {
    Assert.state(tokens.keys != null, "No token keys");
    PropertyTokenHolder getterTokens = new PropertyTokenHolder(tokens.actualName);
    getterTokens.canonicalName = tokens.canonicalName;
    getterTokens.keys = new String[tokens.keys.length - 1];
    System.arraycopy(tokens.keys, 0, getterTokens.keys, 0, tokens.keys.length - 1);

    Object propValue;
    try {
        //获取属性值
        propValue = getPropertyValue(getterTokens);
    }
    catch (NotReadablePropertyException ex) {
        throw new NotWritablePropertyException(getRootClass(), this.nestedPath +
                tokens.canonicalName, "Cannot access indexed value in property referenced " +
                "in indexed property path '" + tokens.canonicalName + "'", ex);
    }

    if (propValue == null) {
        if (isAutoGrowNestedPaths()) {
            int lastKeyIndex = tokens.canonicalName.lastIndexOf('[');
            getterTokens.canonicalName = tokens.canonicalName.substring(0, lastKeyIndex);
            propValue = setDefaultValue(getterTokens);
        }
        else {
            throw new NullValueInNestedPathException(getRootClass(), this.nestedPath +
tokens.canonicalName,
                    "Cannot access indexed value in property referenced " +
                    "in indexed property path '" + tokens.canonicalName + "': returned null");
        }
    }
    return propValue;
}
```

通过对依赖注入代码的分析，我们明白了 Spring IoC 容器是如何将属性值注入 Bean 实例对象上的：

（1）对于集合类型的属性，将属性值解析为目标类型的集合后直接赋值给属性。

（2）对于非集合类型的属性，大量使用 JDK 的反射机制，通过属性的 getter() 方法获取指定属性注入前的值，同时调用属性的 setter() 方法为属性设置注入后的值。

看到这里相信很多人都明白了 Spring 的 setter()注入原理，详细时序图见"资源文件"。

至此，Spring IoC 容器对 Bean 定义资源的定位、载入、解析和依赖注入已经全部分析完毕，Spring IoC 容器中管理了一系列靠依赖关系联系起来的 Bean，不需要手动创建所需的对象，Spring IoC 容器会自动为我们创建，并且为我们注入好相关的依赖。

9.2 Spring IoC 容器中那些鲜为人知的细节

Spring IoC 容器还有一些高级特性，如使用 lazy-init 属性对 Bean 预初始化、使用 FactoryBean 产生或者修饰 Bean 对象的生成、IoC 容器在初始化 Bean 过程中使用 BeanPostProcessor 后置处理器对 Bean 声明周期事件进行管理等。

9.2.1 关于延时加载

我们已经知道，IoC 容器的初始化过程就是对 Bean 定义资源的定位、载入和注册，此时容器对 Bean 的依赖注入并没有发生，依赖注入是在应用程序第一次向容器索取 Bean 时通过 getBean()方法完成的。

当 Bean 定义资源的<bean>元素中配置了 lazy-init=false 属性时，容器将会在初始化时对所配置的 Bean 进行预实例化，Bean 的依赖注入在容器初始化时就已经完成。这样，当应用程序第一次向容器索取被管理的 Bean 时，就不用再初始化和对 Bean 进行依赖注入了，而是直接从容器中获取已经完成依赖注入的 Bean，提高了应用程序第一次向容器获取 Bean 的性能。

1. refresh()方法

IoC 容器读入已经定位的 Bean 定义资源是从 refresh()方法开始的，我们从 AbstractApplicationContext 类的 refresh()方法入手分析，回顾一下源码：

```
@Override
public void refresh() throws BeansException, IllegalStateException {
    ...
    //子类的 refreshBeanFactory()方法启动
    ConfigurableListableBeanFactory beanFactory = obtainFreshBeanFactory();
    ...
}
```

在 refresh()方法中 ConfigurableListableBeanFactory beanFactory = obtainFreshBeanFactory();启动了 Bean 定义资源的载入、注册过程。finishBeanFactoryInitialization()方法是对注册后的 Bean 定义中的预实例化（lazy-init=false，Spring 默认进行预实例化，即为 true）的 Bean 进行处理的地方。

2. 使用 finishBeanFactoryInitialization()处理预实例化的 Bean

当 Bean 定义资源被载入 IoC 容器之后，容器将 Bean 定义资源解析为容器内部的数据结构 BeanDefinition，并注册到容器中，AbstractApplicationContext 类中的 finishBeanFactoryInitialization() 方法对配置了预实例化属性的 Bean 进行预初始化，源码如下：

```
//对配置了 lazy-init 属性的 Bean 进行预实例化处理
protected void finishBeanFactoryInitialization(ConfigurableListableBeanFactory beanFactory)
{
    //这是 Spring 3 新加的代码，为容器指定一个转换服务(ConversionService)
    //在对某些 Bean 属性进行转换时使用
    if (beanFactory.containsBean(CONVERSION_SERVICE_BEAN_NAME) &&
        beanFactory.isTypeMatch(CONVERSION_SERVICE_BEAN_NAME, ConversionService.class)) {
        beanFactory.setConversionService(
            beanFactory.getBean(CONVERSION_SERVICE_BEAN_NAME, ConversionService.class));
    }

    if (!beanFactory.hasEmbeddedValueResolver()) {
        beanFactory.addEmbeddedValueResolver(strVal -> getEnvironment().resolvePlaceholders
(strVal));
    }

    String[] weaverAwareNames = beanFactory.getBeanNamesForType(LoadTimeWeaverAware.class,
false, false);
    for (String weaverAwareName : weaverAwareNames) {
        getBean(weaverAwareName);
    }

    //为了使类型匹配，停止使用临时的类加载器
    beanFactory.setTempClassLoader(null);

    //缓存容器中所有注册的 BeanDefinition 元数据，以防被修改
    beanFactory.freezeConfiguration();

    //对配置了 lazy-init 属性的单例模式的 Bean 进行预实例化处理
    beanFactory.preInstantiateSingletons();
}
```

其中 ConfigurableListableBeanFactory 是一个接口，preInstantiateSingletons()方法由其子类 DefaultListableBeanFactory 提供。

3. 对配置了 lazy-init 属性的单例模式的 Bean 的预实例化

对配置了 lazy-init 属性的单例模式的 Bean 的预实例化相关源码如下：

```
public void preInstantiateSingletons() throws BeansException {
    if (this.logger.isDebugEnabled()) {
        this.logger.debug("Pre-instantiating singletons in " + this);
```

```java
}
List<String> beanNames = new ArrayList<>(this.beanDefinitionNames);

for (String beanName : beanNames) {
    //获取指定名称的 Bean 定义
    RootBeanDefinition bd = getMergedLocalBeanDefinition(beanName);
    //Bean 不是抽象的,是单例模式的,且 lazy-init 属性配置为 false
    if (!bd.isAbstract() && bd.isSingleton() && !bd.isLazyInit()) {
        //如果指定名称的 Bean 是创建容器的 Bean
        if (isFactoryBean(beanName)) {
            //FACTORY_BEAN_PREFIX="&",当 Bean 名称前面加"&"符号
            //时,获取的是容器对象本身,而不是容器产生的 Bean
            //调用 getBean 方法,触发 Bean 实例化和依赖注入
            final FactoryBean<?> factory = (FactoryBean<?>) getBean(FACTORY_BEAN_PREFIX + beanName);
            //标识是否需要预实例化
            boolean isEagerInit;
            if (System.getSecurityManager() != null && factory instanceof SmartFactoryBean) {
                //一个匿名内部类
                isEagerInit = AccessController.doPrivileged((PrivilegedAction<Boolean>) () ->
                    ((SmartFactoryBean<?>) factory).isEagerInit(),
                    getAccessControlContext());
            }
            else {
                isEagerInit = (factory instanceof SmartFactoryBean &&
                    ((SmartFactoryBean<?>) factory).isEagerInit());
            }
            if (isEagerInit) {
                //调用 getBean()方法,触发 Bean 实例化和依赖注入
                getBean(beanName);
            }
        }
        else {
            getBean(beanName);
        }
    }
}
```

通过对 lazy-init 处理源码的分析可以看出,如果设置了 lazy-init 属性,则容器在完成 Bean 定义的注册之后,会通过 getBean()方法触发指定 Bean 的初始化和依赖注入。如前所述,这样当应用程序第一次向容器索取所需的 Bean 时,容器不再需要对 Bean 进行初始化和依赖注入,可直接从已经完成实例化和依赖注入的 Bean 中取一个现成的 Bean,提高了第一次获取 Bean 的性能。

9.2.2 关于 FactoryBean 和 BeanFactory

Spring 中，有两个很容易混淆的类：BeanFactory 和 FactoryBean。

BeanFactory：Bean 工厂，是一个工厂（Factory），Spring IoC 容器的最高层接口就是 BeanFactory，它的作用是管理 Bean，即实例化、定位、配置应用程序中的对象及建立这些对象之间的依赖。

FactoryBean：工厂 Bean，是一个 Bean，作用是产生其他 Bean 实例。这种 Bean 没有什么特别的要求，仅需要提供一个工厂方法，该方法用来返回其他 Bean 实例。在通常情况下，Bean 无须自己实现工厂模式，Spring 容器担任工厂的角色；在少数情况下，容器中的 Bean 本身就是工厂，其作用是产生其他 Bean 实例。

当用户使用容器时，可以使用转义字符"&"来得到 FactoryBean 本身，以区别通过 FactoryBean 产生的实例对象和 FactoryBean 对象本身。在 BeanFactory 中通过如下代码定义了该转义字符：

```
String FACTORY_BEAN_PREFIX = "&";
```

如果 myJndiObject 是一个 FactoryBean，则使用&myJndiObject 得到的是 myJndiObject 对象，而不是 myJndiObject 产生的对象。

1. FactoryBean 源码

```java
//工厂 Bean，用于产生其他对象
public interface FactoryBean<T> {

    //获取容器管理的对象实例
    @Nullable
    T getObject() throws Exception;

    //获取 Bean 工厂创建的对象的类型
    @Nullable
    Class<?> getObjectType();

    //Bean 工厂创建的对象是否是单例模式的，如果是，
    //则整个容器中只有一个实例对象，每次请求都返回同一个实例对象
    default boolean isSingleton() {
        return true;
    }

}
```

2. AbstractBeanFactory 的 getBean()方法

在分析 Spring IoC 容器实例化 Bean 并进行依赖注入的源码时，提到在 getBean()方法触发容器实例化 Bean 时会调用 AbstractBeanFactory 的 doGetBean()方法，其重要源码如下：

```java
protected <T> T doGetBean(final String name, @Nullable final Class<T> requiredType,
    @Nullable final Object[] args, boolean typeCheckOnly) throws BeansException {
    ...
    BeanFactory parentBeanFactory = getParentBeanFactory();
    //当前容器的父容器存在，且当前容器中不存在指定名称的Bean
    if (parentBeanFactory != null && !containsBeanDefinition(beanName)) {
        //解析指定Bean名称的原始名称
        String nameToLookup = originalBeanName(name);
        if (parentBeanFactory instanceof AbstractBeanFactory) {
            return ((AbstractBeanFactory) parentBeanFactory).doGetBean(
                nameToLookup, requiredType, args, typeCheckOnly);
        }
        else if (args != null) {
            //委派父容器根据指定名称和显式的参数查找
            return (T) parentBeanFactory.getBean(nameToLookup, args);
        }
        else {
            //委派父容器根据指定名称和类型查找
            return parentBeanFactory.getBean(nameToLookup, requiredType);
        }
    }
    ...
    return (T) bean;
}

//获取给定Bean的实例对象，主要完成FactoryBean的相关处理
protected Object getObjectForBeanInstance(
    Object beanInstance, String name, String beanName, @Nullable RootBeanDefinition mbd) {

    //容器已经得到了Bean实例对象，这个实例对象可能是一个普通的Bean，
    //也可能是一个工厂Bean，如果是一个工厂Bean，则使用它创建一个Bean实例对象，
    //如果调用本身就想获得一个容器的引用，则返回这个工厂Bean实例对象
    //如果指定的名称是容器的解引用（dereference，即对象本身而非内存地址）
    //且Bean实例也不是创建Bean实例对象的工厂Bean
    if (BeanFactoryUtils.isFactoryDereference(name) && !(beanInstance instanceof FactoryBean)) {
        throw new BeanIsNotAFactoryException(transformedBeanName(name), beanInstance.getClass());
    }

    //如果Bean实例不是工厂Bean，或者指定名称是容器的解引用
    //调用者获取对容器的引用时，直接返回当前的Bean实例
    if (!(beanInstance instanceof FactoryBean) || BeanFactoryUtils.isFactoryDereference(name))
    {
        return beanInstance;
    }

    //处理指定名称不是容器的解引用，或者根据名称获取的Bean实例对象是一个工厂Bean
    //使用工厂Bean创建一个Bean的实例对象
```

```java
        Object object = null;
        if (mbd == null) {
            //从 Bean 工厂缓存中获取指定名称的 Bean 实例对象
            object = getCachedObjectForFactoryBean(beanName);
        }
        //让 Bean 工厂生产指定名称的 Bean 实例对象
        if (object == null) {
            FactoryBean<?> factory = (FactoryBean<?>) beanInstance;
            //如果从 Bean 工厂生产的 Bean 是单例模式的,则缓存
            if (mbd == null && containsBeanDefinition(beanName)) {
                //从容器中获取指定名称的 Bean 定义,如果继承了基类,则合并基类的相关属性
                mbd = getMergedLocalBeanDefinition(beanName);
            }
            //如果从容器得到了 Bean 定义信息,并且 Bean 定义信息不是虚构的,
            //则让工厂 Bean 生产 Bean 实例对象
            boolean synthetic = (mbd != null && mbd.isSynthetic());
            //调用 FactoryBeanRegistrySupport 类的 getObjectFromFactoryBean()方法
            //实现工厂 Bean 生产 Bean 实例对象的过程
            object = getObjectFromFactoryBean(factory, beanName, !synthetic);
        }
        return object;
}
```

在上面获取给定 Bean 的实例对象的 getObjectForBeanInstance()方法中,会调用 FactoryBean-RegistrySupport 类的 getObjectFromFactoryBean()方法,该方法实现了 Bean 工厂生产 Bean 实例对象。

3. AbstractBeanFactory 生产 Bean 实例对象

AbstractBeanFactory 类中生产 Bean 实例对象的主要源码如下:

```java
//Bean 工厂生产 Bean 实例对象
protected Object getObjectFromFactoryBean(FactoryBean<?> factory, String beanName, boolean shouldPostProcess) {
    //Bean 工厂是单例模式,并且 Bean 工厂缓存中存在指定名称的 Bean 实例对象
    if (factory.isSingleton() && containsSingleton(beanName)) {
        //多线程同步,以防止数据不一致
        synchronized (getSingletonMutex()) {
            //直接从 Bean 工厂的缓存中获取指定名称的 Bean 实例对象
            Object object = this.factoryBeanObjectCache.get(beanName);
            //如果 Bean 工厂缓存中没有指定名称的实例对象,则生产该实例对象
            if (object == null) {
                //调用 Bean 工厂的获取对象的方法生产指定 Bean 的实例对象
                object = doGetObjectFromFactoryBean(factory, beanName);
                Object alreadyThere = this.factoryBeanObjectCache.get(beanName);
                if (alreadyThere != null) {
                    object = alreadyThere;
```

```java
            }
            else {
                if (shouldPostProcess) {
                    try {
                        object = postProcessObjectFromFactoryBean(object, beanName);
                    }
                    catch (Throwable ex) {
                        throw new BeanCreationException(beanName,
                                "Post-processing of FactoryBean's singleton object failed", ex);
                    }
                }
                //将生产的实例对象添加到Bean工厂的缓存中
                this.factoryBeanObjectCache.put(beanName, object);
            }
        }
        return object;
    }
}
//调用Bean工厂的获取对象的方法生产指定Bean的实例对象
else {
    Object object = doGetObjectFromFactoryBean(factory, beanName);
    if (shouldPostProcess) {
        try {
            object = postProcessObjectFromFactoryBean(object, beanName);
        }
        catch (Throwable ex) {
            throw new BeanCreationException(beanName, "Post-processing of FactoryBean's object failed", ex);
        }
    }
    return object;
}
}

//调用Bean工厂的方法生产指定Bean的实例对象
private Object doGetObjectFromFactoryBean(final FactoryBean<?> factory, final String beanName)
        throws BeanCreationException {

    Object object;
    try {
        if (System.getSecurityManager() != null) {
            AccessControlContext acc = getAccessControlContext();
            try {
                //实现PrivilegedExceptionAction接口的匿名内部类
                object = AccessController.doPrivileged((PrivilegedExceptionAction<Object>) () ->
                        factory.getObject(), acc);
            }
```

```
            catch (PrivilegedActionException pae) {
                throw pae.getException();
            }
        }
        else {
            //调用 BeanFactory 接口实现类的创建对象方法
            object = factory.getObject();
        }
    }
    catch (FactoryBeanNotInitializedException ex) {
        throw new BeanCurrentlyInCreationException(beanName, ex.toString());
    }
    catch (Throwable ex) {
        throw new BeanCreationException(beanName, "FactoryBean threw exception on object creation", ex);
    }

    //创建出来的实例对象为 null，或者因为单例对象正在创建而返回 null
    if (object == null) {
        if (isSingletonCurrentlyInCreation(beanName)) {
            throw new BeanCurrentlyInCreationException(
                    beanName, "FactoryBean which is currently in creation returned null from getObject");
        }
        object = new NullBean();
    }
    return object;
}
```

从上面的源码分析中可以看出，BeanFactory 接口调用其实现类的获取对象的方法来实现创建 Bean 实例对象的功能。

4. FactoryBean 实现类的获取对象的方法

FactoryBean 接口的实现类非常多，比如 Proxy、RMI、JNDI、ServletContextFactoryBean 等。FactoryBean 接口为 Spring 容器提供了一个很好的封装机制，具体的获取对象的方法由不同的实现类根据不同的实现策略来提供，我们分析一下最简单的 AnnotationTestFactoryBean 类的源码：

```
public class AnnotationTestBeanFactory implements FactoryBean<FactoryCreatedAnnotationTestBean> {
    private final FactoryCreatedAnnotationTestBean instance = new FactoryCreatedAnnotationTestBean();
    public AnnotationTestBeanFactory() {
        this.instance.setName("FACTORY");
    }
    @Override
    public FactoryCreatedAnnotationTestBean getObject() throws Exception {
```

```
      return this.instance;
   }
   //AnnotationTestBeanFactory 产生 Bean 实例对象的实现
   @Override
   public Class<? extends IJmxTestBean> getObjectType() {
      return FactoryCreatedAnnotationTestBean.class;
   }
   @Override
   public boolean isSingleton() {
      return true;
   }
}
```

Proxy、RMI、JNDI 等其他实现类都根据相应的策略提供方法，这里不做一一分析，这已经不是 Spring 的核心功能，感兴趣的"小伙伴"可以自行深入研究。

9.2.3 再述 autowiring

Spring IoC 容器提供了两种管理 Bean 依赖关系的方式：

（1）显式管理：通过 BeanDefinition 的属性值和构造方法实现 Bean 依赖关系管理。

（2）autowiring：Spring IoC 容器有依赖自动装配功能，不需要对 Bean 属性的依赖关系做显式的声明，只需要配置好 autowiring 属性，IoC 容器会自动使用反射查找属性的类型和名称，然后基于属性的类型或者名称来自动匹配容器中的 Bean，从而自动完成依赖注入。

容器对 Bean 的自动装配发生在容器对 Bean 依赖注入的过程中。在对 Spring IoC 容器的依赖注入源码进行分析时，我们已经知道容器对 Bean 实例对象的依赖属性注入发生在 AbstractAutoWireCapableBeanFactory 类的 populateBean() 方法中，下面通过程序流程分析 autowiring 的实现原理。

1. AbstractAutoWireCapableBeanFactory 对 Bean 实例对象进行属性依赖注入

应用程序第一次通过 getBean() 方法（配置了 lazy-init 预实例化属性的除外）向 IoC 容器索取 Bean 时，容器创建 Bean 实例对象，并且对 Bean 实例对象进行属性依赖注入，AbstractAutoWireCapableBeanFactory 的 populateBean() 方法就实现了属性依赖注入的功能，其主要源码如下：

```
//将 Bean 属性设置到生成的实例对象上
protected void populateBean(String beanName, RootBeanDefinition mbd, @Nullable BeanWrapper bw)
{
   …
   //获取容器在解析 Bean 定义时为 BeanDefinition 设置的属性值
   PropertyValues pvs = (mbd.hasPropertyValues() ? mbd.getPropertyValues() : null);
```

```java
//处理依赖注入，首先处理 autowiring 自动装配的依赖注入
if (mbd.getResolvedAutowireMode() == RootBeanDefinition.AUTOWIRE_BY_NAME ||
    mbd.getResolvedAutowireMode() == RootBeanDefinition.AUTOWIRE_BY_TYPE) {
    MutablePropertyValues newPvs = new MutablePropertyValues(pvs);

    //根据 Bean 名称进行 autowiring 自动装配处理
    if (mbd.getResolvedAutowireMode() == RootBeanDefinition.AUTOWIRE_BY_NAME) {
        autowireByName(beanName, mbd, bw, newPvs);
    }

    //根据 Bean 类型进行 autowiring 自动装配处理
    if (mbd.getResolvedAutowireMode() == RootBeanDefinition.AUTOWIRE_BY_TYPE) {
        autowireByType(beanName, mbd, bw, newPvs);
    }

    pvs = newPvs;
}
//对非 autowiring 的属性进行依赖注入处理
...
}
```

2. Spring IoC 容器根据 Bean 名称或者类型进行 autowiring 自动属性依赖注入

Spring IoC 容器根据 Bean 名称或者类型进行 autowiring 自动属性依赖注入的重要代码如下：

```java
//根据类型对属性进行自动依赖注入
protected void autowireByType(
    String beanName, AbstractBeanDefinition mbd, BeanWrapper bw, MutablePropertyValues pvs)
{
    //获取用户定义的类型转换器
    TypeConverter converter = getCustomTypeConverter();
    if (converter == null) {
        converter = bw;
    }

    //存放解析的要注入的属性
    Set<String> autowiredBeanNames = new LinkedHashSet<>(4);
    //对 Bean 对象中非简单属性（不是简单继承的对象，如 8 种原始类型、字符、URL 等都是简单属性）进行处理
    String[] propertyNames = unsatisfiedNonSimpleProperties(mbd, bw);
    for (String propertyName : propertyNames) {
        try {
            //获取指定属性名称的属性描述器
            PropertyDescriptor pd = bw.getPropertyDescriptor(propertyName);
            //不对 Object 类型的属性进行 autowiring 自动依赖注入
            if (Object.class != pd.getPropertyType()) {
```

```
        //获取属性的赋值方法
        MethodParameter MethodParam = BeanUtils.getWriteMethodParameter(pd);
        //检查指定类型是否可以被转换为目标对象的类型
        boolean eager = !PriorityOrdered.class.isInstance(bw.getWrappedInstance());
        //创建一个要被注入的依赖描述
        DependencyDescriptor desc = new AutowireByTypeDependencyDescriptor(MethodParam, eager);
        //根据容器的 Bean 定义解析依赖关系,返回所有要被注入的 Bean 对象
        Object autowiredArgument = resolveDependency(desc, beanName, autowiredBeanNames, converter);
        if (autowiredArgument != null) {
            //将属性赋值为所引用的对象
            pvs.add(propertyName, autowiredArgument);
        }
        for (String autowiredBeanName : autowiredBeanNames) {
            //为指定名称属性注册依赖 Bean 名称,进行属性的依赖注入
            registerDependentBean(autowiredBeanName, beanName);
            if (logger.isDebugEnabled()) {
                logger.debug("Autowiring by type from bean name '" + beanName + "' via property "
                    + propertyName + "' to bean named '" + autowiredBeanName + "'");
            }
        }
        //释放已自动注入的属性
        autowiredBeanNames.clear();
    }
    }
    catch (BeansException ex) {
        throw new UnsatisfiedDependencyException(mbd.getResourceDescription(), beanName, propertyName, ex);
    }
  }
}
```

通过上面的源码分析可以看出,通过属性名进行自动依赖注入相比通过属性类型进行自动依赖注入要稍微简单一些。但是真正实现属性注入的是 DefaultSingletonBeanRegistry 类的 registerDependentBean()方法。

3. DefaultSingletonBeanRegistry 的 registerDependentBean()方法实现属性依赖注入

DefaultSingletonBeanRegistry 的 registerDependentBean()方法实现属性依赖注入的重要代码如下:

```
//为指定的 Bean 注入依赖的 Bean
public void registerDependentBean(String beanName, String dependentBeanName) {
    //处理 Bean 名称,将别名转换为规范的 Bean 名称
    String canonicalName = canonicalName(beanName);
    Set<String> dependentBeans = this.dependentBeanMap.get(canonicalName);
```

```java
if (dependentBeans != null && dependentBeans.contains(dependentBeanName)) {
    return;
}

//多线程同步，保证容器内数据的一致性
//在容器中通过"Bean 名称→全部依赖 Bean 名称集合"查找指定名称 Bean 的依赖 Bean
synchronized (this.dependentBeanMap) {
    //获取指定名称 Bean 的所有依赖 Bean 名称
    dependentBeans = this.dependentBeanMap.get(canonicalName);
    if (dependentBeans == null) {
        //为 Bean 设置依赖 Bean 信息
        dependentBeans = new LinkedHashSet<>(8);
        this.dependentBeanMap.put(canonicalName, dependentBeans);
    }
    //在向容器中通过"Bean 名称→全部依赖 Bean 名称集合"添加 Bean 的依赖信息
    //即，将 Bean 所依赖的 Bean 添加到容器的集合中
    dependentBeans.add(dependentBeanName);
}
//在容器中通过"Bean 名称→指定名称 Bean 的依赖 Bean 集合"查找指定名称 Bean 的依赖 Bean
synchronized (this.dependenciesForBeanMap) {
    Set<String> dependenciesForBean = this.dependenciesForBeanMap.get(dependentBeanName);
    if (dependenciesForBean == null) {
        dependenciesForBean = new LinkedHashSet<>(8);
        this.dependenciesForBeanMap.put(dependentBeanName, dependenciesForBean);
    }
    //在容器中通过"Bean 名称→指定 Bean 的依赖 Bean 名称集合"添加 Bean 的依赖信息
    //即，将 Bean 所依赖的 Bean 添加到容器的集合中
    dependenciesForBean.add(canonicalName);
}
}
```

可以看出，autowiring 的实现过程如下：

（1）对 Bean 的属性调用 getBean()方法，完成依赖 Bean 的初始化和依赖注入。

（2）将依赖 Bean 的属性引用设置到被依赖的 Bean 属性上。

（3）将依赖 Bean 的名称和被依赖 Bean 的名称存储在 IoC 容器的集合中。

Spring IoC 容器的 autowiring 自动属性依赖注入是一个很方便的特性，可以简化开发配置，但是凡事都有两面性，自动属性依赖注入也有不足：首先，Bean 的依赖关系在配置文件中无法很清楚地看出来，会给维护造成一定的困难；其次，由于自动属性依赖注入是 Spring 容器自动执行的，容器是不会智能判断的，如果配置不当，将会带来无法预料的后果。所以在使用自动属性依赖注入时需要综合考虑。

第 10 章
一步一步手绘 Spring AOP 运行时序图

10.1 Spring AOP 初体验

10.1.1 再述 Spring AOP 应用场景

AOP 是 OOP 的延续，是 Aspect Oriented Programming 的缩写，意思是面向切面编程，可以通过预编译和运行时动态代理，实现在不修改源代码的情况下给程序动态统一添加功能。AOP 设计模式孜孜不倦追求的是调用者和被调用者之间的解耦，AOP 也是这个目标的一种实现。

我们现在做的一些非业务，如日志、事务、安全等都会写在业务代码中（也就是说，这些非业务类横切于业务类），但这些代码往往是重复的，复制-粘贴式的代码会给程序的维护带来不便。AOP 就实现了把这些业务需求与系统需求分开，这种解决方式也称为代理机制。

Spring AOP 是一种编程范式，主要目的是将非功能性需求从功能性需求中分离出来，达到解耦的目的。现实生活中也常常使用 AOP 思维来解决实际问题，如飞机组装、汽车组装等，如下两图所示。

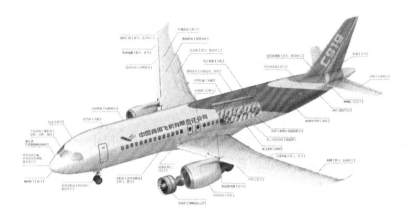

飞机组装示意图

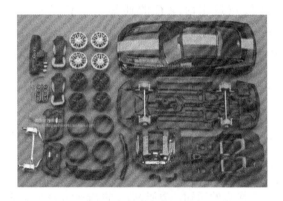

汽车组装示意图

飞机各部件的零件会交给不同的厂家生产,最终由组装工厂将各个部件组装起来变成一个整体。将零件的生产交出去的主要目的是解耦,但是解耦之前必须有统一的标准。

10.1.2 AOP 中必须明白的几个概念

1. 切面(Aspect)

"切面"的官方抽象定义为"一个关注点的模块化,这个关注点可能会横切多个对象"。"切面"由 ApplicationContext 中的<aop:aspect>来配置。

2. 连接点(JoinPoint)

连接点是指程序执行过程中的某一行为,例如 MemberService.get 的调用或者 MemberService.delete 抛出异常等行为。

3. 通知（Advice）

通知是指切面对于某个连接点所产生的动作。其中，一个切面可以包含多个通知。

通知（Advice）有如下几种类型。

（1）前置通知（Before Advice）

前置通知是指在某连接点之前执行的通知，但这个通知不能阻止连接点之前的代码的执行。ApplicationContext 中在<aop:aspect>里面使用<aop:before>元素进行声明，例如 TestAspect 中的 doBefore()方法。

（2）后置通知（After Advice）

后置通知是指当某连接点退出时执行的通知（不论是正常返回还是异常退出）。ApplicationContext 中在<aop:aspect>里面使用<aop:after>元素进行声明。例如，ServiceAspect 中的 returnAfter()方法，Teser 中调用 UserService.delete 抛出异常时，returnAfter()方法仍然执行。

（3）返回后通知（After Return Advice）

返回后通知是指在某连接点正常完成后执行的通知，不包括抛出异常的情况。ApplicationContext 中在<aop:aspect>里面使用<after-returning>元素进行声明。

（4）环绕通知（Around Advice）

环绕通知是指包围一个连接点的通知，类似 Web 的 Servlet 规范中 Filter 的 doFilter()方法。可以在方法的调用前后完成自定义的行为，也可以选择不执行。ApplicationContext 中在<aop:aspect>里面使用<aop:around>元素进行声明，例如 ServiceAspect 中的 around()方法。

（5）异常通知（After Throwing Advice）

异常通知是指在方法抛出异常导致退出时执行的通知。ApplicationContext 中在<aop:aspect>里面使用<aop:after-throwing>元素进行声明，例如 ServiceAspect 中的 returnThrow()方法。

注：可以将多个通知应用到一个目标对象上，即可以将多个切面织入同一个目标对象。

4. 切入点（Pointcut）

切入点是指匹配连接点的断言，在 AOP 中通知和一个切入点表达式关联。切面中的所有通知所关注的连接点都由切入点表达式决定。

5. 目标对象（Target Object）

目标对象是指被一个或者多个切面所通知的对象。例如 AServiceImpl 和 BServiceImpl。当然在实际运行时，Spring AOP 采用代理实现，实际上 AOP 操作的是 TargetObject 的代理对象。

6. AOP 代理（AOP Proxy）

在 Spring AOP 中有两种代理方式：JDK 动态代理和 CGLib 代理。默认情况下，目标对象实现了接口时，采用 JDK 动态代理，例如 AServiceImpl；反之，采用 CGLib 代理，例如 BServiceImpl。强制使用 CGLib 代理需要将 <aop:config> 的 proxy-target-class 属性设置为 true。

10.1.3 使用 Spring AOP 的两种方式

使用 Spring AOP 有两种方式，一种是比较方便和强大的注解方式，另一种则是中规中矩的 XML 配置方式。

使用注解配置 Spring AOP 总体分为两步，第一步是在 XML 文件中激活自动扫描组件功能，同时激活自动代理功能：

```xml
<?xml version="1.0" encoding="UTF-8"?>
<beans xmlns="http://www.springframework.org/schema/beans"
    xmlns:xsi="http://www.w3.org/2001/XMLSchema-instance"
    xmlns:context="http://www.springframework.org/schema/context"
    xmlns:util="http://www.springframework.org/schema/util"
    xsi:schemaLocation="http://www.springframework.org/schema/beans
http://www.springframework.org/schema/beans/spring-beans.xsd
     http://www.springframework.org/schema/util
http://www.springframework.org/schema/util/spring-util-2.0.xsd
     http://www.springframework.org/schema/context
http://www.springframework.org/schema/context/spring-context-3.0.xsd">

   <context:component-scan base-package="com.gupaoedu"/>
   <context:annotation-config />
</beans>
```

第二步是为切面类添加注解：

```java
//声明这是一个组件
@Component
//声明这是一个切面 Bean
@Aspect
@Slf4j
public class AnnotaionAspect {

   //配置切入点,该方法无方法体,主要是为了方便同类中其他方法使用此处配置的切入点
   @Pointcut("execution(* com.gupaoedu.vip.pattern.spring.aop.service..*(..))")
```

```java
public void aspect(){ }

/*
 * 配置前置通知，使用在 aspect()方法上注册的切入点
 * 同时接收 JoinPoint 对象，可以没有该参数
 */
@Before("aspect()")
public void before(JoinPoint joinPoint){

    log.info("before 通知 " + joinPoint);
}

//配置后置通知，使用在 aspect()方法上注册的切入点
@After("aspect()")
public void after(JoinPoint joinPoint){

    log.info("after 通知  " + joinPoint);
}

//配置环绕通知，使用在 aspect()方法上注册的切入点
@Around("aspect()")
public void around(JoinPoint joinPoint){
    long start = System.currentTimeMillis();
    try {
        ((ProceedingJoinPoint) joinPoint).proceed();
        long end = System.currentTimeMillis();
        log.info("around 通知 " + joinPoint + "\tUse time : " + (end - start) + " ms!");
    } catch (Throwable e) {
        long end = System.currentTimeMillis();
        log.info("around 通知 " + joinPoint + "\tUse time : " + (end - start) + " ms with exception : " + e.getMessage());
    }
}

//配置返回后通知，使用在 aspect()方法上注册的切入点
@AfterReturning("aspect()")
public void afterReturn(JoinPoint joinPoint){
    log.info("afterReturn 通知 " + joinPoint);
}

//配置异常通知，使用在 aspect()方法上注册的切入点
@AfterThrowing(pointcut="aspect()", throwing="ex")
public void afterThrow(JoinPoint joinPoint, Exception ex){
    log.info("afterThrow 通知 " + joinPoint + "\t" + ex.getMessage());
}

}
```

测试代码如下：

```java
@ContextConfiguration(locations = {"classpath*:application-context.xml"})
@RunWith(SpringJUnit4ClassRunner.class)
public class AnnotationTester {
    @Autowired
    MemberService annotationService;
    @Autowired
    ApplicationContext app;


    @Test
//  @Ignore
    public void test(){
        System.out.println("=====这是一条华丽的分割线======");

        AnnotaionAspect aspect = app.getBean(AnnotaionAspect.class);
        System.out.println(aspect);
        annotationService.save(new Member());

        System.out.println("=====这是一条华丽的分割线======");
        try {
            annotationService.delete(1L);
        } catch (Exception e) {
            //e.printStackTrace();
        }
    }
}
```

控制台输出如下：

```
=====这是一条华丽的分割线======
com.gupaoedu.aop.aspect.AnnotaionAspect@6ef714a0
[INFO ] [13:04:46] com.gupaoedu.aop.aspect.AnnotaionAspect - before execution(void com.gupaoedu.aop.service.MemberService.save(Member))
[INFO ] [13:04:46] com.gupaoedu.aop.aspect.ArgsAspect - beforeArgUser execution(void com.gupaoedu.aop.service.MemberService.save(Member))
[INFO ] [13:04:46] com.gupaoedu.aop.aspect.AnnotaionAspect - save member Method . . .
[INFO ] [13:04:46] com.gupaoedu.aop.aspect.AnnotaionAspect - around execution(void com.gupaoedu.aop.service.MemberService.save(Member))  Use time : 38 ms!
[INFO ] [13:04:46] com.gupaoedu.aop.aspect.AnnotaionAspect - after execution(void com.gupaoedu.aop.service.MemberService.save(Member))
[INFO ] [13:04:46] com.gupaoedu.aop.aspect.AnnotaionAspect - afterReturn execution(void com.gupaoedu.aop.service.MemberService.save(Member))
=====这是一条华丽的分割线======
[INFO ] [13:04:46] com.gupaoedu.aop.aspect.AnnotaionAspect - before execution(boolean
```

```
com.gupaoedu.aop.service.MemberService.delete(long))
[INFO ] [13:04:46] com.gupaoedu.aop.aspect.ArgsAspect - beforeArgId execution(boolean
com.gupaoedu.aop.service.MemberService.delete(long))  ID:1
[INFO ] [13:04:46] com.gupaoedu.aop.aspect.AnnotaionAspect - delete Method . . .
[INFO ] [13:04:46] com.gupaoedu.aop.aspect.AnnotaionAspect - around execution(boolean
com.gupaoedu.aop.service.MemberService.delete(long))   Use time : 3 ms with exception : spring
aop ThrowAdvice 演示
[INFO ] [13:04:46] com.gupaoedu.aop.aspect.AnnotaionAspect - after execution(boolean
com.gupaoedu.aop.service.MemberService.delete(long))
[INFO ] [13:04:46] com.gupaoedu.aop.aspect.AnnotaionAspect - afterReturn execution(boolean
com.gupaoedu.aop.service.MemberService.delete(long))
```

可以看到，正如预期的那样，虽然我们并没有对 MemberService 类包括其调用方式做任何改变，但是 Spring 仍然拦截到了其中方法的调用，或许这正是 AOP 的魔力所在。

再简单说一下 XML 配置方式，其实也一样简单：

```xml
<bean id="xmlAspect" class="com.gupaoedu.vip.pattern.spring.aop.aspect.XmlAspect"></bean>
<!-- AOP 配置 -->
<aop:config>
    <!-- 声明一个切面,并注入切面 Bean,相当于@Aspect -->
    <aop:aspect ref="xmlAspect">
        <!-- 配置一个切入点,相当于@Pointcut -->
        <aop:pointcut expression="execution(* com.gupaoedu.vip.pattern.spring.aop.service..*
(..))" id="simplePointcut"/>
        <!-- 配置通知,相当于@Before、@After、@AfterReturn、@Around、@AfterThrowing -->
        <aop:before pointcut-ref="simplePointcut" method="before"/>
        <aop:after pointcut-ref="simplePointcut" method="after"/>
        <aop:after-returning pointcut-ref="simplePointcut" method="afterReturn"/>
        <aop:after-throwing pointcut-ref="simplePointcut" method="afterThrow" throwing="ex"/>
    </aop:aspect>
</aop:config>
```

个人觉得 XML 配置方式不如注解方式灵活和强大，你可以不同意这个观点，但是不知道如下的代码会不会让你的想法有所改变呢？

```
//配置切入点，该方法无方法体,主要是为了方便同类中其他方法使用此处配置的切入点
@Pointcut("execution(* com.gupaoedu.aop.service..*(..))")
public void aspect(){  }

//配置前置通知
@Before("execution(com.gupaoedu.model.Member com.gupaoedu.aop.service..*(..))")
public void beforeReturnUser(JoinPoint joinPoint){
    log.info("beforeReturnUser " + joinPoint);
}

//配置前置通知
```

```java
@Before("execution(* com.gupaoedu.aop.service..*(com.gupaoedu.model.Member))")
public void beforeArgUser(JoinPoint joinPoint){
    log.info("beforeArgUser " + joinPoint);
}

//配置前置通知
@Before("aspect()&&args(id)")
public void beforeArgId(JoinPoint joinPoint, long id){
    log.info("beforeArgId " + joinPoint + "\tID:" + id);
}
```

以下是 MemberService 类的代码：

```java
@Service
public class MemberService {

    private final static Logger log = Logger.getLogger(AnnotaionAspect.class);

    public Member get(long id){
        log.info("getMemberById Method . . .");
        return new Member();
    }

    public Member get(){
        log.info("getMember Method . . .");
        return new Member();
    }

    public void save(Member member){
        log.info("save member Method . . .");
    }

    public boolean delete(long id) throws Exception{
        log.info("delete Method . . .");
        throw new Exception("spring aop ThrowAdvice 演示");
    }

}
```

10.1.4 切入点表达式的配置规则

应该说学习 Spring AOP 有两个难点，第一个难点在于理解 AOP 的理念和相关概念，第二个难点在于灵活掌握和使用切入点表达式。概念的理解通常不在一朝一夕，时间长了，自然就明白了。下面简单地介绍一下切入点表达式的配置规则。

通常情况下，表达式中使用"execution"就可以满足大部分要求，格式如下：

```
execution(modifiers-pattern? ret-type-pattern declaring-type-pattern? name-pattern(param-pattern)
throws-pattern?
```

其中各参数的含义如下。

- modifiers-pattern：方法的操作权限。
- ret-type-pattern：返回值。
- declaring-type-pattern：方法所在的包。
- name-pattern：方法名。
- param-pattern：参数名。
- throws-pattern：异常。

除 ret-type-pattern 和 name-pattern 外，其他参数都是可选的。execution(* com.spring.service.*.*(..)) 表示 com.spring.service 包下的返回值为任意类型、方法名任意、参数不做限制的所有方法。

这里再讲一下通知（Advice）参数，可以通过 args 来绑定，这样就可以在通知中访问具体参数了。例如，<aop:aspect>配置如下：

```
<aop:config>
  <aop:aspect ref="xmlAspect">
    <aop:pointcut id="simplePointcut"
              expression="execution(* com.gupaoedu.aop.service..*(..)) and args(msg,..)" />
    <aop:after pointcut-ref="simplePointcut" Method="after"/>
  </aop:aspect>
</aop:config>
```

上面的 args(msg,..) 是指将切入点方法上的第一个 String 类型参数添加到参数名为 msg 的通知的参数上，这样就可以直接使用该参数了。

还可以看到，每个通知方法的第一个参数都是连接点（JoinPoint）。其实，在 Spring 中，任何通知方法都可以将第一个参数定义为 org.aspectj.lang.JoinPoint 类型用于接收当前连接点对象。JoinPoint 接口提供了一系列有用的方法，比如 getArgs()（返回方法参数）、getThis()（返回代理对象）、getTarget()（返回目标）、getSignature()（返回正在被通知的方法的相关信息）和 toString()（打印正在被通知的方法的有用信息）。

10.2 Spring AOP 源码分析

10.2.1 寻找入口

Spring AOP 是由接入 BeanPostProcessor 后置处理器开始的，它是 Spring IoC 容器经常使用的

一个特性，这个 Bean 后置处理器是一个监听器，可以监听容器触发的 Bean 声明周期事件。向容器注册后置处理器以后，容器中管理的 Bean 就具备了接收 IoC 容器回调事件的能力。

BeanPostProcessor 的使用非常简单，只需要提供一个 BeanPostProcessor 接口的实现类，然后在 Bean 的配置文件中设置即可。

1. BeanPostProcessor 的源码

BeanPostProcessor 的源码如下：

```
public interface BeanPostProcessor {

    //在 Bean 的初始化前提供回调入口
    @Nullable
    default Object postProcessBeforeInitialization(Object bean, String beanName) throws BeansException {
        return bean;
    }

    //在 Bean 的初始化之后提供回调入口
    @Nullable
    default Object postProcessAfterInitialization(Object bean, String beanName) throws BeansException {
        return bean;
    }

}
```

这两个回调入口都和容器管理的 Bean 的生命周期事件紧密相关，可以为用户提供在 Spring IoC 容器初始化 Bean 过程中自定义的处理操作。

2. AbstractAutowireCapableBeanFactory 类的 doCreateBean()方法

BeanPostProcessor 后置处理器的调用发生在 Spring IoC 容器完成 Bean 实例对象的创建和属性的依赖注入之后，在对 Spring 依赖注入的源码分析中我们知道，当应用程序第一次调用 getBean()方法（lazy-init 预实例化除外）向 Spring IoC 容器索取指定 Bean 时，触发 Spring IoC 容器创建 Bean 实例对象并进行依赖注入。其实真正实现创建 Bean 对象并进行依赖注入的方法是 AbstractAutowireCapableBeanFactory 类的 doCreateBean()方法，主要源码如下：

```
//真正创建 Bean 的方法
protected Object doCreateBean(final String beanName, final RootBeanDefinition mbd, final @Nullable Object[] args)
        throws BeanCreationException {

    //创建 Bean 实例对象
```

```
...
    Object exposedObject = bean;
    try {
        //对 Bean 属性进行依赖注入
        populateBean(beanName, mbd, instanceWrapper);
        //Bean 实例对象的依赖注入完成以后，开始对 Bean 实例对象进行初始化，为 Bean 实例对象应用
BeanPostProcessor 后置处理器
        exposedObject = initializeBean(beanName, exposedObject, mbd);
    }
    catch (Throwable ex) {
        if (ex instanceof BeanCreationException && beanName.equals(((BeanCreationException) ex).getBeanName())) {
            throw (BeanCreationException) ex;
        }
        else {
            throw new BeanCreationException(
                mbd.getResourceDescription(), beanName, "Initialization of bean failed", ex);
        }
    }
    ...

    //为应用返回所需要的实例对象
    return exposedObject;
}
```

从上面的代码可知，为 Bean 实例对象添加 BeanPostProcessor 后置处理器的入口是 initializeBean()方法。

3. initializeBean()方法

在 AbstractAutowireCapableBeanFactory 类中 initializeBean()方法实现为容器创建的 Bean 实例对象添加 BeanPostProcessor 后置处理器，源码如下：

```
//初始化容器创建的 Bean 实例对象，为其添加 BeanPostProcessor 后置处理器
protected Object initializeBean(final String beanName, final Object bean, @Nullable RootBeanDefinition mbd) {
    //通过 JDK 的安全机制验证权限
    if (System.getSecurityManager() != null) {
        //实现 PrivilegedAction 接口的匿名内部类
        AccessController.doPrivileged((PrivilegedAction<Object>) () -> {
            invokeAwareMethods(beanName, bean);
            return null;
        }, getAccessControlContext());
    }
```

```java
    else {
        //为 Bean 实例对象包装相关属性,如名称、类加载器、所属容器等
        invokeAwareMethods(beanName, bean);
    }

    Object wrappedBean = bean;
    //调用 BeanPostProcessor 后置处理器的回调方法,在 Bean 实例初始化前做一些处理
    if (mbd == null || !mbd.isSynthetic()) {
        wrappedBean = applyBeanPostProcessorsBeforeInitialization(wrappedBean, beanName);
    }

    //调用 Bean 实例初始化方法,这个初始化方法是在 Spring Bean 定义配置文件中通过 init-Method 属性指定的
    try {
        invokeInitMethods(beanName, wrappedBean, mbd);
    }
    catch (Throwable ex) {
        throw new BeanCreationException(
                (mbd != null ? mbd.getResourceDescription() : null),
                beanName, "Invocation of init Method failed", ex);
    }
    //调用 BeanPostProcessor 后置处理器的回调方法,在 Bean 实例初始化之后做一些处理
    if (mbd == null || !mbd.isSynthetic()) {
        wrappedBean = applyBeanPostProcessorsAfterInitialization(wrappedBean, beanName);
    }

    return wrappedBean;
}

@Override
//调用 BeanPostProcessor 后置处理器实例初始化之前的处理方法
public Object applyBeanPostProcessorsBeforeInitialization(Object existingBean, String beanName)
        throws BeansException {
    Object result = existingBean;
    //遍历容器为所创建的 Bean 添加所有 BeanPostProcessor 后置处理器
    for (BeanPostProcessor beanProcessor : getBeanPostProcessors()) {
        //调用 Bean 实例所有后置处理中初始化前前的处理方法,
        //为 Bean 实例对象在初始化之前做一些自定义的处理
        Object current = beanProcessor.postProcessBeforeInitialization(result, beanName);
        if (current == null) {
            return result;
        }
        result = current;
    }
    return result;
}

@Override
```

```
//调用 BeanPostProcessor 后置处理器实例初始化之后的处理方法
public Object applyBeanPostProcessorsAfterInitialization(Object existingBean, String beanName)
        throws BeansException {

    Object result = existingBean;
    //遍历容器为所创建的 Bean 添加所有 BeanPostProcessor 后置处理器
    for (BeanPostProcessor beanProcessor : getBeanPostProcessors()) {
        //调用 Bean 实例所有的后置处理中初始化后的处理方法,
        //为 Bean 实例对象在初始化之后做一些自定义的处理
        Object current = beanProcessor.postProcessAfterInitialization(result, beanName);
        if (current == null) {
            return result;
        }
        result = current;
    }
    return result;
}
```

BeanPostProcessor 初始化前的操作方法和初始化后的操作方法均委派其实现子类实现。Spring 中 BeanPostProcessor 的实现子类非常多,分别完成不同的操作,如 AOP 面向切面编程的注册通知适配器、Bean 对象的数据校验、Bean 继承属性、方法的合并等。下面我们来分析其中一个创建 AOP 代理对象的子类 AbstractAutoProxyCreator,该类重写了 postProcessAfterInitialization()方法。

10.2.2 选择代理策略

下面看一下 postProcessAfterInitialization()方法,它调用了一个非常核心的方法——wrapIfNecessary(),如下所示。

```
@Override
public Object postProcessAfterInitialization(@Nullable Object bean, String beanName) throws
BeansException {
    if (bean != null) {
        Object cacheKey = getCacheKey(bean.getClass(), beanName);
        if (!this.earlyProxyReferences.contains(cacheKey)) {
            return wrapIfNecessary(bean, beanName, cacheKey);
        }
    }
    return bean;
}
...
protected Object wrapIfNecessary(Object bean, String beanName, Object cacheKey) {
    if (StringUtils.hasLength(beanName) && this.targetSourcedBeans.contains(beanName)) {
        return bean;
    }
```

```java
//判断是否应该代理这个 Bean
if (Boolean.FALSE.equals(this.advisedBeans.get(cacheKey))) {
    return bean;
}
/*
 * 判断是否是一些 InfrastructureClass 或者是否应该跳过这个 Bean
 * 所谓 InfrastructureClass 就是指 Advice、PointCut、Advisor 等接口的实现类
 * shouldSkip()方法默认返回 false，由于是 protected 修饰的方法，子类可以覆盖
 */
if (isInfrastructureClass(bean.getClass()) || shouldSkip(bean.getClass(), beanName)) {
    this.advisedBeans.put(cacheKey, Boolean.FALSE);
    return bean;
}

//获取这个 Bean 的通知
Object[] specificInterceptors = getAdvicesAndAdvisorsForBean(bean.getClass(), beanName, null);
if (specificInterceptors != DO_NOT_PROXY) {
    this.advisedBeans.put(cacheKey, Boolean.TRUE);
    //创建代理
    Object proxy = createProxy(
            bean.getClass(), beanName, specificInterceptors, new SingletonTargetSource(bean));
    this.proxyTypes.put(cacheKey, proxy.getClass());
    return proxy;
}

this.advisedBeans.put(cacheKey, Boolean.FALSE);
return bean;
}
...
protected Object createProxy(Class<?> beanClass, @Nullable String beanName,
        @Nullable Object[] specificInterceptors, TargetSource targetSource) {

    if (this.beanFactory instanceof ConfigurableListableBeanFactory) {
        AutoProxyUtils.exposeTargetClass((ConfigurableListableBeanFactory) this.beanFactory,
beanName, beanClass);
    }

    ProxyFactory proxyFactory = new ProxyFactory();
    proxyFactory.copyFrom(this);

    if (!proxyFactory.isProxyTargetClass()) {
        if (shouldProxyTargetClass(beanClass, beanName)) {
            proxyFactory.setProxyTargetClass(true);
        }
        else {
            evaluateProxyInterfaces(beanClass, proxyFactory);
```

```
        }
    }

    Advisor[] advisors = buildAdvisors(beanName, specificInterceptors);
    proxyFactory.addAdvisors(advisors);
    proxyFactory.setTargetSource(targetSource);
    customizeProxyFactory(proxyFactory);

    proxyFactory.setFrozen(this.freezeProxy);
    if (advisorsPreFiltered()) {
        proxyFactory.setPreFiltered(true);
    }

    return proxyFactory.getProxy(getProxyClassLoader());
}
```

整个过程最终调用的是 proxyFactory.getProxy()方法。到这里,proxyFactory 有 JDK 和 CGLib 两种,我们该如何选择呢?使用 DefaultAopProxyFactory 的 createAopProxy()方法:

```
public class DefaultAopProxyFactory implements AopProxyFactory, Serializable {

    @Override
    public AopProxy createAopProxy(AdvisedSupport config) throws AopConfigException {
        if (config.isOptimize() || config.isProxyTargetClass() ||
hasNoUserSuppliedProxyInterfaces(config)) {
            Class<?> targetClass = config.getTargetClass();
            if (targetClass == null) {
                throw new AopConfigException("TargetSource cannot determine target class: " +
                    "Either an interface or a target is required for proxy creation.");
            }
            if (targetClass.isInterface() || Proxy.isProxyClass(targetClass)) {
                return new JdkDynamicAopProxy(config);
            }
            return new ObjenesisCglibAopProxy(config);
        }
        else {
            return new JdkDynamicAopProxy(config);
        }
    }

    private boolean hasNoUserSuppliedProxyInterfaces(AdvisedSupport config) {
        Class<?>[] ifcs = config.getProxiedInterfaces();
        return (ifcs.length == 0 || (ifcs.length == 1 && SpringProxy.class.isAssignableFrom(ifcs[0])));
    }
}
```

10.2.3 调用代理方法

在分析调用逻辑之前先看类图，看看 Spring 中主要的 AOP 组件，如下图所示。

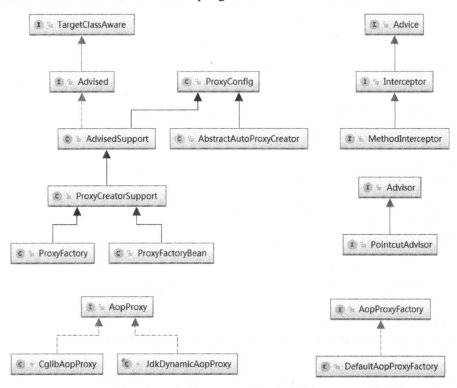

我们已经知道，Spring 提供了两种方式来生成代理：JDK Proxy 和 CGLib。下面来研究一下 Spring 如何使用 JDK 来生成代理对象，具体的生成代码在 JdkDynamicAopProxy 类中，直接看相关代码：

```java
/**
 * 获取代理类要实现的接口，除了 Advised 对象中配置的，还会加上 SpringProxy、Advised(opaque=false)
 * 检查得到的接口中有没有定义 equals 或者 hashcode 的接口
 * 调用 Proxy.newProxyInstance()方法创建代理对象
 */
@Override
public Object getProxy(@Nullable ClassLoader classLoader) {
    if (logger.isDebugEnabled()) {
        logger.debug("Creating JDK dynamic proxy: target source is " +
this.advised.getTargetSource());
    }
    Class<?>[] proxiedInterfaces = AopProxyUtils.completeProxiedInterfaces(this.advised, true);
```

```
    findDefinedEqualsAndHashCodeMethods(proxiedInterfaces);
    return Proxy.newProxyInstance(classLoader, proxiedInterfaces, this);
}
```

代码中的注释已经把代理对象的生成过程解释得非常明白,此处不再赘述。代理对象生成了,那么切面是如何织入的?

InvocationHandler 是 JDK 动态代理的核心,生成的代理对象的方法调用都会委派到 invoke() 方法。从 JdkDynamicAopProxy 的源码可以看到,这个类其实也实现了 InvocationHandler。下面我们分析 Spring AOP 是如何织入切面的,直接看 invoke() 方法:

```
public Object invoke(Object proxy, Method Method, Object[] args) throws Throwable {
    MethodInvocation invocation;
    Object oldProxy = null;
    boolean setProxyContext = false;

    TargetSource targetSource = this.advised.targetSource;
    Object target = null;

    try {
        //目标对象未实现 eqauls()方法
        if (!this.equalsDefined && AopUtils.isEqualsMethod(Method)) {
            return equals(args[0]);
        }
        //目标对象未实现 hashCode()方法
        else if (!this.hashCodeDefined && AopUtils.isHashCodeMethod(Method)) {
            return hashCode();
        }
        else if (Method.getDeclaringClass() == DecoratingProxy.class) {
            return AopProxyUtils.ultimateTargetClass(this.advised);
        }
        //直接反射调用 Advised 接口或者其父接口中定义的方法,不应用通知
        else if (!this.advised.opaque && Method.getDeclaringClass().isInterface() &&
                Method.getDeclaringClass().isAssignableFrom(Advised.class)) {
            return AopUtils.invokeJoinpointUsingReflection(this.advised, Method, args);
        }

        Object retVal;

        if (this.advised.exposeProxy) {
            oldProxy = AopContext.setCurrentProxy(proxy);
            setProxyContext = true;
        }

        //获得目标对象的类
        target = targetSource.getTarget();
```

```java
        Class<?> targetClass = (target != null ? target.getClass() : null);
        //获取可以应用到此方法上的拦截器列表
        List<Object> chain = this.advised.getInterceptorsAndDynamicInterceptionAdvice(Method, targetClass);

        //如果没有可以应用到此方法上的拦截器,则直接反射调用 Method.invoke(target, args)
        if (chain.isEmpty()) {
            Object[] argsToUse = AopProxyUtils.adaptArgumentsIfNecessary(Method, args);
            retVal = AopUtils.invokeJoinpointUsingReflection(target, Method, argsToUse);
        }
        else {
            //创建 ReflectiveMethodInvocation
            invocation = new ReflectiveMethodInvocation(proxy, target, Method, args, targetClass, chain);
            retVal = invocation.proceed();
        }

        Class<?> returnType = Method.getReturnType();
        if (retVal != null && retVal == target &&
                returnType != Object.class && returnType.isInstance(proxy) &&
                !RawTargetAccess.class.isAssignableFrom(Method.getDeclaringClass())) {
            retVal = proxy;
        }
        else if (retVal == null && returnType != Void.TYPE && returnType.isPrimitive()) {
            throw new AopInvocationException(
                    "Null return value from advice does not match primitive return type for: " + Method);
        }
        return retVal;
    }
    finally {
        if (target != null && !targetSource.isStatic()) {
            targetSource.releaseTarget(target);
        }
        if (setProxyContext) {
            AopContext.setCurrentProxy(oldProxy);
        }
    }
}
```

主要实现思路为:先获取应用到此方法上的拦截器链(Interceptor Chain)。如果有拦截器,则应用拦截器,并执行连接点(JoinPoint);如果没有拦截器,则直接反射执行连接点。这里的关键是拦截器链是如何获取的,以及它又是如何执行的。下面来逐一分析。

从上面的代码可以看到,拦截器链是通过 Advised.getInterceptorsAndDynamicInterceptionAdvice() 方法获取的,我们来看这个方法的实现代码:

```java
public List<Object> getInterceptorsAndDynamicInterceptionAdvice(Method Method, @Nullable
Class<?> targetClass) {
    MethodCacheKey cacheKey = new MethodCacheKey(Method);
    List<Object> cached = this.MethodCache.get(cacheKey);
    if (cached == null) {
        cached = this.advisorChainFactory.getInterceptorsAndDynamicInterceptionAdvice(
                this, Method, targetClass);
        this.MethodCache.put(cacheKey, cached);
    }
    return cached;
}
```

可以看到,获取拦截器其实是由 AdvisorChainFactory 的 getInterceptorsAndDynamicInterception-Advice()方法完成的,且获取到的结果会被缓存下来。下面分析 getInterceptorsAndDynamic-InterceptionAdvice()方法的实现:

```java
/**
 * 从提供的配置实例 config 中获取 Advisor 列表,遍历处理这些 Advisor。如果是 IntroductionAdvisor,
 * 则判断此 Advisor 能否应用到目标类 targetClass 上。如果是 PointcutAdvisor,则判断
 * 此 Advisor 能否应用到目标方法 Method 上。将满足条件的 Advisor 通过 AdvisorAdaptor 转化成拦截器列
 * 表返回。
 */
@Override
public List<Object> getInterceptorsAndDynamicInterceptionAdvice(
        Advised config, Method Method, @Nullable Class<?> targetClass) {

    List<Object> interceptorList = new ArrayList<>(config.getAdvisors().length);
    Class<?> actualClass = (targetClass != null ? targetClass : Method.getDeclaringClass());
    //查看是否包含 IntroductionAdvisor
    boolean hasIntroductions = hasMatchingIntroductions(config, actualClass);
    //这里实际上注册了一系列 AdvisorAdapter,用于将 Advisor 转化成 MethodInterceptor
    AdvisorAdapterRegistry registry = GlobalAdvisorAdapterRegistry.getInstance();

    for (Advisor advisor : config.getAdvisors()) {
        if (advisor instanceof PointcutAdvisor) {
            PointcutAdvisor pointcutAdvisor = (PointcutAdvisor) advisor;
            if (config.isPreFiltered() || pointcutAdvisor.getPointcut().getClassFilter().matches
(actualClass)) {
                //这里的两个方法的位置可以互换
                //将 Advisor 转化成 Interceptor
                MethodInterceptor[] interceptors = registry.getInterceptors(advisor);
                //检查当前 Advisor 的切入点是否可以匹配当前方法
                MethodMatcher mm = pointcutAdvisor.getPointcut().getMethodMatcher();
                if (MethodMatchers.matches(mm, Method, actualClass, hasIntroductions)) {
                    if (mm.isRuntime()) {
                        for (MethodInterceptor interceptor : interceptors) {
```

```
                    interceptorList.add(new InterceptorAndDynamicMethodMatcher(interceptor, mm));
                }
            }
            else {
                interceptorList.addAll(Arrays.asList(interceptors));
            }
        }
    }
    else if (advisor instanceof IntroductionAdvisor) {
        IntroductionAdvisor ia = (IntroductionAdvisor) advisor;
        if (config.isPreFiltered() || ia.getClassFilter().matches(actualClass)) {
            Interceptor[] interceptors = registry.getInterceptors(advisor);
            interceptorList.addAll(Arrays.asList(interceptors));
        }
    }
    else {
        Interceptor[] interceptors = registry.getInterceptors(advisor);
        interceptorList.addAll(Arrays.asList(interceptors));
    }
}

return interceptorList;
}
```

这个方法执行完成后，Advised 中配置的能够应用到连接点（JoinPoint）或者目标对象（Target Object）的 Advisor 全部被转化成 MethodInterceptor。接下来我们看得到的拦截器链是怎么起作用的。

```
if (chain.isEmpty()) {
    Object[] argsToUse = AopProxyUtils.adaptArgumentsIfNecessary(Method, args);
    retVal = AopUtils.invokeJoinpointUsingReflection(target, Method, argsToUse);
}
else {
    //创建 MethodInvocation
    invocation = new ReflectiveMethodInvocation(proxy, target, Method, args, targetClass, chain);
    retVal = invocation.proceed();
}
```

从这段代码可以看出，如果得到的拦截器链为空，则直接反射调用目标方法，否则创建 MethodInvocation，调用其 proceed()方法，触发拦截器链的执行，来看具体代码：

```
public Object proceed() throws Throwable {
    //如果拦截器执行完了，则执行连接点
    if (this.currentInterceptorIndex == this.interceptorsAndDynamicMethodMatchers.size() - 1) {
        return invokeJoinpoint();
```

```
        }
        Object interceptorOrInterceptionAdvice =
            this.interceptorsAndDynamicMethodMatchers.get(++this.currentInterceptorIndex);
        InterceptorAndDynamicMethodMatcher dm =
            (InterceptorAndDynamicMethodMatcher) interceptorOrInterceptionAdvice;
        //动态匹配：运行时参数是否满足匹配条件
        if (dm.MethodMatcher.matches(this.Method, this.targetClass, this.arguments)) {
            return dm.interceptor.invoke(this);
        }
        else {
            //动态匹配失败时，略过当前拦截器，调用下一个拦截器
            return proceed();
        }
    }
    else {
        //执行当前拦截器
        return ((MethodInterceptor) interceptorOrInterceptionAdvice).invoke(this);
    }
}
```

至此，拦截器链就完美地形成了。我们再往下看 invokeJoinpointUsingReflection()方法，也就是反射调用。

```
public static Object invokeJoinpointUsingReflection(@Nullable Object target, Method method,
Object[] args)
        throws Throwable {
    try {
        ReflectionUtils.makeAccessible(method);
        return method.invoke(target, args);
    }
    catch (InvocationTargetException ex) {
        throw ex.getTargetException();
    }
    catch (IllegalArgumentException ex) {
        throw new AopInvocationException("AOP configuration seems to be invalid: tried calling
method [" + method + "] on target [" + target + "]", ex);
    }
    catch (IllegalAccessException ex) {
        throw new AopInvocationException("Could not access method [" + method + "]", ex);
    }
}
```

Spring AOP 源码就分析到这儿，相信"小伙伴们"应该有基本思路了，请到"下载资源"中查看运行时序图。

10.2.4 触发通知

在为 AopProxy 代理对象配置拦截器的实现中，有一个取得拦截器的配置过程，这个过程是由 DefaultAdvisorChainFactory 实现的。这个工厂类负责生成拦截器链，在它的 getInterceptorsAnd-DynamicInterceptionAdvice()方法中，有一个适配和注册过程，通过配置 Spring 预先设计好的拦截器，加入了 AOP 实现。

```java
/**
 * 从提供的配置实例 config 中获取 Advisor 列表，遍历处理这些 Advisor。如果是 IntroductionAdvisor，
 * 则判断此 Advisor 能否应用到目标类 targetClass 上。如果是 PointcutAdvisor，则判断
 * 此 Advisor 能否应用到目标方法 Method 上。将满足条件的 Advisor 通过 AdvisorAdaptor 转化成拦截器列
 * 表返回。
 */
@Override
public List<Object> getInterceptorsAndDynamicInterceptionAdvice(
        Advised config, Method Method, @Nullable Class<?> targetClass) {

    List<Object> interceptorList = new ArrayList<>(config.getAdvisors().length);
    Class<?> actualClass = (targetClass != null ? targetClass : Method.getDeclaringClass());
    //查看是否包含 IntroductionAdvisor
    boolean hasIntroductions = hasMatchingIntroductions(config, actualClass);
    //这里实际上注册一系列 AdvisorAdapter，用于将 Advisor 转化成 MethodInterceptor
    AdvisorAdapterRegistry registry = GlobalAdvisorAdapterRegistry.getInstance();

    ...

    return interceptorList;
}
```

GlobalAdvisorAdapterRegistry 类负责拦截器的适配和注册过程。

```java
public abstract class GlobalAdvisorAdapterRegistry {

    private static AdvisorAdapterRegistry instance = new DefaultAdvisorAdapterRegistry();

    public static AdvisorAdapterRegistry getInstance() {
        return instance;
    }

    static void reset() {
        instance = new DefaultAdvisorAdapterRegistry();
    }

}
```

GlobalAdvisorAdapterRegistry 类起到了适配器和单例模式的作用，提供了一个 DefaultAdvisor-

AdapterRegistry 类来完成各种通知的适配和注册过程。

```java
public class DefaultAdvisorAdapterRegistry implements AdvisorAdapterRegistry, Serializable {

    private final List<AdvisorAdapter> adapters = new ArrayList<>(3);

    public DefaultAdvisorAdapterRegistry() {
        registerAdvisorAdapter(new MethodBeforeAdviceAdapter());
        registerAdvisorAdapter(new AfterReturningAdviceAdapter());
        registerAdvisorAdapter(new ThrowsAdviceAdapter());
    }


    @Override
    public Advisor wrap(Object adviceObject) throws UnknownAdviceTypeException {
        if (adviceObject instanceof Advisor) {
            return (Advisor) adviceObject;
        }
        if (!(adviceObject instanceof Advice)) {
            throw new UnknownAdviceTypeException(adviceObject);
        }
        Advice advice = (Advice) adviceObject;
        if (advice instanceof MethodInterceptor) {
            return new DefaultPointcutAdvisor(advice);
        }
        for (AdvisorAdapter adapter : this.adapters) {
            if (adapter.supportsAdvice(advice)) {
                return new DefaultPointcutAdvisor(advice);
            }
        }
        throw new UnknownAdviceTypeException(advice);
    }

    @Override
    public MethodInterceptor[] getInterceptors(Advisor advisor) throws UnknownAdviceTypeException {
        List<MethodInterceptor> interceptors = new ArrayList<>(3);
        Advice advice = advisor.getAdvice();
        if (advice instanceof MethodInterceptor) {
            interceptors.add((MethodInterceptor) advice);
        }
        for (AdvisorAdapter adapter : this.adapters) {
            if (adapter.supportsAdvice(advice)) {
                interceptors.add(adapter.getInterceptor(advisor));
            }
        }
        if (interceptors.isEmpty()) {
```

```
        throw new UnknownAdviceTypeException(advisor.getAdvice());
    }
    return interceptors.toArray(new MethodInterceptor[interceptors.size()]);
}

@Override
public void registerAdvisorAdapter(AdvisorAdapter adapter) {
    this.adapters.add(adapter);
}
}
```

DefaultAdvisorAdapterRegistry 类设置了一系列适配器,正是这些适配器的实现,为 Spring AOP 提供了织入能力。下面以 MethodBeforeAdviceAdapter 类为例,来看具体的实现:

```
class MethodBeforeAdviceAdapter implements AdvisorAdapter, Serializable {

    @Override
    public boolean supportsAdvice(Advice advice) {
        return (advice instanceof MethodBeforeAdvice);
    }

    @Override
    public MethodInterceptor getInterceptor(Advisor advisor) {
        MethodBeforeAdvice advice = (MethodBeforeAdvice) advisor.getAdvice();
        return new MethodBeforeAdviceInterceptor(advice);
    }
}
```

Spring AOP 为了实现 Advice 的织入,设计了特定的拦截器对这些功能进行封装。我们接着看 MethodBeforeAdviceInterceptor 类是如何完成封装的:

```
public class MethodBeforeAdviceInterceptor implements MethodInterceptor, Serializable {

    private MethodBeforeAdvice advice;


    public MethodBeforeAdviceInterceptor(MethodBeforeAdvice advice) {
        Assert.notNull(advice, "Advice must not be null");
        this.advice = advice;
    }

    @Override
    public Object invoke(MethodInvocation mi) throws Throwable {
        this.advice.before(mi.getMethod(), mi.getArguments(), mi.getThis() );
        return mi.proceed();
```

        }
}

可以看到,invoke()方法中首先触发了 Advice 的 before()方法回调,然后才执行 proceed()方法。

AfterReturningAdviceInterceptor 类的源码如下:

```java
public class AfterReturningAdviceInterceptor implements MethodInterceptor, AfterAdvice,
Serializable {

    private final AfterReturningAdvice advice;

    public AfterReturningAdviceInterceptor(AfterReturningAdvice advice) {
        Assert.notNull(advice, "Advice must not be null");
        this.advice = advice;
    }

    @Override
    public Object invoke(MethodInvocation mi) throws Throwable {
        Object retVal = mi.proceed();
        this.advice.afterReturning(retVal, mi.getMethod(), mi.getArguments(), mi.getThis());
        return retVal;
    }
}
```

ThrowsAdviceInterceptor 的源码如下:

```java
public class ThrowsAdviceInterceptor implements MethodInterceptor, AfterAdvice {
// 部分代码被省略
}

public Object invoke(MethodInvocation mi) throws Throwable {
    try {
        return mi.proceed();
    }
    catch (Throwable ex) {
        Method handlerMethod = getExceptionHandler(ex);
        if (handlerMethod != null) {
            invokeHandlerMethod(mi, ex, handlerMethod);
        }
        throw ex;
    }
}

private void invokeHandlerMethod(MethodInvocation mi, Throwable ex, Method method) throws
```

```
Throwable {
    Object[] handlerArgs;
    if (method.getParameterCount() == 1) {
        handlerArgs = new Object[] { ex };
    }
    else {
        handlerArgs = new Object[] {mi.getMethod(), mi.getArguments(), mi.getThis(), ex};
    }
    try {
        method.invoke(this.throwsAdvice, handlerArgs);
    }
    catch (InvocationTargetException targetEx) {
        throw targetEx.getTargetException();
    }
}
```

至此，我们知道了对目标对象的增强是通过拦截器实现的，最后还是请"小伙伴们"到"下载资源"中看一下运行时序图吧。

第 11 章

一步一步手绘 Spring MVC 运行时序图

11.1 初探 Spring MVC 请求处理流程

Spring MVC 相对于前面讲的内容算比较简单的，我们首先引用 *Spring in Action* 一书上的一张图来了解 Spring MVC 的核心组件和大致处理流程，如下图所示。

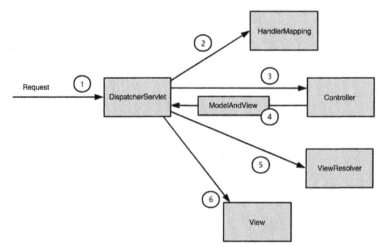

从图中可以看到：

①DispatcherServlet 是 Spring MVC 中的前端控制器（Front Controller），负责接收 Request 并将 Request 转发给对应的处理组件。

②HanlerMapping 是 Spring MVC 中完成 URL 到 Controller 映射的组件。DispatcherServlet 接收 Request，然后从 HandlerMapping 查找处理 Request 的 Controller。

③Controller 处理 Request，并返回 ModelAndView 对象。Controller 是 Spring MVC 中负责处理 Request 的组件（类似于 Struts 2 中的 Action），ModelAndView 是封装结果视图的组件。

④、⑤、⑥是视图解析器解析 ModelAndView 对象并返回对应的视图给客户端的过程。

容器初始化时会建立所有 URL 和 Controller 中方法的对应关系，保存到 Handler Mapping 中，用户请求时根据请求的 URL 快速定位到 Controller 中的某个方法。在 Spring 中先将 URL 和 Controller 的对应关系保存到 Map<url,Controller>中。Web 容器启动时会通知 Spring 初始化容器（加载 Bean 的定义信息和初始化所有单例 Bean），然后 Spring MVC 会遍历容器中的 Bean，获取每一个 Controller 中的所有方法访问的 URL，将 URL 和 Controller 保存到一个 Map 中。这样就可以根据请求快速定位到 Controller，因为最终处理请求的是 Controller 中的方法，Map 中只保留了 URL 和 Controller 的对应关系，所以要根据请求的 URL 进一步确认 Controller 中的方法。其原理就是拼接 Controller 的 URL（Controller 上@RequestMapping 的值）和方法的 URL（Method 上@RequestMapping 的值），与请求的 URL 进行匹配，找到匹配的方法。确定处理请求的方法后，接下来的任务就是参数绑定，把请求中的参数绑定到方法的形式参数上，这是整个请求处理过程中最复杂的一步。

11.2 Spring MVC 九大组件

11.2.1 HandlerMapping

HandlerMapping 是用来查找 Handler 的，也就是处理器，具体的表现形式可以是类，也可以是方法。比如，标注了@RequestMapping 的每个方法都可以看成一个 Handler。Handler 负责实际的请求处理，在请求到达后，HandlerMapping 的作用便是找到请求相应的处理器 Handler 和 Interceptor。

11.2.2 HandlerAdapter

从名字上看，HandlerAdapter 是一个适配器。因为 Spring MVC 中 Handler 可以是任意形式的，只要能够处理请求便可。但是把请求交给 Servlet 的时候，由于 Servlet 的方法结构都是 doService(HttpServletRequest req, HttpServletResponse resp)形式的，要让固定的 Servlet 处理方法调用 Handler 来进行处理，这一步工作便是 HandlerAdapter 要做的事。

11.2.3　HandlerExceptionResolver

从组件的名字上看，HandlerExceptionResolver 是用来处理 Handler 产生的异常情况的组件。具体来说，此组件的作用是根据异常设置 ModelAndView，之后交给渲染方法进行渲染，渲染方法会将 ModelAndView 渲染成页面。不过要注意，HandlerExceptionResolver 只用于解析对请求做处理阶段产生的异常，渲染阶段的异常不归它管，这也是 Spring MVC 组件设计的一大原则——分工明确、互不干涉。

11.2.4　ViewResolver

ViewResolver 即视图解析器，相信大家对这个组件应该很熟悉了。通常在 Spring MVC 的配置文件中，都会配上一个实现类来进行视图解析。这个组件的主要作用是将 String 类型的视图名和 Locale 解析为 View 类型的视图，只有一个 resolveViewName()方法。从方法的定义可以看出，Controller 层返回的 String 类型的视图名 viewName 最终会在这里被解析成 View。View 是用来渲染页面的，也就是说，它会将程序返回的参数和数据填入模板中，生成 HTML 文件。ViewResolver 在这个过程中主要做两件大事：ViewResolver 会找到渲染所用的模板（第一件大事）和所用的技术（第二件大事，其实也就是找到视图的类型，如 JSP）并填入参数。默认情况下，Spring MVC 会为我们自动配置一个 InternalResourceViewResolver，是针对 JSP 类型视图的。

11.2.5　RequestToViewNameTranslator

RequestToViewNameTranslator 组件的作用是从请求中获取 ViewName。因为 ViewResolver 根据 ViewName 查找 View，但有的 Handler 处理完成之后，没有设置 View，也没有设置 ViewName，便要通过这个组件来从请求中查找 ViewName。

11.2.6　LocaleResolver

ViewResolver 组件的 resolveViewName()方法需要两个参数，一个是视图名，另一个就是 Locale。参数 Locale 是从哪来的呢？这就是 LocaleResolver 组件要做的事。LocaleResolver 用于从请求中解析出 Locale，比如在中国 Locale 当然就是 zh-CN，用来表示一个区域。这个组件也是 i18n 的基础。

11.2.7 ThemeResolver

从名字便可看出，ThemeResolver 组件是用来解析主题的。主题就是样式、图片及它们所形成的显示效果的集合。Spring MVC 中一套主题对应一个 properties 文件，里面存放着与当前主题相关的所有资源，如图片、CSS 样式等。创建主题非常简单，只需准备好资源，然后新建一个"主题名.properties"并将资源设置进去，放在 classpath 下，之后便可以在页面中使用了。Spring MVC 中与主题有关的类有 ThemeResolver、ThemeSource 和 Theme。ThemeResolver 负责从请求中解析出主题名，ThemeSource 则根据主题名找到具体的主题，其抽象也就是 Theme，可以通过 Theme 来获取主题和具体的资源。

11.2.8 MultipartResolver

MultipartResolver 是一个大家很熟悉的组件，用于处理上传请求，通过将普通的请求包装成 MultipartHttpServletRequest 来实现。MultipartHttpServletRequest 可以通过 getFile()方法直接获得文件。如果上传多个文件，还可以调用 getFileMap()方法得到 Map<FileName, File> 这样的结构。MultipartResolver 的作用就是封装普通的请求，使其拥有文件上传的功能。

11.2.9 FlashMapManager

说到 FlashMapManager 组件，得先说一下 FlashMap。

FlashMap 用于重定向时的参数传递，比如在处理用户订单时，为了避免重复提交，可以处理完 post 请求后重定向到一个 get 请求，这个 get 请求可以用来显示订单详情之类的信息。这样做虽然可以规避用户重新提交订单的问题，但是在这个页面上要显示订单的信息，这些数据从哪里获取呢？因为重定向是没有传递参数这一功能的，如果不想把参数写进 URL（其实也不推荐这么做，除了 URL 有长度限制，把参数都直接暴露也不安全），那么就可以通过 FlashMap 来传递。只需要在重定向之前将要传递的数据写入请求（可以通过 ServletRequestAttributes.getRequest()方法获得）的属性 OUTPUT_FLASH_MAP_ATTRIBUTE 中，这样在重定向之后的 Handler 中 Spring 就会自动将其设置到 Model 中，在显示订单信息的页面上就可以直接从 Model 中获得数据。

FlashMapManager 就是用来管理 FlashMap 的。

11.3 Spring MVC 源码分析

根据上面分析的 Spring MVC 的九大组件，Spring MVC 的处理过程可分为如下三步：

（1）ApplicationContext 初始化时用 Map 保存所有 URL 和 Controller 类的对应关系。

（2）根据请求 URL 找到对应的 Controller，并从 Controller 中找到处理请求的方法。

（3）将 Request 参数绑定到方法的形参上，执行方法处理请求，并返回结果视图。

11.3.1 初始化阶段

首先找到 DispatcherServlet 类，寻找 init() 方法。我们发现 init() 方法其实在父类 HttpServletBean 中，其源码如下：

```java
public final void init() throws ServletException {
    if (logger.isDebugEnabled()) {
        logger.debug("Initializing servlet '" + getServletName() + "'");
    }

    PropertyValues pvs = new ServletConfigPropertyValues(getServletConfig(), this.requiredProperties);
    if (!pvs.isEmpty()) {
        try {
            //定位资源
            BeanWrapper bw = PropertyAccessorFactory.forBeanPropertyAccess(this);
            //加载配置信息
            ResourceLoader resourceLoader = new ServletContextResourceLoader(getServletContext());
            bw.registerCustomEditor(Resource.class, new ResourceEditor(resourceLoader, getEnvironment()));
            initBeanWrapper(bw);
            bw.setPropertyValues(pvs, true);
        }
        catch (BeansException ex) {
            if (logger.isErrorEnabled()) {
                logger.error("Failed to set bean properties on servlet '" + getServletName() + "'", ex);
            }
            throw ex;
        }
    }

    initServletBean();

    if (logger.isDebugEnabled()) {
        logger.debug("Servlet '" + getServletName() + "' configured successfully");
    }
}
```

在这段代码中调用了一个重要的方法：initServletBean()。initServletBean() 方法的源码如下：

```java
protected final void initServletBean() throws ServletException {
    getServletContext().log("Initializing Spring FrameworkServlet '" + getServletName() + "'");
    if (this.logger.isInfoEnabled()) {
        this.logger.info("FrameworkServlet '" + getServletName() + "': initialization started");
    }
    long startTime = System.currentTimeMillis();

    try {
        this.webApplicationContext = initWebApplicationContext();
        initFrameworkServlet();
    }
    catch (ServletException ex) {
        this.logger.error("Context initialization failed", ex);
        throw ex;
    }
    catch (RuntimeException ex) {
        this.logger.error("Context initialization failed", ex);
        throw ex;
    }

    if (this.logger.isInfoEnabled()) {
        long elapsedTime = System.currentTimeMillis() - startTime;
        this.logger.info("FrameworkServlet '" + getServletName() + "': initialization completed
            in " + elapsedTime + " ms");
    }
}
```

上面这段代码主要就是初始化 IoC 容器，最终会调用 refresh()方法，前面的章节对 IoC 容器的初始化已经讲得很详细，在此不再赘述。我们看到，在 IoC 容器初始化之后，又调用了 onRefresh()方法，它是在 DisptcherServlet 类中实现的，来看源码：

```java
@Override
protected void onRefresh(ApplicationContext context) {
    initStrategies(context);
}

//初始化策略
protected void initStrategies(ApplicationContext context) {
    //多文件上传的组件
    initMultipartResolver(context);
    //初始化本地语言环境
    initLocaleResolver(context);
    //初始化模板处理器
    initThemeResolver(context);
    //初始化 handlerMapping
    initHandlerMappings(context);
```

```
  //初始化参数适配器
  initHandlerAdapters(context);
  //初始化异常拦截器
  initHandlerExceptionResolvers(context);
  //初始化视图预处理器
  initRequestToViewNameTranslator(context);
  //初始化视图转换器
  initViewResolvers(context);
  //初始化 FlashMap 管理器
  initFlashMapManager(context);
}
```

到这就完成了 Spring MVC 的九大组件的初始化。接下来，我们来看 URL 和 Controller 的关系是如何建立的。HandlerMapping 的子类 AbstractDetectingUrlHandlerMapping 实现了 initApplicationContext()方法，我们直接看子类中的初始化容器方法：

```
@Override
public void initApplicationContext() throws ApplicationContextException {
    super.initApplicationContext();
    detectHandlers();
}

/**
 * 建立当前 ApplicationContext 中的所有 Controller 和 URL 的对应关系
 */
protected void detectHandlers() throws BeansException {
    ApplicationContext applicationContext = obtainApplicationContext();
    if (logger.isDebugEnabled()) {
        logger.debug("Looking for URL mappings in application context: " + applicationContext);
    }
    //获取 ApplicationContext 容器中所有 Bean 的名字
    String[] beanNames = (this.detectHandlersInAncestorContexts ?
            BeanFactoryUtils.beanNamesForTypeIncludingAncestors(applicationContext, Object.class) :
            applicationContext.getBeanNamesForType(Object.class));

    //遍历 beanNames，并找到这些 Bean 对应的 URL
    for (String beanName : beanNames) {
        //查找 Bean 上的所有 URL(Controller 上的 URL+方法上的 URL)，该方法由对应的子类实现
        String[] urls = determineUrlsForHandler(beanName);
        if (!ObjectUtils.isEmpty(urls)) {
            //保存 urls 和 beanName 的对应关系，放入 Map<urls,beanName>，
            //该方法在父类 AbstractUrlHandlerMapping 中实现
            registerHandler(urls, beanName);
        }
        else {
            if (logger.isDebugEnabled()) {
```

```
            logger.debug("Rejected bean name '" + beanName + "': no URL paths identified");
        }
    }
  }
}
/** 获取 Controller 中所有方法的 URL，由子类实现，典型的模板模式 **/
protected abstract String[] determineUrlsForHandler(String beanName);
```

determineUrlsForHandler(String beanName)方法的作用是获取每个 Controller 中的 URL，不同的子类有不同的实现，这是典型的模板模式。因为开发中用得最多的就是用注解来配置 Controller 中的 URL，BeanNameUrlHandlerMapping 是 AbstractDetectingUrlHandlerMapping 的子类，用于处理注解形式的 URL 映射。我们这里以 BeanNameUrlHandlerMapping 为例来进行分析，看看如何查找 beanName 上所有映射的 URL。

```
/**
 * 获取 Controller 中所有的 URL
 */
@Override
protected String[] determineUrlsForHandler(String beanName) {
  List<String> urls = new ArrayList<>();
  if (beanName.startsWith("/")) {
     urls.add(beanName);
  }
  String[] aliases = obtainApplicationContext().getAliases(beanName);
  for (String alias : aliases) {
     if (alias.startsWith("/")) {
        urls.add(alias);
     }
  }
  return StringUtils.toStringArray(urls);
}
```

到这里 HandlerMapping 组件已经建立了所有 URL 和 Controller 的对应关系。

11.3.2 运行调用阶段

运行调用是由请求触发的，所以入口为 DispatcherServlet 的核心方法 doService()，doService() 中的核心由 doDispatch()实现，源代码如下：

```
/** 中央控制器，控制请求的转发 **/
protected void doDispatch(HttpServletRequest request, HttpServletResponse response) throws Exception {
  HttpServletRequest processedRequest = request;
  HandlerExecutionChain mappedHandler = null;
```

```java
boolean multipartRequestParsed = false;

WebAsyncManager asyncManager = WebAsyncUtils.getAsyncManager(request);

try {
    ModelAndView mv = null;
    Exception dispatchException = null;

    try {
        // 1.检查是否是文件上传的请求
        processedRequest = checkMultipart(request);
        multipartRequestParsed = (processedRequest != request);

        // 2.取得处理当前请求的 Controller,这里也称为 Hanlder,即处理器,
        //    第一步的意义就在这里体现了。这里并不是直接返回 Controller,
        //    而是返回 HandlerExecutionChain 请求处理器链对象,
        //    该对象封装了 Handler 和 interceptor
        mappedHandler = getHandler(processedRequest);
        // 如果 Handler 为空,则返回 404
        if (mappedHandler == null) {
            noHandlerFound(processedRequest, response);
            return;
        }

        //3. 获取处理请求的处理器适配器 HandlerAdapter
        HandlerAdapter ha = getHandlerAdapter(mappedHandler.getHandler());

        // 处理 last-modified 请求头
        String Method = request.getMethod();
        boolean isGet = "GET".equals(Method);
        if (isGet || "HEAD".equals(Method)) {
            long lastModified = ha.getLastModified(request, mappedHandler.getHandler());
            if (logger.isDebugEnabled()) {
                logger.debug("Last-Modified value for [" + getRequestUri(request) + "] is: " + lastModified);
            }
            if (new ServletWebRequest(request, response).checkNotModified(lastModified) && isGet) {
                return;
            }
        }

        if (!mappedHandler.applyPreHandle(processedRequest, response)) {
            return;
        }

        // 4.实际处理器处理请求,返回结果视图对象
```

```java
            mv = ha.handle(processedRequest, response, mappedHandler.getHandler());

            if (asyncManager.isConcurrentHandlingStarted()) {
              return;
            }

            //结果视图对象的处理
            applyDefaultViewName(processedRequest, mv);
            mappedHandler.applyPostHandle(processedRequest, response, mv);
        }
        catch (Exception ex) {
            dispatchException = ex;
        }
        catch (Throwable err) {
            dispatchException = new NestedServletException("Handler dispatch failed", err);
        }
        processDispatchResult(processedRequest, response, mappedHandler, mv, dispatchException);
    }
    catch (Exception ex) {
        triggerAfterCompletion(processedRequest, response, mappedHandler, ex);
    }
    catch (Throwable err) {
        triggerAfterCompletion(processedRequest, response, mappedHandler,
            new NestedServletException("Handler processing failed", err));
    }
    finally {
        if (asyncManager.isConcurrentHandlingStarted()) {
          if (mappedHandler != null) {
            //请求成功响应之后的方法
            mappedHandler.applyAfterConcurrentHandlingStarted(processedRequest, response);
          }
        }
        else {
          if (multipartRequestParsed) {
            cleanupMultipart(processedRequest);
          }
        }
    }
}
```

getHandler(processedRequest)方法实际上从 HandlerMapping 中找到 URL 和 Controller 的对应关系，也就是 Map<url,Controller>。我们知道，最终处理请求的是 Controller 中的方法，现在只是知道了 Controller，如何确认 Controller 中处理请求的方法呢？继续往下看。

从 Map<urls,beanName>中取得 Controller 后，经过拦截器的预处理方法，再通过反射获取该方法上的注解和参数，解析方法和参数上的注解，然后反射调用方法获取 ModelAndView 结果视图。最后调用 RequestMappingHandlerAdapter 的 handle()中的核心代码，由 handleInternal (request, response, handler)实现：

```
@Override
protected ModelAndView handleInternal(HttpServletRequest request,
    HttpServletResponse response, HandlerMethod handlerMethod) throws Exception {

  ModelAndView mav;
  checkRequest(request);

  if (this.synchronizeOnSession) {
     HttpSession session = request.getSession(false);
     if (session != null) {
        Object mutex = WebUtils.getSessionMutex(session);
        synchronized (mutex) {
           mav = invokeHandlerMethod(request, response, handlerMethod);
        }
     }
     else {
        mav = invokeHandlerMethod(request, response, handlerMethod);
     }
  }
  else {
     mav = invokeHandlerMethod(request, response, handlerMethod);
  }

  if (!response.containsHeader(HEADER_CACHE_CONTROL)) {
     if (getSessionAttributesHandler(handlerMethod).hasSessionAttributes()) {
        applyCacheSeconds(response, this.cacheSecondsForSessionAttributeHandlers);
     }
     else {
        prepareResponse(response);
     }
  }

  return mav;
}
```

整个处理过程中最核心的步骤其实就是拼接 Controller 的 URL 和方法的 URL，与 Request 的 URL 进行匹配，找到匹配的方法。

```
/** 根据 URL 获取处理请求的方法 **/
@Override
protected HandlerMethod getHandlerInternal(HttpServletRequest request) throws Exception {
  //如果请求 URL 为 http://localhost:8080/web/hello.json，则 lookupPath=web/hello.json
  String lookupPath = getUrlPathHelper().getLookupPathForRequest(request);
  if (logger.isDebugEnabled()) {
```

```java
        logger.debug("Looking up handler method for path " + lookupPath);
    }
    this.mappingRegistry.acquireReadLock();
    try {
        //遍历 Controller 上的所有方法，获取 URL 匹配的方法
        HandlerMethod handlerMethod = lookupHandlerMethod(lookupPath, request);
        if (logger.isDebugEnabled()) {
            if (handlerMethod != null) {
                logger.debug("Returning handler method [" + handlerMethod + "]");
            }
            else {
                logger.debug("Did not find handler method for [" + lookupPath + "]");
            }
        }
        return (handlerMethod != null ? handlerMethod.createWithResolvedBean() : null);
    }
    finally {
        this.mappingRegistry.releaseReadLock();
    }
}
```

通过上面的代码分析，已经找到处理请求的 Controller 中的方法了，下面看如何解析该方法上的参数，并反射调用该方法。

```java
/** 获取处理请求的方法，执行并返回结果视图 **/
@Nullable
protected ModelAndView invokeHandlerMethod(HttpServletRequest request,
        HttpServletResponse response, HandlerMethod handlerMethod) throws Exception {

    ServletWebRequest webRequest = new ServletWebRequest(request, response);
    try {
        WebDataBinderFactory binderFactory = getDataBinderFactory(handlerMethod);
        ModelFactory modelFactory = getModelFactory(handlerMethod, binderFactory);

        ServletInvocableHandlerMethod invocableMethod = createInvocableHandlerMethod(handlerMethod);
        if (this.argumentResolvers != null) {
            invocableMethod.setHandlerMethodArgumentResolvers(this.argumentResolvers);
        }
        if (this.returnValueHandlers != null) {
            invocableMethod.setHandlerMethodReturnValueHandlers(this.returnValueHandlers);
        }
        invocableMethod.setDataBinderFactory(binderFactory);
        invocableMethod.setParameterNameDiscoverer(this.parameterNameDiscoverer);

        ModelAndViewContainer mavContainer = new ModelAndViewContainer();
        mavContainer.addAllAttributes(RequestContextUtils.getInputFlashMap(request));
        modelFactory.initModel(webRequest, mavContainer, invocableMethod);
        mavContainer.setIgnoreDefaultModelOnRedirect(this.ignoreDefaultModelOnRedirect);
```

```java
AsyncWebRequest asyncWebRequest = WebAsyncUtils.createAsyncWebRequest(request, response);
asyncWebRequest.setTimeout(this.asyncRequestTimeout);

WebAsyncManager asyncManager = WebAsyncUtils.getAsyncManager(request);
asyncManager.setTaskExecutor(this.taskExecutor);
asyncManager.setAsyncWebRequest(asyncWebRequest);
asyncManager.registerCallableInterceptors(this.callableInterceptors);
asyncManager.registerDeferredResultInterceptors(this.deferredResultInterceptors);

if (asyncManager.hasConcurrentResult()) {
    Object result = asyncManager.getConcurrentResult();
    mavContainer = (ModelAndViewContainer) asyncManager.getConcurrentResultContext()[0];
    asyncManager.clearConcurrentResult();
    if (logger.isDebugEnabled()) {
        logger.debug("Found concurrent result value [" + result + "]");
    }
    invocableMethod = invocableMethod.wrapConcurrentResult(result);
}

invocableMethod.invokeAndHandle(webRequest, mavContainer);
if (asyncManager.isConcurrentHandlingStarted()) {
    return null;
}

return getModelAndView(mavContainer, modelFactory, webRequest);
}
finally {
    webRequest.requestCompleted();
}
}
```

invocableMethod.invokeAndHandle()最终要实现的目的是：完成请求中的参数和方法参数上数据的绑定。Spring MVC 中提供两种从请求参数到方法中参数的绑定方式：

（1）通过注解进行绑定，@RequestParam。

（2）通过参数名称进行绑定。

通过注解进行绑定，只要在方法的参数前面声明@RequestParam("name")，就可以将请求中参数 name 的值绑定到方法的该参数上。

通过参数名称进行绑定的前提是必须获取方法中参数的名称，Java 反射只提供了获取方法参数类型的方法，并没有提供获取参数名称的方法。Spring MVC 解决这个问题的方法是用 asm 框架读取字节码文件。asm 框架是一个字节码操作框架，更多介绍可以参考其官网。个人建议通过注解进行绑定，如下代码所示，这样就可以省去 asm 框架的读取字节码的操作。

```java
@Nullable
public Object invokeForRequest(NativeWebRequest request, @Nullable ModelAndViewContainer
mavContainer, Object... providedArgs) throws Exception {

    Object[] args = getMethodArgumentValues(request, mavContainer, providedArgs);
    if (logger.isTraceEnabled()) {
        logger.trace("Invoking '" + ClassUtils.getQualifiedMethodName(getMethod(), getBeanType())
                + "' with arguments " + Arrays.toString(args));
    }
    Object returnValue = doInvoke(args);
    if (logger.isTraceEnabled()) {
        logger.trace("Method [" + ClassUtils.getQualifiedMethodName(getMethod(), getBeanType())
                + "] returned [" + returnValue + "]");
    }
    return returnValue;
}

private Object[] getMethodArgumentValues(NativeWebRequest request, @Nullable
ModelAndViewContainer mavContainer,
        Object... providedArgs) throws Exception {
    MethodParameter[] parameters = getMethodParameters();
    Object[] args = new Object[parameters.length];
    for (int i = 0; i < parameters.length; i++) {
        MethodParameter parameter = parameters[i];
        parameter.initParameterNameDiscovery(this.parameterNameDiscoverer);
        args[i] = resolveProvidedArgument(parameter, providedArgs);
        if (args[i] != null) {
            continue;
        }
        if (this.argumentResolvers.supportsParameter(parameter)) {
            try {
                args[i] = this.argumentResolvers.resolveArgument(
                        parameter, mavContainer, request, this.dataBinderFactory);
                continue;
            }
            catch (Exception ex) {
                if (logger.isDebugEnabled()) {
                    logger.debug(getArgumentResolutionErrorMessage("Failed to resolve", i), ex);
                }
                throw ex;
            }
        }
        if (args[i] == null) {
            throw new IllegalStateException("Could not resolve method parameter at index " +
                    parameter.getParameterIndex() + " in " + parameter.getExecutable().toGenericString()
                    + ": " + getArgumentResolutionErrorMessage("No suitable resolver for", i));
        }
    }
}
```

```
    return args;
}
```

关于 asm 框架获取方法参数的内容这里就不再进行分析了,感兴趣的"小伙伴"可以继续自行深入了解。

到这里,方法的参数列表也获取到了,可以直接进行方法的调用了。最后我们再来梳理一下 Spring MVC 核心组件的关联关系,如下图所示。

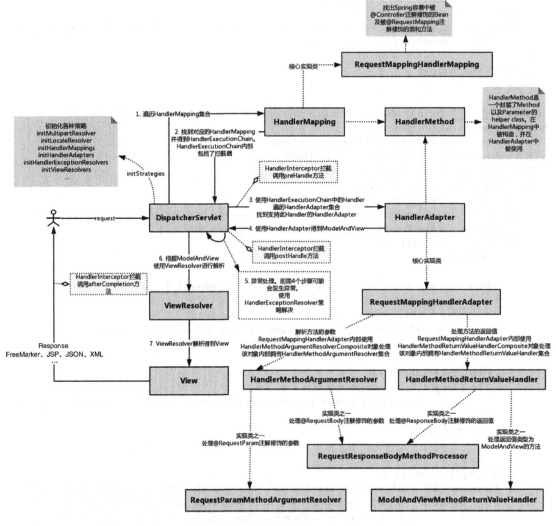

再来看看时序图,请到"下载资源"中查看。

11.4 Spring MVC 优化建议

前面我们已经对 Spring MVC 的工作原理和源码进行了分析，在这个过程中有几个优化点。

1. Controller 如果能保持单例模式，尽量使用单例模式

这样可以减小创建对象和回收对象的开销。也就是说，如果 Controller 的类变量和实例变量可以以方法形参声明就尽量以方法形参声明，不要以类变量和实例变量声明，这样可以避免线程安全问题。

2. 处理请求的方法中的形参务必加上 @RequestParam 注解

这样可以避免 Spring MVC 使用 asm 框架读取 .class 文件获取方法参数名。即便 Spring MVC 对读取出的方法参数名进行了缓存，如果能不读取 .class 文件当然更好。

3. 缓存 URL

在阅读源码的过程中，我们发现 Spring MVC 并没有对处理 URL 的方法进行缓存，也就是说，每次都要根据请求 URL 去匹配 Controller 中的方法的 URL，如果把 URL 和方法的关系缓存起来，会不会带来性能上的提升呢？不幸的是，负责解析 URL 和方法对应关系的 ServletHandlerMethodResolver 是一个私有的内部类，不能直接通过继承该类增强代码，必须在代码后重新编译。当然，如果将 URL 缓存起来，必须考虑缓存的线程安全问题。

第 4 篇
Spring 手写实战

第 12 章　环境准备
第 13 章　IoC 顶层结构设计
第 14 章　完成 DI 模块的功能
第 15 章　完成 MVC 模块的功能
第 16 章　完成 AOP 代码织入

第 12 章
环境准备

12.1　IDEA 集成 Lombok 插件

12.1.1　安装插件

　　IntelliJ IDEA 是一款非常优秀的集成开发工具，功能强大，而且插件众多。Lombok 是开源的代码生成库，是一款非常实用的小工具，我们在编辑实体类时可以通过 Lombok 注解减少 getter、setter 等方法的编写，在更改实体类时只需要修改属性即可，减少了很多重复代码的编写工作。

　　首先需要安装 IntelliJ IDEA 中的 Lombok 插件，打开 IntelliJ IDEA 后单击菜单栏中的 File→Settings（如右图所示），或者使用快捷键 Ctrl+Alt+S 进入设置界面。

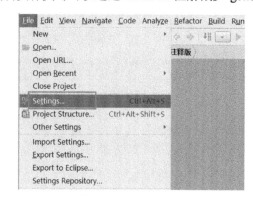

单击设置界面中的 Plugins 进行插件的安装，在右侧单击 Browse repositories 按钮，如下图所示。

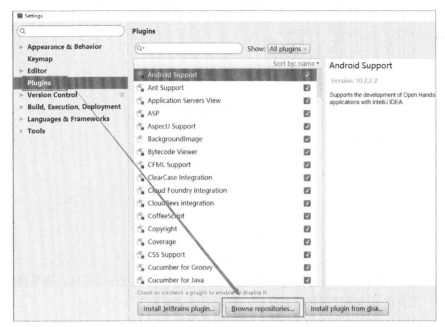

然后在搜索界面中输入 lombok 可以查询到下方的 Lombok Plugin，单击 Lombok Plugin 可在右侧看到 Install 按钮，单击该按钮便可开始安装，如下图所示。

我们在如下安装界面可以看到 Lombok 支持的所有注解。

在安装过程中有 Downloading Plugins 的提示，安装过程中进度条会变化。需要提醒的是，在安装过程中一定要保证网络连接可用且良好，否则可能会安装失败。

安装成功后可以看到下图右侧的 Restart IntelliJ IDEA 按钮，此时可先不操作，因为还有后续的配置工作。

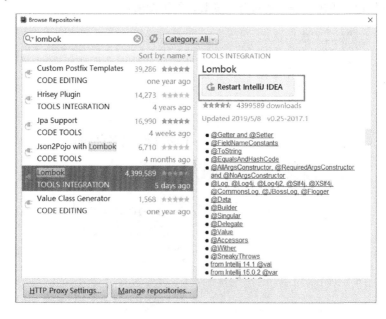

回到 Plugins，此时在下图右侧可以搜索到 Lombok（而安装前搜索不到）。

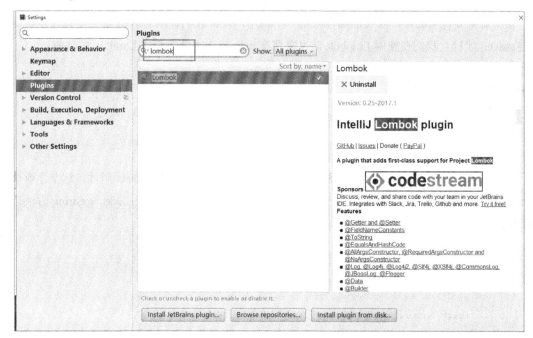

12.1.2 配置注解处理器

在如下设置界面单击 Build, Execution, Deployment→Compiler→Annotation Processors，然后在右侧勾选 Enable annotation processing 复选项即可。

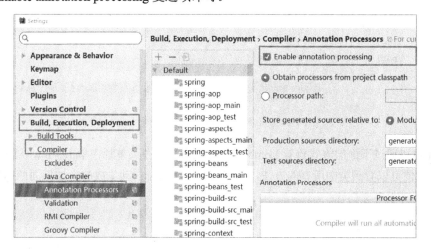

12.1.3 使用插件

使用前需要说明的是，安装的插件只是一个调用，就像我们使用 maven 插件一样，本机需要安装 maven 才行。我们在使用 Lombok 前也需要在 pom.xml 文件中添加 Lombok 的依赖。

```
#托管的类扫描包路径#
<dependency>
    <groupId>org.projectlombok</groupId>
    <artifactId>lombok</artifactId>
    <version>1.16.10</version>
</dependency>
```

接下来编辑一个 Config 测试类，添加两个属性，最后在类上添加@Data 属性，这个注解可以帮我们在.class 文件中生成类中所有属性的 get/set、equals、canEqual、hashCode、toString 方法等，如下图所示。

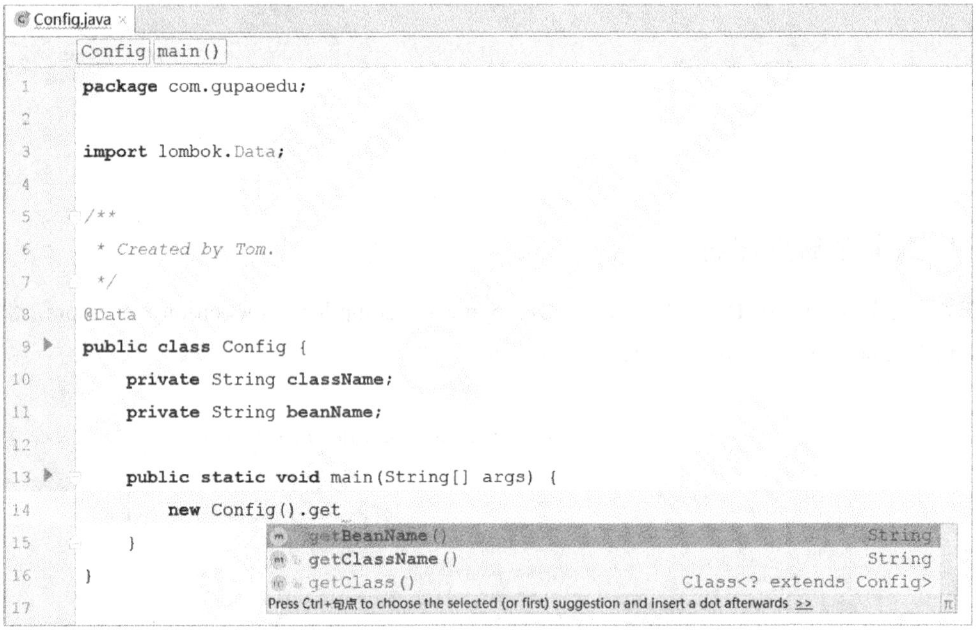

Config.java
Config main()
package com.gupaoedu;
import lombok.Data;
/**
 * Created by Tom.
 */
@Data
public class Config {
    private String className;
    private String beanName;
    public static void main(String[] args) {
        new Config().get
    }
}
getBeanName() String
getClassName() String
getClass() Class<? extends Config>
Press Ctrl+句点 to choose the selected (or first) suggestion and insert a dot afterwards >>

我们还可以通过下面的方式查看 Lombok 生成的方法。在菜单栏中单击 View→Tool Windows→Structure，便可以看到类中所有的方法，这些都是 Lombok 自动生成的，如下面两图所示。

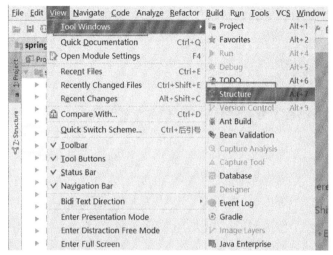

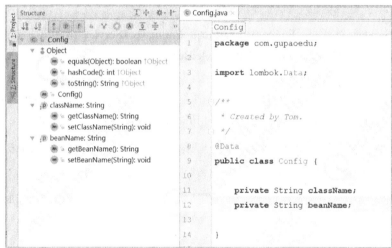

12.2 从 Servlet 到 ApplicationContext

在"用 300 行代码手写提炼 Spring 核心原理"一章中，我们已经了解 Spring MVC 的入口是 DispatcherSerlvet，并实现了 DispatcherServlet 的 init()方法，在 init()方法中完成了 IoC 容器的初始化。而在使用 Spring 的过程中，见得最多的是 ApplicationContext，似乎 Spring 托管的所有实例 Bean 都可以通过调用 getBean()方法来获得。那么 ApplicationContext 又是从何而来的呢？从 Spring 源码中可以看到，DispatcherServlet 的类图如下图所示。

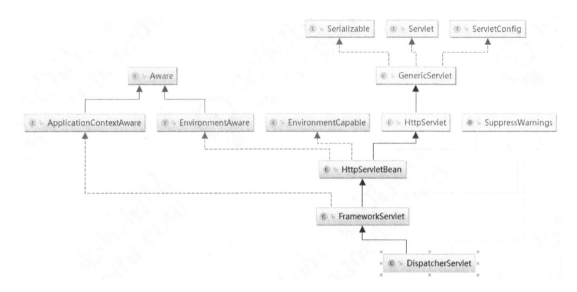

DispatcherServlet 继承了 FrameworkServlet，FrameworkServlet 继承了 HttpServletBean，HttpServletBean 继承了 HttpServlet。在 HttpServletBean 的 init()方法中调用了 FrameworkServlet 的 initServletBean()方法，在 initServletBean()方法中初始化 WebApplicationContext 实例。在 initServletBean()方法中调用了 DispatcherServlet 重写的 onRefresh()方法。在 DispatcherServlet 的 onRefresh()方法中又调用了 initStrategies()方法，初始化 Spring MVC 的九大组件。

其实，通过上面复杂的调用关系可以得出一个结论：在 Servlet 的 init()方法中初始化了 IoC 容器和 Spring MVC 所依赖的九大组件。

在手写之前先将框架类关系画出来，如下图所示，顺便也回顾一下我们之前讲过的 IoC 容器结构。

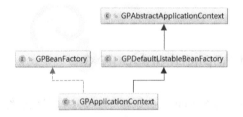

12.3　准备基础配置

在开始手写之前我们先做一个约定，所有的手写类都以 GP 开头，以区别于 Spring 框架中的原生类，方便对比理解。如 DispatcherServlet 在这个 Mini 版本中会命名为 GPDispatcherServlet，所有的方法名尽量和原生 Spring 保持一致，可能有些参数列表会做一些微调，主要目的是理解设计思路。

12.3.1 application.properties 配置

还是先从 application.properties 文件开始，用 application.properties 来代替 application.xml，具体配置如下：

```
#托管的类扫描包路径#
scanPackage=com.gupaoedu.vip.spring.demo
```

12.3.2 pom.xml 配置

接下来看 pom.xml 的配置，主要关注 jar 包依赖：

```xml
<properties>
    <!-- dependency versions -->
    <servlet.api.version>2.4</servlet.api.version>
</properties>

<dependencies>
    <!-- requied start -->
    <dependency>
        <groupId>javax.servlet</groupId>
        <artifactId>servlet-api</artifactId>
        <version>${servlet.api.version}</version>
        <scope>provided</scope>
    </dependency>
    <!-- requied end -->

    <dependency>
        <groupId>org.projectlombok</groupId>
        <artifactId>lombok</artifactId>
        <version>1.16.10</version>
    </dependency>

    <dependency>
        <groupId>org.slf4j</groupId>
        <artifactId>slf4j-api</artifactId>
        <version>1.7.25</version>
    </dependency>

    <dependency>
```

```xml
      <groupId>ch.qos.logback</groupId>
      <artifactId>logback-classic</artifactId>
      <version>1.2.3</version>
   </dependency>

</dependencies>
```

12.3.3　web.xml 配置

web.xml 配置如下：

```xml
<?xml version="1.0" encoding="UTF-8"?>
<web-app xmlns:xsi="http://www.w3.org/2001/XMLSchema-instance"
   xmlns="http://java.sun.com/xml/ns/j2ee" xmlns:javaee="http://java.sun.com/xml/ns/javaee"
   xmlns:web="http://java.sun.com/xml/ns/javaee/web-app_2_5.xsd"
   xsi:schemaLocation="http://java.sun.com/xml/ns/j2ee
http://java.sun.com/xml/ns/j2ee/web-app_2_4.xsd"
   version="2.4">
   <display-name>Gupao Spring Application</display-name>

   <servlet>
      <servlet-name>gupaomvc</servlet-name>
      <servlet-class>com.gupaoedu.vip.spring.formework.webmvc.servlet.GPDispatcherServlet</servlet-class>
      <init-param>
         <param-name>contextConfigLocation</param-name>
         <param-value>classpath:application.properties</param-value>
      </init-param>
      <load-on-startup>1</load-on-startup>
   </servlet>

   <servlet-mapping>
      <servlet-name>gupaomvc</servlet-name>
      <url-pattern>/*</url-pattern>
   </servlet-mapping>

</web-app>
```

12.3.4　GPDispatcherServlet

GPDispatcherServlet 代码如下：

```java
package com.gupaoedu.vip.spring.formework.webmvc.servlet;
import javax.servlet.ServletConfig;
import javax.servlet.ServletException;
import javax.servlet.http.HttpServlet;
```

```java
import javax.servlet.http.HttpServletRequest;
import javax.servlet.http.HttpServletResponse;
import java.io.IOException;
//Servlet 只是作为一个 MVC 的启动入口
public class GPDispatcherServlet extends HttpServlet {
    @Override
    public void init(ServletConfig config) throws ServletException {
    }
    @Override
    protected void doGet(HttpServletRequest req, HttpServletResponse resp) throws ServletException, IOException {
        this.doPost(req,resp);
    }
    @Override
    protected void doPost(HttpServletRequest req, HttpServletResponse resp) throws ServletException, IOException {
    }
}
```

第 13 章
IoC 顶层结构设计

13.1 Annotation（自定义配置）模块

Annotation 的代码实现我们还是沿用 Mini 版本的，保持不变，复制过来便可。

13.1.1 @GPService

@GPService 代码如下：

```
package com.gupaoedu.vip.spring.formework.annotation;
import java.lang.annotation.Documented;
import java.lang.annotation.ElementType;
import java.lang.annotation.Retention;
import java.lang.annotation.RetentionPolicy;
import java.lang.annotation.Target;
/**
 * 业务逻辑，注入接口
 */
@Target({ElementType.TYPE})
@Retention(RetentionPolicy.RUNTIME)
@Documented
```

```
public @interface GPService {
    String value() default "";
}
```

13.1.2 @GPAutowired

@GPAutowired 代码如下：

```
package com.gupaoedu.vip.spring.formework.annotation;
import java.lang.annotation.Documented;
import java.lang.annotation.ElementType;
import java.lang.annotation.Retention;
import java.lang.annotation.RetentionPolicy;
import java.lang.annotation.Target;
/**
 * 自动注入
 */
@Target({ElementType.FIELD})
@Retention(RetentionPolicy.RUNTIME)
@Documented
public @interface GPAutowired {
    String value() default "";
}
```

13.1.3 @GPController

@GPController 代码如下：

```
package com.gupaoedu.vip.spring.formework.annotation;
import java.lang.annotation.Documented;
import java.lang.annotation.ElementType;
import java.lang.annotation.Retention;
import java.lang.annotation.RetentionPolicy;
import java.lang.annotation.Target;
/**
 * 页面交互
 */
@Target({ElementType.TYPE})
@Retention(RetentionPolicy.RUNTIME)
@Documented
public @interface GPController {
    String value() default "";
}
```

13.1.4 @GPRequestMapping

@GPRequestMapping 代码如下：

```java
package com.gupaoedu.vip.spring.formework.annotation;
import java.lang.annotation.Documented;
import java.lang.annotation.ElementType;
import java.lang.annotation.Retention;
import java.lang.annotation.RetentionPolicy;
import java.lang.annotation.Target;
/**
 * 请求 URL
 */
@Target({ElementType.METHOD,ElementType.TYPE})
@Retention(RetentionPolicy.RUNTIME)
@Documented
public @interface GPRequestMapping {
    String value() default "";
}
```

13.1.5 @GPRequestParam

@GPRequestParam 代码如下：

```java
package com.gupaoedu.vip.spring.formework.annotation;
import java.lang.annotation.Documented;
import java.lang.annotation.ElementType;
import java.lang.annotation.Retention;
import java.lang.annotation.RetentionPolicy;
import java.lang.annotation.Target;
/**
 * 请求参数映射
 */
@Target(ElementType.PARAMETER)
@Retention(RetentionPolicy.RUNTIME)
@Documented
public @interface GPRequestParam {
    String value() default "";
}
```

13.2 core（顶层接口）模块

13.2.1 GPFactoryBean

关于顶层接口设计，通过前面的学习我们了解了 FactoryBean 的基本作用，在此不做过多解释。

```
package com.gupaoedu.vip.spring.formework.core;
public interface GPFactoryBean {
}
```

13.2.2　GPBeanFactory

作为所有 IoC 容器的顶层设计，前面也已经详细介绍了 BeanFactory 的作用。

```
package com.gupaoedu.vip.spring.formework.core;

/**
 * 单例工厂的顶层设计
 */
public interface GPBeanFactory {
    /**
     * 根据 beanName 从 IoC 容器中获得一个实例 Bean
     * @param beanName
     * @return
     */
    Object getBean(String beanName) throws Exception;

    public Object getBean(Class<?> beanClass) throws Exception;
}
```

13.3　beans（配置封装）模块

13.3.1　GPBeanDefinition

BeanDefinition 主要用于保存 Bean 相关的配置信息。

```
package com.gupaoedu.vip.spring.formework.beans.config;
//用来存储配置文件中的信息
//相当于保存在内存中的配置
public class GPBeanDefinition {
    private String beanClassName;    //原生 Bean 的全类名
    private boolean lazyInit = false; //标记是否延时加载
    private String factoryBeanName;  //保存 beanName，在 IoC 容器中存储的 key
    public String getBeanClassName() {
        return beanClassName;
    }
    public void setBeanClassName(String beanClassName) {
        this.beanClassName = beanClassName;
```

```java
    }
    public boolean isLazyInit() {
        return lazyInit;
    }
    public void setLazyInit(boolean lazyInit) {
        this.lazyInit = lazyInit;
    }
    public String getFactoryBeanName() {
        return factoryBeanName;
    }
    public void setFactoryBeanName(String factoryBeanName) {
        this.factoryBeanName = factoryBeanName;
    }
}
```

13.3.2　GPBeanWrapper

BeanWrapper 主要用于封装创建后的对象实例，代理对象（Proxy Object）或者原生对象（Original Object）都由 BeanWrapper 来保存。

```java
package com.gupaoedu.vip.spring.formework.beans;

public class GPBeanWrapper {

    private Object wrappedInstance;
    private Class<?> wrappedClass;

    public GPBeanWrapper(Object wrappedInstance){
        this.wrappedInstance = wrappedInstance;
    }

    public Object getWrappedInstance(){
        return this.wrappedInstance;
    }

    //返回代理以后的Class
    //可能会是这个 $Proxy0
    public Class<?> getWrappedClass(){
        return this.wrappedInstance.getClass();
    }
}
```

13.4 context（IoC 容器）模块

13.4.1 GPAbstractApplicationContext

IoC 容器实现类的顶层抽象类，实现 IoC 容器相关的公共逻辑。为了尽可能地简化，在这个 Mini 版本中，暂时只设计了一个 refresh()方法。

```java
package com.gupaoedu.vip.spring.formework.context.support;

/**
 * IoC 容器实现的顶层设计
 */
public abstract class GPAbstractApplicationContext {
    //受保护，只提供给子类重写
    public void refresh() throws Exception {}
}
```

13.4.2 GPDefaultListableBeanFactory

DefaultListableBeanFactory 是众多 IoC 容器子类的典型代表。在 Mini 版本中我只做了一个简单的设计，就是定义顶层的 IoC 缓存，也就是一个 Map，属性名字也和原生 Spring 保持一致，定义为 beanDefinitionMap，以方便大家对比理解。

```java
package com.gupaoedu.vip.spring.formework.beans.support;

import com.gupaoedu.vip.spring.formework.beans.config.GPBeanDefinition;
import com.gupaoedu.vip.spring.formework.context.support.GPAbstractApplicationContext;

import java.util.Map;
import java.util.concurrent.ConcurrentHashMap;

public class GPDefaultListableBeanFactory extends GPAbstractApplicationContext{
    //存储注册信息的 BeanDefinition
    protected final Map<String, GPBeanDefinition> beanDefinitionMap = new ConcurrentHashMap<String, GPBeanDefinition>();
}
```

13.4.3 GPApplicationContext

ApplicationContext 是直接接触用户的入口，主要实现 DefaultListableBeanFactory 的 refresh()

方法和 BeanFactory 的 getBean()方法，完成 IoC、DI、AOP 的衔接。

```java
package com.gupaoedu.vip.spring.formework.context;

import com.gupaoedu.vip.spring.formework.annotation.GPAutowired;
import com.gupaoedu.vip.spring.formework.annotation.GPController;
import com.gupaoedu.vip.spring.formework.annotation.GPService;
import com.gupaoedu.vip.spring.formework.beans.GPBeanWrapper;
import com.gupaoedu.vip.spring.formework.beans.config.GPBeanPostProcessor;
import com.gupaoedu.vip.spring.formework.core.GPBeanFactory;
import com.gupaoedu.vip.spring.formework.beans.config.GPBeanDefinition;
import com.gupaoedu.vip.spring.formework.beans.support.GPBeanDefinitionReader;
import com.gupaoedu.vip.spring.formework.beans.support.GPDefaultListableBeanFactory;

import java.lang.reflect.Field;
import java.util.List;
import java.util.Map;
import java.util.Properties;
import java.util.concurrent.ConcurrentHashMap;

/**
 * 按之前源码分析的套路，IoC、DI、MVC、AOP
 */
public class GPApplicationContext extends GPDefaultListableBeanFactory implements GPBeanFactory {

    private String [] configLoactions;
    private GPBeanDefinitionReader reader;

    //单例的 IoC 容器缓存
    private Map<String,Object> factoryBeanObjectCache = new ConcurrentHashMap<String, Object>();
    //通用的 IoC 容器
    private Map<String,GPBeanWrapper> factoryBeanInstanceCache = new ConcurrentHashMap<String, GPBeanWrapper>();

    public GPApplicationContext(String... configLoactions){
        this.configLoactions = configLoactions;
        try {
            refresh();
        } catch (Exception e) {
            e.printStackTrace();
        }
    }

    @Override
    public void refresh() throws Exception{
        //1. 定位，定位配置文件
```

```java
        reader = new GPBeanDefinitionReader(this.configLoactions);

        //2. 加载配置文件，扫描相关的类，把它们封装成BeanDefinition
        List<GPBeanDefinition> beanDefinitions = reader.loadBeanDefinitions();

        //3. 注册，把配置信息放到容器里面（伪IoC容器）
        doRegisterBeanDefinition(beanDefinitions);

        //4. 把不是延时加载的类提前初始化
        doAutowrited();
    }

    //只处理非延时加载的情况
    private void doAutowrited() {
        for (Map.Entry<String, GPBeanDefinition> beanDefinitionEntry :
super.beanDefinitionMap.entrySet()) {
            String beanName = beanDefinitionEntry.getKey();
            if(!beanDefinitionEntry.getValue().isLazyInit()) {
                try {
                    getBean(beanName);
                } catch (Exception e) {
                    e.printStackTrace();
                }
            }
        }
    }

    private void doRegisterBeanDefinition(List<GPBeanDefinition> beanDefinitions) throws
Exception {

        for (GPBeanDefinition beanDefinition: beanDefinitions) {
            if(super.beanDefinitionMap.containsKey(beanDefinition.getFactoryBeanName())){
                throw new Exception("The "" + beanDefinition.getFactoryBeanName() + "" is
exists!!");
            }

super.beanDefinitionMap.put(beanDefinition.getFactoryBeanName(),beanDefinition);
        }
        //到这里为止，容器初始化完毕
    }
    public Object getBean(Class<?> beanClass) throws Exception {
        return getBean(beanClass.getName());
    }

    //依赖注入，从这里开始，读取BeanDefinition中的信息
    //然后通过反射机制创建一个实例并返回
    //Spring做法是，不会把最原始的对象放出去，会用一个BeanWrapper来进行一次包装
```

```java
//装饰器模式:
//1. 保留原来的 OOP 关系
//2. 需要对它进行扩展、增强(为了以后的 AOP 打基础)
public Object getBean(String beanName) throws Exception {

    return null;
}

public String[] getBeanDefinitionNames() {
    return this.beanDefinitionMap.keySet().toArray(new String[this.beanDefinitionMap.size()]);
}

public int getBeanDefinitionCount(){
    return this.beanDefinitionMap.size();
}

public Properties getConfig(){
    return this.reader.getConfig();
}
}
```

13.4.4　GPBeanDefinitionReader

根据约定，BeanDefinitionReader 主要完成对 application.properties 配置文件的解析工作，实现逻辑非常简单。通过构造方法获取从 ApplicationContext 传过来的 locations 配置文件路径，然后解析、扫描并保存所有相关的类并提供统一的访问入口。

```java
package com.gupaoedu.vip.spring.formework.beans.support;

import com.gupaoedu.vip.spring.formework.beans.config.GPBeanDefinition;

import java.io.File;
import java.io.IOException;
import java.io.InputStream;
import java.net.URL;
import java.util.ArrayList;
import java.util.List;
import java.util.Properties;

//对配置文件进行查找、读取、解析
public class GPBeanDefinitionReader {

    private List<String> registyBeanClasses = new ArrayList<String>();
```

```java
    private Properties config = new Properties();

    //固定配置文件中的key，相对于XML的规范
    private final String SCAN_PACKAGE = "scanPackage";

    public GPBeanDefinitionReader(String... locations){
        //通过URL定位找到其所对应的文件，然后转换为文件流
        InputStream is = this.getClass().getClassLoader().getResourceAsStream(locations[0].replace("classpath:",""));
        try {
            config.load(is);
        } catch (IOException e) {
            e.printStackTrace();
        }finally {
            if(null != is){
                try {
                    is.close();
                } catch (IOException e) {
                    e.printStackTrace();
                }
            }
        }

        doScanner(config.getProperty(SCAN_PACKAGE));
    }

    private void doScanner(String scanPackage) {
        //转换为文件路径，实际上就是把.替换为/
        URL url = this.getClass().getClassLoader().getResource("/" + scanPackage.replaceAll("\\.","/"));
        File classPath = new File(url.getFile());
        for (File file : classPath.listFiles()) {
            if(file.isDirectory()){
                doScanner(scanPackage + "." + file.getName());
            }else{
                if(!file.getName().endsWith(".class")){ continue;}
                String className = (scanPackage + "." + file.getName().replace(".class",""));
                registyBeanClasses.add(className);
            }
        }
    }

    public Properties getConfig(){
        return this.config;
    }
```

```java
//把配置文件中扫描到的所有配置信息转换为GPBeanDefinition对象，以便于之后的IoC操作
public List<GPBeanDefinition> loadBeanDefinitions(){
    List<GPBeanDefinition> result = new ArrayList<GPBeanDefinition>();
    try {
        for (String className : registyBeanClasses) {
            Class<?> beanClass = Class.forName(className);
            if(beanClass.isInterface()) { continue; }

            result.add(doCreateBeanDefinition(toLowerFirstCase(beanClass.getSimpleName()),beanClass.getName()));

            Class<?> [] interfaces = beanClass.getInterfaces();
            for (Class<?> i : interfaces) {
                result.add(doCreateBeanDefinition(i.getName(),beanClass.getName()));
            }

        }
    }catch (Exception e){
        e.printStackTrace();
    }
    return result;
}


//把每一个配置信息解析成一个BeanDefinition
private GPBeanDefinition doCreateBeanDefinition(String factoryBeanName,String beanClassName){
    GPBeanDefinition beanDefinition = new GPBeanDefinition();
    beanDefinition.setBeanClassName(beanClassName);
    beanDefinition.setFactoryBeanName(factoryBeanName);
    return beanDefinition;
}

//将类名首字母改为小写
//为了简化程序逻辑，就不做其他判断了，大家了解就好
private String toLowerFirstCase(String simpleName) {
    char [] chars = simpleName.toCharArray();
    //因为大小写字母的ASCII码相差32
    //而且大写字母的ASCII码要小于小写字母的ASCII码
    //在Java中，对char做算术运算，实际上就是对ASCII码做算术运算
    chars[0] += 32;
    return String.valueOf(chars);
}

}
```

13.4.5　GPApplicationContextAware

相信很多"小伙伴"都用过 ApplicationContextAware 接口，主要是通过实现侦听机制得到一个回调方法，从而得到 IoC 容器的上下文，即 ApplicationContext。在这个 Mini 版本中只是做了一个顶层设计，告诉大家这样一种现象，并没有做具体实现。这不是本书的重点，感兴趣的"小伙伴"可以自行尝试。

```java
package com.gupaoedu.vip.spring.formework.context;
/**
 * 通过解耦方式获得 IoC 容器的顶层设计
 * 后面将通过一个监听器去扫描所有的类，只要实现了此接口，
 * 将自动调用 setApplicationContext()方法，从而将 IoC 容器注入目标类中
 */
public interface GPApplicationContextAware {
    void setApplicationContext(GPApplicationContext applicationContext);
}
```

第 14 章
完成 DI 模块的功能

在之前的源码分析中我们已经了解到,依赖注入(DI)的入口是 getBean()方法,前面的 IoC 手写部分基本流程已通。先在 GPApplicationContext 中定义好 IoC 容器,然后将 GPBeanWrapper 对象保存到 Map 中。在 GPApplicationContext 中设计两个 Map:factoryBeanObjectCache 保存单例对象的缓存,factoryBeanInstanceCache 保存 GPBeanWrapper 的缓存,变量命名也和原生 Spring 一致,这两个对象的设计其实就是注册式单例模式的经典应用。

```
public class GPApplicationContext extends GPDefaultListableBeanFactory implements GPBeanFactory {
    private String [] configLocations;

    private GPBeanDefinitionReader reader;

    //用来保证注册式单例的容器
    private Map<String,Object> factoryBeanObjectCache = new HashMap<String, Object>();

    //用来存储所有的被代理过的对象
    private Map<String,GPBeanWrapper> factoryBeanInstanceCache = new ConcurrentHashMap<String,GPBeanWrapper>();
    ...
}
```

14.1 从 getBean()方法开始

下面我们从完善 getBean()方法开始：

```java
@Override
public Object getBean(String beanName) {

    GPBeanDefinition beanDefinition = super.beanDefinitionMap.get(beanName);

    try{

        //生成通知事件
        GPBeanPostProcessor beanPostProcessor = new GPBeanPostProcessor();

        Object instance = instantiateBean(beanDefinition);
        if(null == instance){ return null;}

        //在实例初始化以前调用一次
        beanPostProcessor.postProcessBeforeInitialization(instance,beanName);

        GPBeanWrapper beanWrapper = new GPBeanWrapper(instance);

        this.factoryBeanInstanceCache.put(beanName,beanWrapper);

        //在实例初始化以后调用一次
        beanPostProcessor.postProcessAfterInitialization(instance,beanName);

        populateBean(beanName,instance);

        //通过这样调用，相当于给我们自己留有了可操作的空间
        return this.factoryBeanInstanceCache.get(beanName).getWrappedInstance();
    }catch (Exception e){
//        e.printStackTrace();
        return null;
    }
}

private void populateBean(String beanName,Object instance){

    Class clazz = instance.getClass();

    if(!(clazz.isAnnotationPresent(GPController.class) ||
            clazz.isAnnotationPresent(GPService.class))){
        return;
    }
}
```

```java
        Field [] fields = clazz.getDeclaredFields();

        for (Field field : fields) {
            if (!field.isAnnotationPresent(GPAutowired.class)){ continue; }

            GPAutowired autowired = field.getAnnotation(GPAutowired.class);

            String autowiredBeanName = autowired.value().trim();

            if("".equals(autowiredBeanName)){
                autowiredBeanName = field.getType().getName();
            }

            field.setAccessible(true);

            try {

                field.set(instance,this.factoryBeanInstanceCache.get(autowiredBeanName).getWrappedInstance());

            } catch (IllegalAccessException e) {
//                e.printStackTrace();
            }

        }

    }

    //传一个BeanDefinition，就返回一个实例Bean
    private Object instantiateBean(GPBeanDefinition beanDefinition){
        Object instance = null;
        String className = beanDefinition.getBeanClassName();
        try{

            //因为根据Class才能确定一个类是否有实例
            if(this.factoryBeanObjectCache.containsKey(className)){
                instance = this.factoryBeanObjectCache.get(className);
            }else{
                Class<?> clazz = Class.forName(className);
                instance = clazz.newInstance();

                this.factoryBeanObjectCache.put(beanDefinition.getFactoryBeanName(),instance);
            }

            return instance;
        }catch (Exception e){
            e.printStackTrace();
```

```
            }
            return null;
        }
```

14.2 GPBeanPostProcessor

原生 Spring 中的 BeanPostProcessor 是为对象初始化事件设置的一种回调机制。这个 Mini 版本中只做说明，不做具体实现，感兴趣的"小伙伴"可以继续深入研究 Spring 源码。

```
package com.gupaoedu.vip.spring.formework.beans.config;

public class GPBeanPostProcessor {

    //为在 Bean 的初始化之前提供回调入口
    public Object postProcessBeforeInitialization(Object bean, String beanName) throws Exception {
        return bean;
    }

    //为在 Bean 的初始化之后提供回调入口
    public Object postProcessAfterInitialization(Object bean, String beanName) throws Exception {
        return bean;
    }
}
```

至此，DI 部分就手写完成了，也就是说完成了 Spring 的核心部分。"小伙伴们"是不是发现其实还是很简单的？

第 15 章 完成 MVC 模块的功能

接下来我们来完成 MVC 模块的功能，应该不需要再做说明。Spring MVC 的入口就是从 DispatcherServlet 开始的，而前面的章节中已完成了 web.xml 的基础配置。下面就从 DispatcherServlet 开始添砖加瓦。

15.1 MVC 顶层设计

15.1.1 GPDispatcherServlet

我们已经了解到 Servlet 的生命周期由 init()到 service()再到 destory()组成，destory()方法我们不做实现。前面我们讲过，这是 J2EE 中模板模式的典型应用。下面先定义好全局变量：

```
package com.gupaoedu.vip.spring.formework.webmvc.servlet;

import com.gupaoedu.vip.spring.formework.annotation.GPController;
import com.gupaoedu.vip.spring.formework.annotation.GPRequestMapping;
import com.gupaoedu.vip.spring.formework.context.GPApplicationContext;
import com.gupaoedu.vip.spring.formework.webmvc.*;
import lombok.extern.slf4j.Slf4j;
```

```java
import javax.servlet.ServletConfig;
import javax.servlet.ServletException;
import javax.servlet.http.HttpServlet;
import javax.servlet.http.HttpServletRequest;
import javax.servlet.http.HttpServletResponse;
import java.io.File;
import java.io.IOException;
import java.lang.reflect.Method;
import java.util.*;
import java.util.regex.Matcher;
import java.util.regex.Pattern;

//Servlet 只是作为一个 MVC 的启动入口
@Slf4j
public class GPDispatcherServlet extends HttpServlet {

    private final String LOCATION = "contextConfigLocation";

    //读者可以思考一下这样设计的经典之处
    //GPHandlerMapping 最核心的设计，也是最经典的
    //它直接干掉了 Struts、Webwork 等 MVC 框架
    private List<GPHandlerMapping> handlerMappings = new ArrayList<GPHandlerMapping>();

    private Map<GPHandlerMapping,GPHandlerAdapter> handlerAdapters = new HashMap<GPHandlerMapping, GPHandlerAdapter>();

    private List<GPViewResolver> viewResolvers = new ArrayList<GPViewResolver>();

    private GPApplicationContext context;

}
```

下面实现 init()方法，我们主要完成 IoC 容器的初始化和 Spring MVC 九大组件的初始化。

```java
@Override
public void init(ServletConfig config) throws ServletException {
    //相当于把 IoC 容器初始化了
    context = new GPApplicationContext(config.getInitParameter(LOCATION));
    initStrategies(context);
}

protected void initStrategies(GPApplicationContext context) {

    //有九种策略
    //针对每个用户请求，都会经过一些处理策略处理，最终才能有结果输出
    //每种策略可以自定义干预，但是最终的结果都一致
```

```
        // ============   这里说的就是传说中的九大组件 ================
        initMultipartResolver(context);//文件上传解析,如果请求类型是multipart,将通过
MultipartResolver进行文件上传解析
        initLocaleResolver(context);//本地化解析
        initThemeResolver(context);//主题解析

        /** 我们自己会实现 */
        //GPHandlerMapping 用来保存Controller中配置的RequestMapping和Method的对应关系
        initHandlerMappings(context);//通过HandlerMapping将请求映射到处理器
        /** 我们自己会实现 */
        //HandlerAdapters 用来动态匹配Method参数,包括类转换、动态赋值
        initHandlerAdapters(context);//通过HandlerAdapter进行多类型的参数动态匹配

        initHandlerExceptionResolvers(context);//如果执行过程中遇到异常,将交给
HandlerExceptionResolver来解析
        initRequestToViewNameTranslator(context);//直接将请求解析到视图名

        /** 我们自己会实现 */
        //通过ViewResolvers实现动态模板的解析
        //自己解析一套模板语言
        initViewResolvers(context);//通过viewResolver将逻辑视图解析到具体视图实现

        initFlashMapManager(context);//Flash映射管理器
    }

    private void initFlashMapManager(GPApplicationContext context) {}
    private void initRequestToViewNameTranslator(GPApplicationContext context) {}
    private void initHandlerExceptionResolvers(GPApplicationContext context) {}
    private void initThemeResolver(GPApplicationContext context) {}
    private void initLocaleResolver(GPApplicationContext context) {}
    private void initMultipartResolver(GPApplicationContext context) {}

    //将Controller中配置的RequestMapping和Method进行一一对应
    private void initHandlerMappings(GPApplicationContext context) {
        //按照我们通常的理解应该是一个Map
        //Map<String,Method> map;
        //map.put(url,Method)

        //首先从容器中获取所有的实例
        String [] beanNames = context.getBeanDefinitionNames();
        try {
            for (String beanName : beanNames) {
                //到了MVC层,对外提供的方法只有一个getBean()方法
                //返回的对象不是BeanWrapper,怎么办?
                Object controller = context.getBean(beanName);
                //Object controller = GPAopUtils.getTargetObject(proxy);
                Class<?> clazz = controller.getClass();
```

```java
            if (!clazz.isAnnotationPresent(GPController.class)) {
                continue;
            }

            String baseUrl = "";

            if (clazz.isAnnotationPresent(GPRequestMapping.class)) {
                GPRequestMapping requestMapping = clazz.getAnnotation(GPRequestMapping.class);
                baseUrl = requestMapping.value();
            }

            //扫描所有的 public 类型的方法
            Method[] methods = clazz.getMethods();
            for (Method method : methods) {
                if (!method.isAnnotationPresent(GPRequestMapping.class)) {
                    continue;
                }

                GPRequestMapping requestMapping = method.getAnnotation(GPRequestMapping.class);
                String regex = ("/" + baseUrl + requestMapping.value().replaceAll("\\*", ".*")).replaceAll("/+", "/");
                Pattern pattern = Pattern.compile(regex);
                this.handlerMappings.add(new GPHandlerMapping(pattern, controller, method));
                log.info("Mapping: " + regex + " , " + method);

            }

        }
    }catch (Exception e){
        e.printStackTrace();
    }

}

private void initHandlerAdapters(GPApplicationContext context) {
    //在初始化阶段,我们能做的就是,将这些参数的名字或者类型按一定的顺序保存下来
    //因为后面用反射调用的时候,传的形参是一个数组
    //可以通过记录这些参数的位置 index,逐个从数组中取值,这样就和参数的顺序无关了
    for (GPHandlerMapping handlerMapping : this.handlerMappings){
        //每个方法有一个参数列表,这里保存的是形参列表
        this.handlerAdapters.put(handlerMapping,new GPHandlerAdapter());
    }

}

private void initViewResolvers(GPApplicationContext context) {
```

```
//在页面中输入 http://localhost/first.html
//解决页面名字和模板文件关联的问题
String templateRoot = context.getConfig().getProperty("templateRoot");
String templateRootPath = this.getClass().getClassLoader().getResource
(templateRoot).getFile();

File templateRootDir = new File(templateRootPath);

for (File template : templateRootDir.listFiles()) {
    this.viewResolvers.add(new GPViewResolver(templateRoot));
}

}
```

在上面的代码中,我们只实现了九大组件中的三大核心组件的基本功能,分别是 HandlerMapping、HandlerAdapter、ViewResolver,完成 MVC 最核心的调度功能。其中 HandlerMapping 就是策略模式的应用,用输入 URL 间接调用不同的 Method 已达到获取结果的目的。顾名思义,HandlerAdapter 应用的是适配器模式,将 Request 的字符型参数自动适配为 Method 的 Java 实参,主要实现参数列表自动适配和类型转换功能。ViewResolver 也算一种策略,根据不同的请求选择不同的模板引擎来进行页面的渲染。

接下来看 service()方法,它主要负责接收请求,得到 Request 和 Response 对象。在 Servlet 子类中 service()方法被拆分成 doGet()方法和 doPost()方法。我们在 doGet()方法中直接调用 doPost()方法,在 doPost()方法中调用 doDispatch()方法,真正的调用逻辑由 doDispatch()来执行。

```
@Override
    protected void doGet(HttpServletRequest req, HttpServletResponse resp) throws
ServletException, IOException {
        this.doPost(req,resp);
    }

    @Override
    protected void doPost(HttpServletRequest req, HttpServletResponse resp) throws
ServletException, IOException {
        try {
            doDispatch(req, resp);
        }catch (Exception e){
            resp.getWriter().write("<font size='25' color='blue'>500 Exception</font><br/>Details:
<br/>" + Arrays.toString(e.getStackTrace()).replaceAll("\\[|\\]","")
                    .replaceAll("\\s","\r\n") + "<font color='green'><i>Copyright@GupaoEDU
</i></font>");
            e.printStackTrace();
        }
    }
```

```java
    private void doDispatch(HttpServletRequest req, HttpServletResponse resp) throws Exception{

        //根据用户请求的URL来获得一个Handler
        GPHandlerMapping handler = getHandler(req);
        if(handler == null){
            processDispatchResult(req,resp,new GPModelAndView("404"));
            return;
        }

        GPHandlerAdapter ha = getHandlerAdapter(handler);

        //这一步只是调用方法，得到返回值
        GPModelAndView mv = ha.handle(req, resp, handler);

        //这一步才是真的输出
        processDispatchResult(req,resp, mv);

    }

    private void processDispatchResult(HttpServletRequest request,HttpServletResponse response, GPModelAndView mv) throws Exception {
        //调用viewResolver的resolveViewName()方法
        if(null == mv){ return;}

        if(this.viewResolvers.isEmpty()){ return;}

        if (this.viewResolvers != null) {
            for (GPViewResolver viewResolver : this.viewResolvers) {
                GPView view = viewResolver.resolveViewName(mv.getViewName(), null);
                if (view != null) {
                    view.render(mv.getModel(),request,response);
                    return;
                }
            }
        }

    }

    private GPHandlerAdapter getHandlerAdapter(GPHandlerMapping handler) {
        if(this.handlerAdapters.isEmpty()){return null;}
        GPHandlerAdapter ha = this.handlerAdapters.get(handler);
        if (ha.supports(handler)) {
            return ha;
        }
        return null;
```

```
}

private GPHandlerMapping getHandler(HttpServletRequest req) {

    if(this.handlerMappings.isEmpty()){ return null;}

    String url = req.getRequestURI();
    String contextPath = req.getContextPath();
    url = url.replace(contextPath,"").replaceAll("/+","/");

    for (GPHandlerMapping handler : this.handlerMappings) {
        Matcher matcher = handler.getPattern().matcher(url);
        if(!matcher.matches()){ continue;}
        return handler;
    }

    return null;
}
```

下面补充实现上面的代码中缺失的依赖类。

15.1.2 GPHandlerMapping

我们已经知道 HandlerMapping 主要用来保存 URL 和 Method 的对应关系，这里其实使用的是策略模式。

```
package com.gupaoedu.vip.spring.formework.webmvc;

import java.lang.reflect.Method;
import java.util.regex.Pattern;

public class GPHandlerMapping {
    private Object controller;  //目标方法所在的contrller对象
    private Method method;  //URL 对应的目标方法
    private Pattern pattern;    //URL 的封装

    public GPHandlerMapping(Pattern pattern,Object controller, Method method) {
        this.controller = controller;
        this.method = method;
        this.pattern = pattern;
    }

    public Object getController() {
        return controller;
    }
```

```java
    public void setController(Object controller) {
        this.controller = controller;
    }

    public Method getMethod() {
        return method;
    }

    public void setMethod(Method method) {
        this.method = method;
    }

    public Pattern getPattern() {
        return pattern;
    }

    public void setPattern(Pattern pattern) {
        this.pattern = pattern;
    }
}
```

15.1.3　GPHandlerAdapter

原生 Spring 的 HandlerAdapter 主要完成请求传递到服务端的参数列表与 Method 实参列表的对应关系，完成参数值的类型转换工作。核心方法是 handle()，在 handle()方法中用反射来调用被适配的目标方法，并将转换包装好的参数列表传递过去。

```java
package com.gupaoedu.vip.spring.formework.webmvc;

import com.gupaoedu.vip.spring.formework.annotation.GPRequestParam;

import javax.servlet.http.HttpServletRequest;
import javax.servlet.http.HttpServletResponse;
import java.lang.annotation.Annotation;
import java.util.Arrays;
import java.util.HashMap;
import java.util.Map;

//专人干专事
public class GPHandlerAdapter {

    public boolean supports(Object handler){
        return (handler instanceof GPHandlerMapping);
    }
```

```java
public GPModelAndView handle(HttpServletRequest req, HttpServletResponse resp, Object handler) throws Exception{
    GPHandlerMapping handlerMapping = (GPHandlerMapping)handler;

    //每个方法有一个参数列表，这里保存的是形参列表
    Map<String,Integer> paramMapping = new HashMap<String, Integer>();

    //这里只是给出命名参数
    Annotation[][] pa = handlerMapping.getMethod().getParameterAnnotations();
    for (int i = 0; i < pa.length ; i ++) {
        for (Annotation a : pa[i]) {
            if(a instanceof GPRequestParam){
                String paramName = ((GPRequestParam) a).value();
                if(!"".equals(paramName.trim())){
                    paramMapping.put(paramName,i);
                }
            }
        }
    }

    //根据用户请求的参数信息，跟Method中的参数信息进行动态匹配
    //resp 传进来的目的只有一个：将其赋值给方法参数，仅此而已

    //只有当用户传过来的ModelAndView为空的时候，才会新建一个默认的

    //1. 要准备好这个方法的形参列表
    //方法重载时形参的决定因素：参数的个数、参数的类型、参数顺序、方法的名字
    //只处理Request和Response
    Class<?>[] paramTypes = handlerMapping.getMethod().getParameterTypes();
    for (int i = 0;i < paramTypes.length; i ++) {
        Class<?> type = paramTypes[i];
        if(type == HttpServletRequest.class ||
               type == HttpServletResponse.class){
            paramMapping.put(type.getName(),i);
        }
    }

    //2. 得到自定义命名参数所在的位置
    //用户通过URL传过来的参数列表
    Map<String,String[]> reqParameterMap = req.getParameterMap();

    //3. 构造实参列表
    Object [] paramValues = new Object[paramTypes.length];

    for (Map.Entry<String,String[]> param : reqParameterMap.entrySet()) {
```

```java
        String value = Arrays.toString(param.getValue()).replaceAll("\\[|\\]","").replaceAll("\\s","");

            if(!paramMapping.containsKey(param.getKey())){continue;}

            int index = paramMapping.get(param.getKey());

            //因为页面传过来的值都是 String 类型的,而在方法中定义的类型是千变万化的
            //所以要针对我们传过来的参数进行类型转换
            paramValues[index] = caseStringValue(value,paramTypes[index]);
        }

        if(paramMapping.containsKey(HttpServletRequest.class.getName())) {
            int reqIndex = paramMapping.get(HttpServletRequest.class.getName());
            paramValues[reqIndex] = req;
        }

        if(paramMapping.containsKey(HttpServletResponse.class.getName())) {
            int respIndex = paramMapping.get(HttpServletResponse.class.getName());
            paramValues[respIndex] = resp;
        }

        //4. 从 handler 中取出 Controller、Method,然后利用反射机制进行调用

        Object result = handlerMapping.getMethod().invoke(handlerMapping.getController(),paramValues);

        if(result == null){ return null; }

        boolean isModelAndView = handlerMapping.getMethod().getReturnType() == GPModelAndView.class;
        if(isModelAndView){
            return (GPModelAndView)result;
        }else{
            return null;
        }
    }

    private Object caseStringValue(String value,Class<?> clazz){
        if(clazz == String.class){
            return value;
        }else if(clazz == Integer.class){
            return  Integer.valueOf(value);
        }else if(clazz == int.class){
            return Integer.valueOf(value).intValue();
        }else {
            return null;
        }
```

    }
}

15.1.4 GPModelAndView

原生 Spring 中 ModelAndView 类主要用于封装页面模板和要往页面传送的参数的对应关系。

```java
package com.gupaoedu.vip.spring.formework.webmvc;

import java.util.Map;

public class GPModelAndView {

    private String viewName; //页面模板的名称
    private Map<String,?> model; //往页面传送的参数

    public GPModelAndView(String viewName) {
        this(viewName,null);
    }
    public GPModelAndView(String viewName, Map<String, ?> model) {
        this.viewName = viewName;
        this.model = model;
    }

    public String getViewName() {
        return viewName;
    }

    public void setViewName(String viewName) {
        this.viewName = viewName;
    }

    public Map<String, ?> getModel() {
        return model;
    }

    public void setModel(Map<String, ?> model) {
        this.model = model;
    }
}
```

15.1.5 GPViewResolver

原生 Spring 中的 ViewResolver 主要完成模板名称和模板解析引擎的匹配。通过在 Serlvet 中调

用 resolveViewName()方法来获得模板所对应的 View。在这个 Mini 版本中简化了实现，只实现了一套默认的模板引擎，语法也是完全自定义的。

```java
package com.gupaoedu.vip.spring.formework.webmvc;

import java.io.File;
import java.util.Locale;

//设计这个类的主要目的是:
//1. 将一个静态文件变为一个动态文件
//2. 根据用户传送不同的参数，产生不同的结果
//最终输出字符串，交给 Response 输出
public class GPViewResolver {
    private final String DEFAULT_TEMPLATE_SUFFIX = ".html";

    private File templateRootDir;
    private String viewName;

    public GPViewResolver(String templateRoot){
        String templateRootPath = this.getClass().getClassLoader().getResource(templateRoot).getFile();
        this.templateRootDir = new File(templateRootPath);
    }

    public GPView resolveViewName(String viewName, Locale locale) throws Exception {
        this.viewName = viewName;
        if(null == viewName || "".equals(viewName.trim())){ return null;}
        viewName = viewName.endsWith(DEFAULT_TEMPLATE_SUFFIX) ? viewName : (viewName + DEFAULT_TEMPLATE_SUFFIX);
        File templateFile = new File((templateRootDir.getPath() + "/" + viewName).replaceAll("/+", "/"));
        return new GPView(templateFile);
    }

    public String getViewName() {
        return viewName;
    }
}
```

15.1.6 GPView

这里的 GPView 就是前面所说的自定义模板解析引擎，其核心方法是 render()。在 render()方法中完成对模板的渲染，最终返回浏览器能识别的字符串，通过 Response 输出。

```java
package com.gupaoedu.vip.spring.formework.webmvc;

import javax.servlet.http.HttpServletRequest;
import javax.servlet.http.HttpServletResponse;
import java.io.RandomAccessFile;
import java.util.Map;
import java.io.File;
import java.util.regex.Matcher;
import java.util.regex.Pattern;

public class GPView {

    public static final String DEFAULT_CONTENT_TYPE = "text/html;charset=utf-8";

    private File viewFile;

    public GPView(File viewFile){
        this.viewFile = viewFile;
    }

    public String getContentType(){
        return DEFAULT_CONTENT_TYPE;
    }

    public void render(Map<String, ?> model,HttpServletRequest request, HttpServletResponse response) throws Exception{
        StringBuffer sb = new StringBuffer();
        RandomAccessFile ra = new RandomAccessFile(this.viewFile,"r");

        try {
            String line = null;
            while (null != (line = ra.readLine())) {
                line = new String(line.getBytes("ISO-8859-1"),"utf-8");
                Pattern pattern = Pattern.compile("¥\\{[^\\}]+\\}",Pattern.CASE_INSENSITIVE);
                Matcher matcher = pattern.matcher(line);

                while (matcher.find()) {

                    String paramName = matcher.group();
                    paramName = paramName.replaceAll("¥\\{|\\}","");
                    Object paramValue = model.get(paramName);
                    if (null == paramValue) { continue; }
                    //要把¥{}中间的这个字符串取出来
                    line = matcher.replaceFirst(makeStringForRegExp(paramValue.toString()));
                    matcher = pattern.matcher(line);
```

```
            }
            sb.append(line);
        }
    }finally {
        ra.close();
    }
    response.setCharacterEncoding("utf-8");
    //response.setContentType(DEFAULT_CONTENT_TYPE);
    response.getWriter().write(sb.toString());
}

//处理特殊字符
public static String makeStringForRegExp(String str) {
    return str.replace("\\", "\\\\").replace("*", "\\*")
    .replace("+", "\\+").replace("|", "\\|")
    .replace("{", "\\{").replace("}", "\\}")
    .replace("(", "\\(").replace(")", "\\)")
    .replace("^", "\\^").replace("$", "\\$")
    .replace("[", "\\[").replace("]", "\\]")
    .replace("?", "\\?").replace(",", "\\,")
    .replace(".", "\\.").replace("&", "\\&");
}
}
```

从上面的代码可以看出，GPView 是基于 HTML 文件来对页面进行渲染的。但是加入了一些自定义语法，例如在模板页面中扫描到￥{name}这样的表达式，就会从 ModelAndView 的 Model 中找到 name 所对应的值，并且用正则表达式将其替换（外国人喜欢用美元符号$，我们的模板引擎就用人民币符号￥）。

15.2 业务代码实现

15.2.1 IQueryService

定义一个负责查询业务的顶层接口 IQueryService，提供一个 query()方法：

```
package com.gupaoedu.vip.spring.demo.service;

/**
 * 查询业务
 *
 */
public interface IQueryService {
```

```java
/**
 * 查询
 */
public String query(String name);
}
```

15.2.2 QueryService

查询业务的实现 QueryService 也非常简单，就是打印一下调用时间和传入的参数，并封装为 JSON 格式返回：

```java
package com.gupaoedu.vip.spring.demo.service.impl;

import java.text.SimpleDateFormat;
import java.util.Date;

import com.gupaoedu.vip.spring.demo.service.IQueryService;
import com.gupaoedu.vip.spring.formework.annotation.GPService;
import lombok.extern.slf4j.Slf4j;

/**
 * 查询业务
 *
 */
@GPService
@Slf4j
public class QueryService implements IQueryService {
    /**
     * 查询
     */
    public String query(String name) {
        SimpleDateFormat sdf = new SimpleDateFormat("yyyy-MM-dd HH:mm:ss");
        String time = sdf.format(new Date());
        String json = "{name:\"" + name + "\",time:\"" + time + "\"}";
        log.info("这是在业务方法中打印的：" + json);
        return json;
    }
}
```

15.2.3 IModifyService

定义一个增、删、改业务的顶层接口 IModifyService：

```java
package com.gupaoedu.vip.spring.demo.service;
```

```java
/**
 * 增、删、改业务
 */
public interface IModifyService {
    /**
     * 增加
     */
    public String add(String name, String addr) ;
    /**
     * 修改
     */
    public String edit(Integer id, String name);
    /**
     * 删除
     */
    public String remove(Integer id);
}
```

15.2.4 ModifyService

增、删、改业务的实现 ModifyService 也非常简单,主要是打印传过来的参数:

```java
package com.gupaoedu.vip.spring.demo.service.impl;
import com.gupaoedu.vip.spring.demo.service.IModifyService;
import com.gupaoedu.vip.spring.formework.annotation.GPService;
/**
 * 增、删、改业务
 */
@GPService
public class ModifyService implements IModifyService {
    /**
     * 增加
     */
    public String add(String name,String addr) {
        return "modifyService add,name=" + name + ",addr=" + addr;
    }
    /**
     * 修改
     */
    public String edit(Integer id,String name) {
        return "modifyService edit,id=" + id + ",name=" + name;
    }
    /**
     * 删除
     */
```

```
    public String remove(Integer id) {
        return "modifyService id=" + id;
    }
}
```

15.2.5 MyAction

Controller 的主要功能是负责调度，不做业务实现。业务实现方法全部在 Service 层，一般我们会将 Service 实例注入 Controller。MyAction 中主要实现对 IQueryService 和 IModifyService 的调度，统一返回结果：

```
package com.gupaoedu.vip.spring.demo.action;

import java.io.IOException;
import javax.servlet.http.HttpServletRequest;
import javax.servlet.http.HttpServletResponse;
import com.gupaoedu.vip.spring.demo.service.IModifyService;
import com.gupaoedu.vip.spring.demo.service.IQueryService;
import com.gupaoedu.vip.spring.formework.annotation.GPAutowired;
import com.gupaoedu.vip.spring.formework.annotation.GPController;
import com.gupaoedu.vip.spring.formework.annotation.GPRequestMapping;
import com.gupaoedu.vip.spring.formework.annotation.GPRequestParam;
import com.gupaoedu.vip.spring.formework.webmvc.GPModelAndView;

/**
 * 公布接口 URL
 */
@GPController
@GPRequestMapping("/web")
public class MyAction {

    @GPAutowired IQueryService queryService;
    @GPAutowired IModifyService modifyService;

    @GPRequestMapping("/query.json")
    public GPModelAndView query(HttpServletRequest request, HttpServletResponse response,
                    @GPRequestParam("name") String name){
        String result = queryService.query(name);
        return out(response,result);
    }
    @GPRequestMapping("/add*.json")
    public GPModelAndView add(HttpServletRequest request,HttpServletResponse response,
            @GPRequestParam("name") String name,@GPRequestParam("addr") String addr){
        String result = modifyService.add(name,addr);
        return out(response,result);
```

```java
    }
    @GPRequestMapping("/remove.json")
    public GPModelAndView remove(HttpServletRequest request,HttpServletResponse response,
            @GPRequestParam("id") Integer id){
        String result = modifyService.remove(id);
        return out(response,result);
    }
    @GPRequestMapping("/edit.json")
    public GPModelAndView edit(HttpServletRequest request,HttpServletResponse response,
            @GPRequestParam("id") Integer id,
            @GPRequestParam("name") String name){
        String result = modifyService.edit(id,name);
        return out(response,result);
    }

    private GPModelAndView out(HttpServletResponse resp,String str){
        try {
            resp.getWriter().write(str);
        } catch (IOException e) {
            e.printStackTrace();
        }
        return null;
    }
}
```

15.2.6 PageAction

专门设计 PageAction 是为了演示 Mini 版 Spring 对模板引擎的支持，实现从 Controller 层到 View 层的传参，以及对模板的渲染进行最终输出：

```java
package com.gupaoedu.vip.spring.demo.action;

import java.util.HashMap;
import java.util.Map;
import com.gupaoedu.vip.spring.demo.service.IQueryService;
import com.gupaoedu.vip.spring.formework.annotation.GPAutowired;
import com.gupaoedu.vip.spring.formework.annotation.GPController;
import com.gupaoedu.vip.spring.formework.annotation.GPRequestMapping;
import com.gupaoedu.vip.spring.formework.annotation.GPRequestParam;
import com.gupaoedu.vip.spring.formework.webmvc.GPModelAndView;

/**
 * 公布接口 URL
 */
@GPController
@GPRequestMapping("/")
```

```java
public class PageAction {

    @GPAutowired IQueryService queryService;

    @GPRequestMapping("/first.html")
    public GPModelAndView query(@GPRequestParam("teacher") String teacher){
        String result = queryService.query(teacher);
        Map<String,Object> model = new HashMap<String,Object>();
        model.put("teacher", teacher);
        model.put("data", result);
        model.put("token", "123456");
        return new GPModelAndView("first.html",model);
    }

}
```

15.3 定制模板页面

为了更全面地演示页面渲染效果,分别定义了 first.html 对应 PageAction 中的 first.html 请求、404.html 默认页和 500.html 异常默认页。

15.3.1 first.html

first.html 定义如下:

```html
<!DOCTYPE html>
<html lang="zh-cn">
<head>
    <meta charset="utf-8">
    <title>咕泡学院 SpringMVC 模板引擎演示</title>
</head>
<center>
    <h1>大家好,我是¥{teacher}老师<br/>欢迎大家一起来探索 Spring 的世界</h1>
    <h3>Hello,My name is ¥{teacher}</h3>
    <div>¥{data}</div>
    Token 值:¥{token}
</center>
</html>
```

15.3.2 404.html

404.html 定义如下:

```html
<!DOCTYPE html>
<html lang="zh-cn">
```

```
<head>
    <meta charset="utf-8">
    <title>页面去火星了</title>
</head>
<body>
    <font size='25' color='red'>404 Not Found</font><br/><font color='green'><i>Copyright
@GupaoEDU</i></font>
</body>
</html>
```

15.3.3　500.html

500.html 定义如下：

```
<!DOCTYPE html>
<html lang="zh-cn">
<head>
    <meta charset="utf-8">
    <title>服务器好像累了</title>
</head>
<body>
    <font size='25' color='blue'>500 服务器好像有点累了，需要休息一下</font><br/>
    <b>Message:￥{detail}</b><br/>
    <b>StackTrace:￥{stackTrace}</b><br/>
    <font color='green'><i>Copyright@GupaoEDU</i></font>
</body>
</html>
```

15.4　运行效果演示

在浏览器中输入 http://localhost/web/query.json?name=Tom，就会映射到 MyAction 中的 @GPRequestMapping("query.json")对应的 query()方法，得到如下图所示结果。

```
{name:"Tom",time:"2019-02-15
16:01:10"}
```

在浏览器中输入 http://localhost/web/addTom.json?name=tom&addr=HunanChangsha，就会映射到 MyAction 中的@GPRequestMapping("add*.json")对应的 add()方法，得到如下图所示结果。

```
modifyService
add, name=tom, addr=HunanChangsha
```

在浏览器中输入 http://localhost/web/remove.json?id=66，就会映射到 MyAction 中的 @GPRequestMapping("remove.json")对应的 remove()方法，并将 id 自动转换为 int 类型，得到如下图所示结果。

> modifyService id=66

在浏览器中输入 http://localhost/web/edit.json?id=666&name=Tom，就会映射到 MyAction 中的 @GPRequestMapping("edit.json")对应的 edit()方法，并将 id 自动转换为 int 类型，得到如下图所示结果。

> modifyService edit, id=666, name=Tom

在浏览器中输入 http://localhost/first.html?teacher=Tom，就会映射到 PageAction 中的 @GPRequestMapping("first.html")对应的 query()方法，得到如下图所示结果。

> **大家好，我是Tom老师**
> **欢迎大家一起来探索Spring的世界**
> Hello,My name is Tom
> {name:"Tom",time:"2019-02-15 16:05:03"}
> Token值：123456

到这里，已经实现了 Spring 从 IoC、ID 到 MVC 的完整功能。虽然忽略了一些细节，但是我们已经了解到，Spring 的核心设计思想其实并没有我们想象得那么神秘。我们已经巧妙地用到了工厂模式、静态代理模式、适配器模式、模板模式、策略模式、委派模式等，使得代码变得非常优雅。

第 16 章
完成 AOP 代码织入

前面我们已经完成了 Spring IoC、DI、MVC 三大核心模块的功能，并保证了功能可用。接下来要完成 Spring 的另一个核心模块——AOP，这也是最难的部分。

16.1 基础配置

首先，在 application.properties 中增加如下自定义配置，作为 Spring AOP 的基础配置：

```
#多切面配置可以在 key 前面加前缀
#例如 aspect.logAspect.

#切面表达式#
pointCut=public .* com.gupaoedu.vip.spring.demo.service..*Service..*(.*)
#切面类#
aspectClass=com.gupaoedu.vip.spring.demo.aspect.LogAspect
#切面前置通知#
aspectBefore=before
#切面后置通知#
aspectAfter=after
#切面异常通知#
aspectAfterThrow=afterThrowing
```

```
#切面异常类型#
aspectAfterThrowingName=java.lang.Exception
```

为了加强理解，我们对比一下 Spring AOP 的原生配置：

```xml
<bean id="xmlAspect" class="com.gupaoedu.aop.aspect.XmlAspect"></bean>

<!-- AOP 配置 -->
<aop:config>

    <!-- 声明一个切面，并注入切面 Bean，相当于@Aspect -->
    <aop:aspect ref="xmlAspect">
       <!-- 配置一个切入点，相当于@Pointcut -->
       <aop:pointcut expression="execution(* com.gupaoedu.aop.service..*(..))" id="simplePointcut"/>
       <!-- 配置通知，相当于@Before、@After、@AfterReturn、@Around、@AfterThrowing -->
       <aop:before pointcut-ref="simplePointcut" method="before"/>
       <aop:after pointcut-ref="simplePointcut" method="after"/>
       <aop:after-returning pointcut-ref="simplePointcut" method="afterReturn"/>
       <aop:after-throwing pointcut-ref="simplePointcut" method="afterThrow" throwing="ex"/>
    </aop:aspect>

</aop:config>
```

为了方便，我们用 properties 文件来代替 XML，以简化操作。

16.2 完成 AOP 顶层设计

16.2.1 GPJoinPoint

定义一个切点的抽象，这是 AOP 的基础组成单元。我们可以理解为这是某一个业务方法的附加信息。可想而知，切点应该包含业务方法本身、实参列表和方法所属的实例对象，还可以在 GPJoinPoint 中添加自定义属性，看下面的代码：

```java
package com.gupaoedu.vip.spring.formework.aop.aspect;

import java.lang.reflect.Method;

/**
 * 回调连接点，通过它可以获得被代理的业务方法的所有信息
 */
public interface GPJoinPoint {

    Method getMethod();  //业务方法本身
```

```
    Object[] getArguments();    //该方法的实参列表

    Object getThis();   //该方法所属的实例对象

    //在 JoinPoint 中添加自定义属性
    void setUserAttribute(String key, Object value);
    //从已添加的自定义属性中获取一个属性值
    Object getUserAttribute(String key);

}
```

16.2.2 GPMethodInterceptor

方法拦截器是 AOP 代码增强的基本组成单元，其子类主要有 **GPMethodBeforeAdvice**、**GPAfterReturningAdvice** 和 **GPAfterThrowingAdvice**。

```
package com.gupaoedu.vip.spring.formework.aop.intercept;

/**
 * 方法拦截器顶层接口
 */
public interface GPMethodInterceptor{
    Object invoke(GPMethodInvocation mi) throws Throwable;
}
```

16.2.3 GPAopConfig

定义 AOP 的配置信息的封装对象，以方便在之后的代码中相互传递。

```
package com.gupaoedu.vip.spring.formework.aop;
import lombok.Data;

/**
 * AOP 配置封装
 */
@Data
public class GPAopConfig {
//以下配置与 properties 文件中的属性一一对应
    private String pointCut;    //切面表达式
    private String aspectBefore;    //前置通知方法名
    private String aspectAfter;    //后置通知方法名
    private String aspectClass;    //要织入的切面类
    private String aspectAfterThrow;    //异常通知方法名
    private String aspectAfterThrowingName;    //需要通知的异常类型
}
```

16.2.4　GPAdvisedSupport

GPAdvisedSupport 主要完成对 AOP 配置的解析。其中 pointCutMatch()方法用来判断目标类是否符合切面规则，从而决定是否需要生成代理类，对目标方法进行增强。而 getInterceptorsAndDynamic-InterceptionAdvice()方法主要根据 AOP 配置，将需要回调的方法封装成一个拦截器链并返回提供给外部获取。

```java
package com.gupaoedu.vip.spring.formework.aop.support;

import com.gupaoedu.vip.spring.formework.aop.GPAopConfig;
import com.gupaoedu.vip.spring.formework.aop.aspect.GPAfterReturningAdvice;
import com.gupaoedu.vip.spring.formework.aop.aspect.GPAfterThrowingAdvice;
import com.gupaoedu.vip.spring.formework.aop.aspect.GPMethodBeforeAdvice;

import java.lang.reflect.Method;
import java.util.*;
import java.util.regex.Matcher;
import java.util.regex.Pattern;

/**
 * 主要用来解析和封装 AOP 配置
 */
public class GPAdvisedSupport {
    private Class targetClass;
    private Object target;
    private Pattern pointCutClassPattern;

    private transient Map<Method, List<Object>> methodCache;

    private GPAopConfig config;

    public GPAdvisedSupport(GPAopConfig config){
        this.config = config;
    }

    public Class getTargetClass() {
        return targetClass;
    }

    public void setTargetClass(Class targetClass) {
        this.targetClass = targetClass;
        parse();
    }

    public Object getTarget() {
        return target;
```

```java
    }

    public void setTarget(Object target) {
        this.target = target;
    }

    public List<Object> getInterceptorsAndDynamicInterceptionAdvice(Method method, Class<?> targetClass) throws Exception {
        List<Object> cached = methodCache.get(method);

        //缓存未命中，则进行下一步处理
        if (cached == null) {
            Method m = targetClass.getMethod(method.getName(),method.getParameterTypes());
            cached = methodCache.get(m);
            //存入缓存
            this.methodCache.put(m, cached);
        }
        return cached;
    }

    public boolean pointCutMatch(){
        return pointCutClassPattern.matcher(this.targetClass.toString()).matches();
    }

    private void parse(){
        //pointCut 表达式
        String pointCut = config.getPointCut()
                .replaceAll("\\.","\\\\.")
                .replaceAll("\\\\.\\*",".*")
                .replaceAll("\\(","\\\\(")
                .replaceAll("\\)","\\\\)");

        String pointCutForClass = pointCut.substring(0,pointCut.lastIndexOf("\\(") - 4);
        pointCutClassPattern = Pattern.compile("class " + pointCutForClass.substring(pointCutForClass.lastIndexOf(" ")+1));

        methodCache = new HashMap<Method, List<Object>>();
        Pattern pattern = Pattern.compile(pointCut);

        try {
            Class aspectClass = Class.forName(config.getAspectClass());
            Map<String,Method> aspectMethods = new HashMap<String,Method>();
            for (Method m : aspectClass.getMethods()){
                aspectMethods.put(m.getName(),m);
            }

            //在这里得到的方法都是原生方法
```

```java
            for (Method m : targetClass.getMethods()){
                String methodString = m.toString();
                if(methodString.contains("throws")){
                    methodString = methodString.substring(0,methodString.lastIndexOf("throws")).trim();
                }
                Matcher matcher = pattern.matcher(methodString);
                if(matcher.matches()){
                    //能满足切面规则的类,添加到AOP配置中
                    List<Object> advices = new LinkedList<Object>();
                    //前置通知
                    if(!(null == config.getAspectBefore() || "".equals(config.getAspectBefore().trim()))) {
                        advices.add(new GPMethodBeforeAdvice(aspectMethods.get(config.getAspectBefore()), aspectClass.newInstance()));
                    }
                    //后置通知
                    if(!(null == config.getAspectAfter() || "".equals(config.getAspectAfter().trim()))) {
                        advices.add(new GPAfterReturningAdvice(aspectMethods.get(config.getAspectAfter()), aspectClass.newInstance()));
                    }
                    //异常通知
                    if(!(null == config.getAspectAfterThrow() || "".equals(config.getAspectAfterThrow().trim()))) {
                        GPAfterThrowingAdvice afterThrowingAdvice = new GPAfterThrowingAdvice(aspectMethods.get(config.getAspectAfterThrow()), aspectClass.newInstance());
                        afterThrowingAdvice.setThrowingName(config.getAspectAfterThrowingName());
                        advices.add(afterThrowingAdvice);
                    }
                    methodCache.put(m,advices);
                }
            }
        } catch (Exception e) {
            e.printStackTrace();
        }
    }
}
```

16.2.5 GPAopProxy

GPAopProxy是代理工厂的顶层接口,其子类主要有两个:GPCglibAopProxy和GPJdkDynamicAopProxy,分别实现CGlib代理和JDK Proxy代理。

```java
package com.gupaoedu.vip.spring.formework.aop;
/**
```

```
 * 代理工厂的顶层接口，提供获取代理对象的顶层入口
 */
//默认就用 JDK 动态代理
public interface GPAopProxy {
//获得一个代理对象
    Object getProxy();
//通过自定义类加载器获得一个代理对象
    Object getProxy(ClassLoader classLoader);
}
```

16.2.6 GPCglibAopProxy

本书未实现 CglibAopProxy，感兴趣的"小伙伴"可以自行尝试。

```
package com.gupaoedu.vip.spring.formework.aop;

import com.gupaoedu.vip.spring.formework.aop.support.GPAdvisedSupport;

/**
 * 使用 CGlib API 生成代理类，在此不举例
 * 感兴趣的"小伙伴"可以自行实现
 */
public class GPCglibAopProxy implements GPAopProxy {
    private GPAdvisedSupport config;

    public GPCglibAopProxy(GPAdvisedSupport config){
        this.config = config;
    }

    @Override
    public Object getProxy() {
        return null;
    }

    @Override
    public Object getProxy(ClassLoader classLoader) {
        return null;
    }
}
```

16.2.7 GPJdkDynamicAopProxy

下面来看 GPJdkDynamicAopProxy 的实现，主要功能在 invoke()方法中。从代码量来看其实不多，主要是调用了 GPAdvisedSupport 的 getInterceptorsAndDynamicInterceptionAdvice()方法获得拦截器链。在目标类中，每一个被增强的目标方法都对应一个拦截器链。

```java
package com.gupaoedu.vip.spring.formework.aop;

import com.gupaoedu.vip.spring.formework.aop.intercept.GPMethodInvocation;
import com.gupaoedu.vip.spring.formework.aop.support.GPAdvisedSupport;

import java.lang.reflect.InvocationHandler;
import java.lang.reflect.Method;
import java.lang.reflect.Proxy;
import java.util.List;

/**
 * 使用 JDK Proxy API 生成代理类
 */
public class GPJdkDynamicAopProxy implements GPAopProxy,InvocationHandler {
    private GPAdvisedSupport config;

    public GPJdkDynamicAopProxy(GPAdvisedSupport config){
        this.config = config;
    }

    //把原生的对象传进来
    public Object getProxy(){
        return getProxy(this.config.getTargetClass().getClassLoader());
    }

    @Override
    public Object getProxy(ClassLoader classLoader) {
        return Proxy.newProxyInstance(classLoader,this.config.getTargetClass().getInterfaces(),this);
    }

    //invoke()方法是执行代理的关键入口
    @Override
    public Object invoke(Object proxy, Method method, Object[] args) throws Throwable {
//将每一个 JoinPoint 也就是被代理的业务方法（Method）封装成一个拦截器，组合成一个拦截器链
        List<Object> interceptorsAndDynamicMethodMatchers =
config.getInterceptorsAndDynamicInterceptionAdvice(method,this.config.getTargetClass());
//交给拦截器链 MethodInvocation 的 proceed()方法执行
        GPMethodInvocation invocation = new GPMethodInvocation(proxy,this.config.getTarget(),method,args,this.config.getTargetClass(),interceptorsAndDynamicMethodMatchers);
        return invocation.proceed();
    }
}
```

从代码中可以看出，从 GPAdvisedSupport 中获得的拦截器链又被当作参数传入 GPMethodInvocation 的构造方法中。那么 GPMethodInvocation 中到底又对方法链做了什么呢？

16.2.8 GPMethodInvocation

GPMethodInvocation 的代码如下：

```java
package com.gupaoedu.vip.spring.formework.aop.intercept;

import com.gupaoedu.vip.spring.formework.aop.aspect.GPJoinPoint;

import java.lang.reflect.Method;
import java.util.List;

/**
 * 执行拦截器链，相当于 Spring 中 ReflectiveMethodInvocation 的功能
 */
public class GPMethodInvocation implements GPJoinPoint {

    private Object proxy; //代理对象
    private Method method; //代理的目标方法
    private Object target; //代理的目标对象
    private Class<?> targetClass; //代理的目标类
    private Object[] arguments; //代理的方法的实参列表
    private List<Object> interceptorsAndDynamicMethodMatchers; //回调方法链

//保存自定义属性
private Map<String, Object> userAttributes;


    private int currentInterceptorIndex = -1;

    public GPMethodInvocation(Object proxy, Object target, Method method, Object[] arguments,
                              Class<?> targetClass, List<Object> interceptorsAndDynamicMethodMatchers) {
        this.proxy = proxy;
        this.target = target;
        this.targetClass = targetClass;
        this.method = method;
        this.arguments = arguments;
        this.interceptorsAndDynamicMethodMatchers = interceptorsAndDynamicMethodMatchers;
    }

    public Object proceed() throws Throwable {
//如果 Interceptor 执行完了，则执行 joinPoint
        if (this.currentInterceptorIndex == this.interceptorsAndDynamicMethodMatchers.size() - 1) {
            return this.method.invoke(this.target,this.arguments);
        }
        Object interceptorOrInterceptionAdvice =
this.interceptorsAndDynamicMethodMatchers.get(++this.currentInterceptorIndex);
```

```java
//如果要动态匹配joinPoint
if (interceptorOrInterceptionAdvice instanceof GPMethodInterceptor) {
        GPMethodInterceptor mi = (GPMethodInterceptor) interceptorOrInterceptionAdvice;
        return mi.invoke(this);
    } else {
//执行当前Intercetpor

        return proceed();
    }
}

    @Override
    public Method getMethod() {
        return this.method;
    }

    @Override
    public Object[] getArguments() {
        return this.arguments;
    }

    @Override
    public Object getThis() {
        return this.target;
    }

public void setUserAttribute(String key, Object value) {
    if (value != null) {
        if (this.userAttributes == null) {
            this.userAttributes = new HashMap<String,Object>();
        }
        this.userAttributes.put(key, value);
    }
    else {
        if (this.userAttributes != null) {
            this.userAttributes.remove(key);
        }
    }
}

    public Object getUserAttribute(String key) {
        return (this.userAttributes != null ? this.userAttributes.get(key) : null);
    }

}
```

从代码中可以看出，proceed()方法才是 MethodInvocation 的关键所在。在 proceed()中，先进行判断，如果拦截器链为空，则说明目标方法无须增强，直接调用目标方法并返回。如果拦截器链不为空，则将拦截器链中的方法按顺序执行，直到拦截器链中所有方法全部执行完毕。

16.3 设计 AOP 基础实现

16.3.1 GPAdvice

GPAdvice 作为所有回调通知的顶层接口设计，在 Mini 版本中为了尽量和原生 Spring 保持一致，只是被设计成了一种规范，并没有实现任何功能。

```java
/**
 * 回调通知顶层接口
 */
public interface GPAdvice {

}
```

16.3.2 GPAbstractAspectJAdvice

使用模板模式设计 GPAbstractAspectJAdvice 类，封装拦截器回调的通用逻辑，主要封装反射动态调用方法，其子类只需要控制调用顺序即可。

```java
package com.gupaoedu.vip.spring.formework.aop.aspect;

import java.lang.reflect.Method;

/**
 * 封装拦截器回调的通用逻辑，在 Mini 版本中主要封装了反射动态调用方法
 */
public abstract class GPAbstractAspectJAdvice implements GPAdvice {

    private Method aspectMethod;
    private Object aspectTarget;

    public GPAbstractAspectJAdvice(
            Method aspectMethod, Object aspectTarget) {
        this.aspectMethod = aspectMethod;
        this.aspectTarget = aspectTarget;
    }

    //反射动态调用方法
```

```
    protected Object invokeAdviceMethod(GPJoinPoint joinPoint,Object returnValue,Throwable ex)
            throws Throwable {
        Class<?> [] paramsTypes = this.aspectMethod.getParameterTypes();
        if(null == paramsTypes || paramsTypes.length == 0) {
            return this.aspectMethod.invoke(aspectTarget);
        }else {
            Object[] args = new Object[paramsTypes.length];
            for (int i = 0; i < paramsTypes.length; i++) {
                if(paramsTypes[i] == GPJoinPoint.class){
                    args[i] = joinPoint;
                }else if(paramsTypes[i] == Throwable.class){
                    args[i] = ex;
                }else if(paramsTypes[i] == Object.class){
                    args[i] = returnValue;
                }
            }
            return this.aspectMethod.invoke(aspectTarget,args);
        }
    }
}
```

16.3.3 GPMethodBeforeAdvice

GPMethodBeforeAdvice 继承 GPAbstractAspectJAdvice，实现 GPAdvice 和 GPMethodInterceptor 接口，在 invoke()中控制前置通知的调用顺序。

```
package com.gupaoedu.vip.spring.formework.aop.aspect;

import com.gupaoedu.vip.spring.formework.aop.intercept.GPMethodInterceptor;
import com.gupaoedu.vip.spring.formework.aop.intercept.GPMethodInvocation;

import java.lang.reflect.Method;

/**
 * 前置通知具体实现
 */
public class GPMethodBeforeAdvice extends GPAbstractAspectJAdvice implements GPAdvice,
GPMethodInterceptor {

    private GPJoinPoint joinPoint;

    public GPMethodBeforeAdvice(Method aspectMethod, Object target) {
        super(aspectMethod, target);
    }

    public void before(Method method, Object[] args, Object target) throws Throwable {
```

```java
        invokeAdviceMethod(this.joinPoint,null,null);
    }

    public Object invoke(GPMethodInvocation mi) throws Throwable {
        this.joinPoint = mi;
        this.before(mi.getMethod(), mi.getArguments(), mi.getThis());
        return mi.proceed();
    }
}
```

16.3.4 GPAfterReturningAdvice

GPAfterReturningAdvice 继承 GPAbstractAspectJAdvice，实现 GPAdvice 和 GPMethodInterceptor 接口，在 invoke() 中控制后置通知的调用顺序。

```java
package com.gupaoedu.vip.spring.formework.aop.aspect;

import com.gupaoedu.vip.spring.formework.aop.intercept.GPMethodInterceptor;
import com.gupaoedu.vip.spring.formework.aop.intercept.GPMethodInvocation;

import java.lang.reflect.Method;

/**
 * 后置通知具体实现
 */
public class GPAfterReturningAdvice extends GPAbstractAspectJAdvice implements GPAdvice,
GPMethodInterceptor {

    private GPJoinPoint joinPoint;
    public GPAfterReturningAdvice(Method aspectMethod, Object target) {
        super(aspectMethod, target);
    }

    @Override
    public Object invoke(GPMethodInvocation mi) throws Throwable {
        Object retVal = mi.proceed();
        this.joinPoint = mi;
        this.afterReturning(retVal, mi.getMethod(), mi.getArguments(), mi.getThis());
        return retVal;
    }

    public void afterReturning(Object returnValue, Method method, Object[] args,Object target)
throws Throwable{
        invokeAdviceMethod(joinPoint,returnValue,null);
    }

}
```

16.3.5　GPAfterThrowingAdvice

GPAfterThrowingAdvice 继承 GPAbstractAspectJAdvice，实现 GPAdvice 和 GPMethodInterceptor 接口，在 invoke()中控制异常通知的调用顺序。

```java
package com.gupaoedu.vip.spring.formework.aop.aspect;

import com.gupaoedu.vip.spring.formework.aop.intercept.GPMethodInterceptor;
import com.gupaoedu.vip.spring.formework.aop.intercept.GPMethodInvocation;

import java.lang.reflect.Method;

/**
 * 异常通知具体实现
 */
public class GPAfterThrowingAdvice extends GPAbstractAspectJAdvice implements GPAdvice,
GPMethodInterceptor {

    private String throwingName;
    private GPMethodInvocation mi;

    public GPAfterThrowingAdvice(Method aspectMethod, Object target) {
        super(aspectMethod, target);
    }

    public void setThrowingName(String name) {
        this.throwingName = name;
    }

    @Override
    public Object invoke(GPMethodInvocation mi) throws Throwable {
        try {
            return mi.proceed();
        }catch (Throwable ex) {
            invokeAdviceMethod(mi,null,ex.getCause());
            throw ex;
        }
    }
}
```

感兴趣的"小伙伴"可以参看 Spring 源码，自行实现环绕通知的调用逻辑。

16.3.6　接入 getBean()方法

在上面的代码中，我们已经完成了 Spring AOP 模块的核心功能，那么接下如何集成到 IoC 容器中去呢？找到 GPApplicationContext 的 getBean()方法，我们知道 getBean()中负责 Bean 初始化的方法其实就是 instantiateBean()，在初始化时就可以确定是否返回原生 Bean 或 Proxy Bean。代码实现如下：

```java
//传一个 BeanDefinition，返回一个实例 Bean
private Object instantiateBean(GPBeanDefinition beanDefinition){
    Object instance = null;
    String className = beanDefinition.getBeanClassName();
    try{

        //因为根据 Class 才能确定一个类是否有实例
        if(this.singletonBeanCacheMap.containsKey(className)){
            instance = this.singletonBeanCacheMap.get(className);
        }else{
            Class<?> clazz = Class.forName(className);
            instance = clazz.newInstance();

            GPAdvisedSupport config = instantionAopConfig(beanDefinition);
            config.setTargetClass(clazz);
            config.setTarget(instance);

            if(config.pointCutMatch()) {
                instance = createProxy(config).getProxy();
            }
            this.factoryBeanObjectCache.put(className,instance);
            this.factoryBeanObjectCache.put(beanDefinition.getFactoryBeanName(),instance);
        }

        return instance;
    }catch (Exception e){
        e.printStackTrace();
    }

    return null;
}

private GPAdvisedSupport instantionAopConfig(GPBeanDefinition beanDefinition) throws Exception{

    GPAopConfig config = new GPAopConfig();
    config.setPointCut(reader.getConfig().getProperty("pointCut"));
    config.setAspectClass(reader.getConfig().getProperty("aspectClass"));
    config.setAspectBefore(reader.getConfig().getProperty("aspectBefore"));
    config.setAspectAfter(reader.getConfig().getProperty("aspectAfter"));
    config.setAspectAfterThrow(reader.getConfig().getProperty("aspectAfterThrow"));
    config.setAspectAfterThrowingName(reader.getConfig().getProperty("aspectAfterThrowingName"));

    return new GPAdvisedSupport(config);
}
```

```java
private GPAopProxy createProxy(GPAdvisedSupport config) {
    Class targetClass = config.getTargetClass();
    if (targetClass.getInterfaces().length > 0) {
        return new GPJdkDynamicAopProxy(config);
    }
    return new GPCglibAopProxy(config);
}
```

从上面的代码中可以看出，在 instantiateBean()方法中调用 createProxy()决定代理工厂的调用策略，然后调用代理工厂的 proxy()方法创建代理对象。最终代理对象将被封装到 BeanWrapper 中并保存到 IoC 容器。

16.4 织入业务代码

通过前面的代码编写，所有的核心模块和底层逻辑都已经实现，"万事俱备，只欠东风。"接下来，该是"见证奇迹的时刻了"。我们来织入业务代码，做一个测试。创建 LogAspect 类，实现对业务方法的监控。主要记录目标方法的调用日志，获取目标方法名、实参列表、每次调用所消耗的时间。

16.4.1 LogAspect

LogAspect 的代码如下：

```java
package com.gupaoedu.vip.spring.demo.aspect;

import com.gupaoedu.vip.spring.formework.aop.aspect.GPJoinPoint;
import lombok.extern.slf4j.Slf4j;

import java.util.Arrays;

/**
 * 定义一个织入的切面逻辑，也就是要针对目标代理对象增强的逻辑
 * 本类主要完成对方法调用的监控，监听目标方法每次执行所消耗的时间
 */
@Slf4j
public class LogAspect {

    //在调用一个方法之前，执行 before()方法
    public void before(GPJoinPoint joinPoint){
        joinPoint.setUserAttribute("startTime_" + joinPoint.getMethod().getName(),System.currentTimeMillis());
        //这个方法中的逻辑是由我们自己写的
        log.info("Invoker Before Method!!!" +
```

```java
            "\nTargetObject:" + joinPoint.getThis() +
            "\nArgs:" + Arrays.toString(joinPoint.getArguments()));
}

//在调用一个方法之后，执行 after()方法
public void after(GPJoinPoint joinPoint){
    log.info("Invoker After Method!!!" +
            "\nTargetObject:" + joinPoint.getThis() +
            "\nArgs:" + Arrays.toString(joinPoint.getArguments()));
    long     startTime    =    (Long)   joinPoint.getUserAttribute("startTime_"   +
joinPoint.getMethod().getName());
    long endTime = System.currentTimeMillis();
    System.out.println("use time :" + (endTime - startTime));
}

public void afterThrowing(GPJoinPoint joinPoint, Throwable ex){
    log.info("出现异常" +
            "\nTargetObject:" + joinPoint.getThis() +
            "\nArgs:" + Arrays.toString(joinPoint.getArguments()) +
            "\nThrows:" + ex.getMessage());
}
}
```

通过上面的代码可以发现，每一个回调方法都加了一个参数 GPJoinPoint，还记得 GPJoinPoint 为何物吗？事实上，GPMethodInvocation 就是 GPJoinPoint 的实现类。而 GPMethodInvocation 又是在 GPJdkDynamicAopPorxy 的 invoke()方法中实例化的，即每个被代理对象的业务方法会对应一个 GPMethodInvocation 实例。也就是说，MethodInvocation 的生命周期是被代理对象中业务方法的生命周期的对应。前面我们已经了解，调用 GPJoinPoint 的 setUserAttribute()方法可以在 GPJoinPoint 中自定义属性，调用 getUserAttribute()方法可以获取自定义属性的值。

在 LogAspect 的 before()方法中，在 GPJoinPoint 中设置了 startTime 并赋值为系统时间，即记录方法开始调用时间到 MethodInvocation 的上下文。在 LogAspect 的 after()方法中获取 startTime，再次获取的系统时间保存到 endTime。在 AOP 拦截器链回调中，before()方法肯定在 after()方法之前调用，因此两次获取的系统时间会形成一个时间差，这个时间差就是业务方法执行所消耗的时间。通过这个时间差，就可以判断业务方法在单位时间内的性能消耗，是不是设计得非常巧妙？事实上，市面上几乎所有的系统监控框架都是基于这样一种思想来实现的，可以高度解耦并减少代码侵入。

16.4.2　IModifyService

为了演示异常回调通知，我们给之前定义的 IModifyService 接口的 add()方法添加了抛出异常的功能，看下面的代码实现：

```java
package com.gupaoedu.vip.spring.demo.service;

/**
 * 增、删、改业务
 */
public interface IModifyService {

    /**
     * 增加
     */
    String add(String name, String addr) throws Exception;

    /**
     * 修改
     */
    String edit(Integer id, String name);

    /**
     * 删除
     */
    String remove(Integer id);

}
```

16.4.3 ModifyService

ModifyService 的代码如下：

```java
package com.gupaoedu.vip.spring.demo.service.impl;

import com.gupaoedu.vip.spring.demo.service.IModifyService;
import com.gupaoedu.vip.spring.formework.annotation.GPService;

/**
 * 增、删、改业务
 */
@GPService
public class ModifyService implements IModifyService {

    /**
     * 增加
     */
    public String add(String name,String addr) throws Exception {
        throw new Exception("故意抛出异常，测试切面通知是否生效");
//        return "modifyService add,name=" + name + ",addr=" + addr;
    }

    /**
     * 修改
     */
```

```java
public String edit(Integer id,String name) {
    return "modifyService edit,id=" + id + ",name=" + name;
}
/**
 * 删除
 */
public String remove(Integer id) {
    return "modifyService id=" + id;
}
}
```

16.5 运行效果演示

在浏览器中输入 http://localhost/web/add.json?name=Tom&addr=HunanChangsha，就可以直观明了地看到 Service 层抛出的异常信息，如下图所示。

```
500 服务器好像有点累了，需要休息一下
Message:故意抛出异常，测试切面通知是否生效
StackTrace:[com.gupaoedu.vip.spring.demo.service.impl.ModifyService.add(ModifyService.java:18),
sun.reflect.NativeMethodAccessorImpl.invoke0(Native Method),
sun.reflect.NativeMethodAccessorImpl.invoke(NativeMethodAccessorImpl.java:62),
sun.reflect.DelegatingMethodAccessorImpl.invoke(DelegatingMethodAccessorImpl.java:43),
java.lang.reflect.Method.invoke(Method.java:498),
com.gupaoedu.vip.spring.formework.aop.intercept.GPMethodInvocation.proceed(GPMethodInvocation.java:34),
com.gupaoedu.vip.spring.formework.aop.aspect.GPAfterThrowingAdvice.invoke(GPAfterThrowingAdvice.java:27),
com.gupaoedu.vip.spring.formework.aop.intercept.GPMethodInvocation.proceed(GPMethodInvocation.java:40),
com.gupaoedu.vip.spring.formework.aop.aspect.GPAfterReturningAdvice.invoke(GPAfterReturningAdvice.java:20),
com.gupaoedu.vip.spring.formework.aop.intercept.GPMethodInvocation.proceed(GPMethodInvocation.java:40),
com.gupaoedu.vip.spring.formework.aop.aspect.GPMethodBeforeAdvice.invoke(GPMethodBeforeAdvice.java:26),
com.gupaoedu.vip.spring.formework.aop.intercept.GPMethodInvocation.proceed(GPMethodInvocation.java:40),
com.gupaoedu.vip.spring.formework.aop.GPJdkDynamicAopProxy.invoke(GPJdkDynamicAopProxy.java:38),
com.sun.proxy.$Proxy27.add(Unknown Source),
com.gupaoedu.vip.spring.demo.action.MyAction.add(MyAction.java:43),
sun.reflect.NativeMethodAccessorImpl.invoke0(Native Method),
sun.reflect.NativeMethodAccessorImpl.invoke(NativeMethodAccessorImpl.java:62),
sun.reflect.DelegatingMethodAccessorImpl.invoke(DelegatingMethodAccessorImpl.java:43),
java.lang.reflect.Method.invoke(Method.java:498),
com.gupaoedu.vip.spring.formework.webmvc.GPHandlerAdapter.handle(GPHandlerAdapter.java:91),
com.gupaoedu.vip.spring.formework.webmvc.servlet.GPDispatcherServlet.doDispatch(GPDispatcherServlet.java:19,
com.gupaoedu.vip.spring.formework.webmvc.servlet.GPDispatcherServlet.doPost(GPDispatcherServlet.java:172),
com.gupaoedu.vip.spring.formework.webmvc.servlet.GPDispatcherServlet.doGet(GPDispatcherServlet.java:166),
javax.servlet.http.HttpServlet.service(HttpServlet.java:707),
javax.servlet.http.HttpServlet.service(HttpServlet.java:820),
org.mortbay.jetty.servlet.ServletHolder.handle(ServletHolder.java:511),
org.mortbay.jetty.servlet.ServletHandler.handle(ServletHandler.java:401),
org.mortbay.jetty.security.SecurityHandler.handle(SecurityHandler.java:216),
org.mortbay.jetty.servlet.SessionHandler.handle(SessionHandler.java:182),
org.mortbay.jetty.handler.ContextHandler.handle(ContextHandler.java:766),
org.mortbay.jetty.webapp.WebAppContext.handle(WebAppContext.java:450),
org.mortbay.jetty.handler.ContextHandlerCollection.handle(ContextHandlerCollection.java:230),
org.mortbay.jetty.handler.HandlerCollection.handle(HandlerCollection.java:114),
org.mortbay.jetty.handler.HandlerWrapper.handle(HandlerWrapper.java:152),
org.mortbay.jetty.Server.handle(Server.java:326),
org.mortbay.jetty.HttpConnection.handleRequest(HttpConnection.java:542),
org.mortbay.jetty.HttpConnection$RequestHandler.headerComplete(HttpConnection.java:928),
org.mortbay.jetty.HttpParser.parseNext(HttpParser.java:549),
org.mortbay.jetty.HttpParser.parseAvailable(HttpParser.java:212),
org.mortbay.jetty.HttpConnection.handle(HttpConnection.java:404),
org.mortbay.io.nio.SelectChannelEndPoint.run(SelectChannelEndPoint.java:410),
org.mortbay.thread.QueuedThreadPool$PoolThread.run(QueuedThreadPool.java:582)]
Copyright@GupaoEDU
```

控制台输出如下图所示。

```
13:44:03.129 [1848373999@qtp-1916269505-2] INFO com.gupaoedu.vip.spring.demo.aspect.LogAspect - Invoker Before Method!!!
TargetObject:com.gupaoedu.vip.spring.demo.service.impl.ModifyService@4e438732
Args:[Tom, HunanChangsha]
13:44:03.129 [1848373999@qtp-1916269505-2] INFO com.gupaoedu.vip.spring.demo.aspect.LogAspect - 出现异常
TargetObject:com.gupaoedu.vip.spring.demo.service.impl.ModifyService@4e438732
Args:[Tom, HunanChangsha]
Throws:这是Tom老师故意抛的异常！！
```

通过控制台输出，可以看到异常通知成功捕获异常信息，触发了 GPMethodBeforeAdvice 和 GPAfterThrowingAdvice，而并未触发 GPAfterReturningAdvice，符合我们的预期。

下面再做一个测试，输入 http://localhost/web/query.json?name=Tom，结果如下图所示：

{name:"Tom",time:"2019-04-14 16:45:08"}

控制台输出如下图所示：

```
13:46:28.200 [1848373999@qtp-1916269505-2] INFO com.gupaoedu.vip.spring.demo.aspect.LogAspect - Invoker Before Method!!!
TargetObject:com.gupaoedu.vip.spring.demo.service.impl.QueryService@374d4d8e
Args:[Tom]
13:46:28.200 [1848373999@qtp-1916269505-2] INFO com.gupaoedu.vip.spring.demo.service.impl.QueryService - 这是在业务方法中打印的：{name:"Tom",time:"2019-05-27 13:46:28"}
13:46:28.200 [1848373999@qtp-1916269505-2] INFO com.gupaoedu.vip.spring.demo.aspect.LogAspect - Invoker After Method!!!
TargetObject:com.gupaoedu.vip.spring.demo.service.impl.QueryService@374d4d8e
Args:[Tom]
use time :1
```

通过控制台输出可以看到，分别捕获了前置通知、后置通知，并打印了相关信息，符合我们的预期。

至此 AOP 模块大功告成，是不是有一种小小的成就感，跃跃欲试？在整个 Mini 版本实现中有些细节没有过多考虑，更多的是希望给"小伙伴们"提供一种学习源码的思路。手写源码不是为了重复造轮子，也不是为了装"高大上"，其实只是我们推荐给大家的一种学习方式。

第 5 篇
Spring 数据访问

第 17 章　数据库事务原理详解
第 18 章　Spring JDBC 源码初探
第 19 章　基于 Spring JDBC 手写 ORM 框架

第 17 章
数据库事务原理详解

17.1 从 Spring 事务配置说起

先回顾一下 Spring 事务的基础配置：

```xml
<aop:aspectj-autoproxy proxy-target-class="true"/>

<bean id="transactionManager" class="org.springframework.jdbc.datasource.DataSourceTransactionManager">
    <property name="dataSource" ref="dataSource"/>
</bean>

<tx:annotation-driven transaction-manager="transactionManager"/>

<!-- 配置事务传播特性 -->
<tx:advice id="transactionAdvice" transaction-manager="transactionManager">
    <tx:attributes>
        <tx:method name="add*" propagation="REQUIRED" rollback-for="Exception,RuntimeException,SQLException"/>
        <tx:method name="remove*" propagation="REQUIRED" rollback-for="Exception,RuntimeException,SQLException"/>
        <tx:method name="modify*" propagation="REQUIRED" rollback-for="Exception,RuntimeException,SQLException"/>
```

```xml
    <tx:method name="login" propagation="NOT_SUPPORTED"/>
    <tx:method name="query*" read-only="true"/>
  </tx:attributes>
</tx:advice>

<aop:config>
    <aop:pointcut expression="execution(public * com.gupaoedu.vip..*.service..*Service.*(..))" id="transactionPointcut"/>
    <aop:advisor pointcut-ref="transactionPointcut" advice-ref="transactionAdvice"/>
</aop:config>
```

Spring 事务管理基于 AOP 实现，主要作用是统一封装非功能性需求。

17.2 事务的基本概念

事务（Transaction）是访问并可能更新数据库中各种数据项的一个程序执行单元（unit）。

特点：事务是恢复和并发控制的基本单位。事务应该具有 4 个属性：原子性、一致性、隔离性、持久性。这 4 个属性通常称为 ACID 特性。

- **原子性（Automicity）**：一个事务是一个不可分割的工作单位，事务中包括的诸多操作，要么都做，要么都不做。
- **一致性（Consistency）**：事务必须使数据库从一个一致性状态变到另一个一致性状态。一致性与原子性是密切相关的。
- **隔离性（Isolation）**：一个事务的执行不能被其他事务干扰，即一个事务内部的操作及使用的数据对并发的其他事务是隔离的，并发执行的各个事务之间不能互相干扰。
- **持久性（Durability）**：持久性也称永久性（Permanence），指一个事务一旦提交，它对数据库中数据的改变就应该是永久性的，接下来的其他操作或故障不应该对其有任何影响。

17.3 事务的基本原理

Spring 事务的本质其实就是数据库对事务的支持，没有数据库的事务支持，Spring 是无法提供事务功能的。对于纯 JDBC 操作数据库，想要用到事务，可以按照以下步骤进行：

（1）获取连接 Connection con = DriverManager.getConnection()。

（2）开启事务 con.setAutoCommit(true/false)。

（3）执行 CRUD 操作。

（4）提交事务/回滚事务 con.commit() / con.rollback()。

（5）关闭连接 con.close()。

使用 Spring 的事务管理功能后，我们可以不再写步骤（2）和（4）的代码，而是由 Spring 自动完成。那么 Spring 是如何在我们书写的 CRUD 操作之前和之后开启事务和关闭事务的呢？解决了这个问题，也就可以从整体上理解 Spring 的事务管理实现原理了。下面简单地介绍一下，以注解方式为例。

在配置文件中开启注解驱动，在相关的类和方法上通过注解@Transactional 标识。

Spring 在启动的时候会解析生成相关的 Bean，这时候会查看拥有相关注解的类和方法，并且为这些类和方法生成代理，根据@Transactional 的相关参数进行相关配置注入，这样就在代理中为我们把相关的事务处理掉了（开启正常提交事务、异常回滚事务）。

真正的数据库层的事务提交和回滚是通过 binlog 或者 redo log 实现的。

17.4 Spring 事务的传播属性

所谓 Spring 事务的传播属性，就是定义在多个事务同时存在的时候，Spring 应该如何处理这些事务的行为。这些属性在 TransactionDefinition 中定义，具体常量的解释如下表所示。

常量	解释
PROPAGATION_REQUIRED	支持当前事务，如果当前没有事务，就新建一个事务。这是最常见的选择，也是 Spring 默认的事务的传播属性
PROPAGATION_REQUIRES_NEW	新建事务，如果当前存在事务，把当前事务挂起。新建的事务将和被挂起的事务没有任何关系，是两个独立的事务。外部事务失败回滚之后，不能回滚内部事务执行的结果，内部事务失败抛出异常，被外部事务捕获，也可以不处理回滚操作
PROPAGATION_SUPPORTS	支持当前事务，如果当前没有事务，就以非事务方式执行
PROPAGATION_MANDATORY	支持当前事务，如果当前没有事务，就抛出异常
PROPAGATION_NOT_SUPPORTED	以非事务方式执行操作，如果当前存在事务，就把当前事务挂起
PROPAGATION_NEVER	以非事务方式执行，如果当前存在事务，则抛出异常

续表

常量	解释
PROPAGATION_NESTED	如果有活动事务，则运行在一个嵌套的事务中。如果没有活动事务，则按REQUIRED属性执行。它使用了一个单独的事务，这个事务拥有多个可以回滚的保存点。内部事务的回滚不会对外部事务造成影响。它只对DataSourceTransactionManager事务管理器有效

17.5 数据库事务隔离级别

数据库事务隔离级别如下表所示。

隔离级别	值	导致的问题
Read-Uncommitted	0	导致脏读
Read-Committed	1	避免脏读，允许不可重复读和幻读
Repeatable-Read	2	避免脏读，不可重复读，允许幻读
Serializable	3	串行化读，事务只能一个一个执行，避免了脏读、不可重复读、幻读。执行速度慢，使用时要慎重

脏读：一个事务对数据进行了增、删、改，但未提交，另一个事务可以读取到未提交的数据。如果第一个事务这时候回滚了，那么第二个事务就读到了脏数据。

不可重复读：一个事务中发生了两次读操作，在第一次读操作和第二次读操作之间，另外一个事务对数据进行了修改，这时候两次读取的数据是不一致的。

幻读：第一个事务对一定范围的数据进行了批量修改，第二个事务在这个范围内增加了一条数据，这时候第一个事务就会丢失对新增数据的修改。

数据库事务隔离级别越高，越能保证数据的完整性和一致性，但是对并发性能的影响也越大。大多数数据库（比如SQLServer和Oracle）事务默认隔离级别为Read-Commited，少数数据库（比如MySQL InnoDB）事务默认隔离级别为Repeatable-Read。

17.6 Spring中的事务隔离级别

Spring中的事务隔离级别如下表所示。

常量	解释
ISOLATION_DEFAULT	这是PlatformTransactionManager默认的事务隔离级别，使用数据库默认的事务隔离级别。另外4个与JDBC的事务隔离级别相对应
ISOLATION_READ_UNCOMMITTED	这是最低的事务隔离级别，它允许另外一个事务看到这个事务未提交的数据。这种隔离级别会产生脏读、不可重复读和幻读
ISOLATION_READ_COMMITTED	保证一个事务修改的数据提交后才能被另一个事务读取。另一个事务不能读取该事务未提交的数据
ISOLATION_REPEATABLE_READ	这种事务隔离级别可以防止脏读、不可重复读，但是可能出现幻读
ISOLATION_SERIALIZABLE	这是花费最高代价但是最可靠的事务隔离级别，事务被处理为顺序执行

17.7 事务的嵌套

通过前面的理论知识的铺垫，我们大致知道了数据库事务和Spring事务的一些属性和特点，接下来通过分析一些嵌套事务场景来深入理解Spring事务传播机制。

假设外部事务Service A的Method A()调用内部事务ServiceB的MethodB()。

1. PROPAGATION_REQUIRED（Spring 默认事务属性）

如果ServiceB.MethodB()的事务属性定义为PROPAGATION_REQUIRED，那么执行ServiceA.MethodA()的时候Spring已经发起了事务,这时调用ServiceB.MethodB()，ServiceB.MethodB()看到自己已经运行在ServiceA.MethodA()的事务内部，就不再发起新的事务。

假如ServiceB.MethodB()运行的时候发现自己没有在事务中，它就会为自己分配一个事务。

这样，在ServiceA.MethodA()或者ServiceB.MethodB()内的任何地方出现异常，事务都会被回滚。

2. PROPAGATION_REQUIRES_NEW

如果设计ServiceA.MethodA()的事务属性为PROPAGATION_REQUIRED，ServiceB.MethodB()的事务属性为PROPAGATION_REQUIRES_NEW，那么当执行到ServiceB.MethodB()的时候，ServiceA.MethodA()所在的事务就会挂起，ServiceB.MethodB()会发起一个新的事务，等待ServiceB.MethodB()的事务完成以后，挂起的事务才会继续执行。

它与PROPAGATION_REQUIRED的区别在于事务的回滚程度。因为ServiceB.MethodB()新发起一个事务，存在两个不同的事务。如果ServiceB.MethodB()已经提交，那么ServiceA.MethodA()回滚失败时ServiceB.MethodB()是不会回滚的。如果ServiceB.MethodB()回滚失败，它抛出的异常被 ServiceA.MethodA() 捕获，ServiceA.MethodA() 的事务仍然可能提交（主要看 ServiceB MethodB() 抛出的异常是不是 ServiceA.MethodA() 会回滚的异常）。

3. PROPAGATION_SUPPORTS

假设 ServiceB.MethodB() 的事务属性为 PROPAGATION_SUPPORTS，那么当执行到 ServiceB.MethodB() 时，如果发现 ServiceA.MethodA() 已经开启了一个事务，则加入当前的事务。如果发现 ServiceA.MethodA() 没有开启事务，则自己也不开启事务。对于这种事务属性，内部方法的事务完全依赖于最外部的事务。

4. PROPAGATION_NESTED

这种情况比较复杂，ServiceB.MethodB() 的事务属性被配置为 PROPAGATION_NESTED，此时两者之间将如何协作呢？ServiceB.MethodB() 如果回滚，那么内部事务（即 ServiceB.MethodB()）将回滚到它执行前的 SavePoint，而外部事务（即 ServiceA.MethodA()）可以有以下两种处理方式。

（1）捕获异常，执行异常分支逻辑

```
void MethodA() {
  try {
    ServiceB.MethodB();
  } catch (SomeException) {
    // 执行其他事务，如 ServiceC.MethodC();
  }
}
```

这种方式也是嵌套事务最有价值的地方，它起到了分支执行的效果，如果 ServiceB.MethodB() 失败，那么执行 ServiceC.MethodC()，而 ServiceB.MethodB() 已经回滚到它执行之前的 SavePoint，所以不会产生脏数据（相当于此方法从未执行过），这种特性可以用在某些特殊的业务中，而 PROPAGATION_REQUIRED 和 PROPAGATION_REQUIRES_NEW 都没有办法做到这一点。

（2）外部事务回滚/提交

代码不做任何修改，如果内部事务（ServiceB.MethodB()）回滚，首先 ServiceB.MethodB() 回滚到它执行之前的 SavePoint（在任何情况下都会如此），外部事务（即 ServiceA.MethodA()）将根据具体的配置决定自己是提交还是回滚。

另外三种事务传播属性基本用不到，在此不做分析。

17.8　Spring 事务 API 架构图

Spring 事务 API 架构图如下图所示。

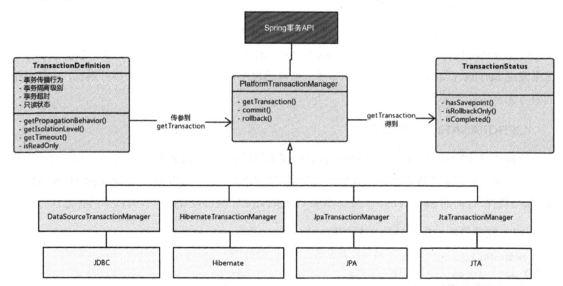

使用 Spring 进行基本的 JDBC 数据库访问有多种选择。Spring 提供了不同的工作模式：JdbcTemplate，在 Spring 2.5 中新提供的 SimpleJdbc 类能够更好地处理数据库元数据；还有一种称为 RDBMS Object 风格的面向对象封装模式，有点类似于 JDO 的查询设计。请注意，即使你选择了其中的一种工作模式，依然可以在代码中混用其他任何一种模式以获取其带来的好处。所有的工作模式都需要 JDBC 2.0 以上数据库驱动的支持，其中一些高级功能可能需要 JDBC 3.0 以上数据库驱动的支持。

17.9　浅谈分布式事务

在现今互联网界，分布式系统和微服务架构盛行。一个简单操作，在服务端很可能是由多个服务和数据库实例协同完成的。在一致性要求较高的场景下，多个独立操作之间的一致性问题显得格外棘手。

基于水平扩容能力和成本考虑，传统的强一致的解决方案（如单机事务）纷纷被抛弃。其理论依据就是响当当的 CAP 理论，即 Consistency（一致性）、Availability（可用性）和 Partition Tolerance（分区容错性）。为了可用性和分区容错性，往往忍痛放弃强一致性，转而追求最终一致性。

1. 分布式系统的特性

在分布式系统中，同时满足 CAP 理论中的一致性、可用性和分区容错性是不可能的。在绝大多数场景，都需要牺牲强一致性来换取系统的高可用性，系统往往只需要保证最终一致性。

分布式事务服务（Distributed Transaction Service，DTS）是一个分布式事务框架，用来保障在大规模分布式环境下事务的最终一致性。

CAP 理论告诉我们，在分布式存储系统中，最多只能实现两点。由于当前的网络硬件原因肯定会出现延迟、丢包等问题，所以分区容错性是必须实现的，我们只能在一致性和可用性之间进行权衡。

为了保障系统的可用性，互联网系统大都将强一致性需求转换成最终一致性需求，并通过系统执行幂等性，保证数据的最终一致性。

2. 理解数据的一致性

强一致性：当更新操作完成之后，任何多个后续进程或者线程的访问都会返回最新的值。这是对用户最友好的，就是用户上一次写什么，下一次就保证能读到什么。根据 CAP 理论，这种实现需要牺牲可用性。

弱一致性：系统并不保证后续进程或者线程的访问都会返回最新的值。在数据写入成功之后，系统不承诺立即可以读到最新写入的值，也不会承诺多久之后可以读到。

最终一致性：弱一致性的特定形式。系统保证在没有后续更新的前提下，最终返回上一次更新操作的值。在没有故障发生的前提下，不一致窗口的时间主要受通信延迟、系统负载和副本的个数影响。DNS 是一个典型的最终一致性系统。

第 18 章
Spring JDBC 源码初探

JdbcTemplate：这个类是经典的、也是最常用的 Spring 对于 JDBC 访问的方案。这也是最低级别的封装，其他的工作模式事实上在底层使用了 JdbcTemplate 作为实现基础。JdbcTemplate 在 JDK 1.4 以上的环境中工作得很好。

NamedParameterJdbcTemplate：这个类对 JdbcTemplate 做了封装，提供了更加便捷的基于命名参数的使用方式，而不是传统的 JDBC 所使用的 "?" 作为参数的占位符。这种方式在你需要为某个 SQL 指定许多个参数时，显得更加直观易用。它必须工作在 JDK 1.4 以上的环境中。

SimpleJdbcTemplate：这个类结合了 JdbcTemplate 和 NamedParameterJdbcTemplate 的最常用功能，同时也利用了一些 Java 5 的特性所带来的优势，例如泛型、varargs 和 autoboxing 等，从而提供了更加简便的 API 访问方式。它需要工作在 Java 5 以上的环境中。

SimpleJdbcInsert 和 SimpleJdbcCall：这两个类可以充分利用数据库元数据的特性来简化配置。使用这两个类进行编程时，可以仅提供数据库表名或者存储过程的名称及一个 Map 作为参数。其中 Map 的 key 需要与数据库表中的字段保持一致。这两个类通常和 SimpleJdbcTemplate 配合使用，需要工作在 JDK 5 以上的环境中，同时数据库需要提供足够的元数据信息。

RDBMS 对象包括 MappingSqlQuery、SqlUpdate 和 StoredProcedure：这种方式允许你在初始

化数据访问层时创建可重用并且线程安全的对象。该对象在定义查询语句、声明查询参数并编译相应的 Query 之后被模型化。一旦模型化完成，任何执行函数都可以传入不同的参数对其进行多次调用。这种方式需要工作在 JDK 1.4 以上的环境中。

18.1 异常处理

Spring JDBC 中的异常结构如下图所示，可到"下载资源"中详细查看。

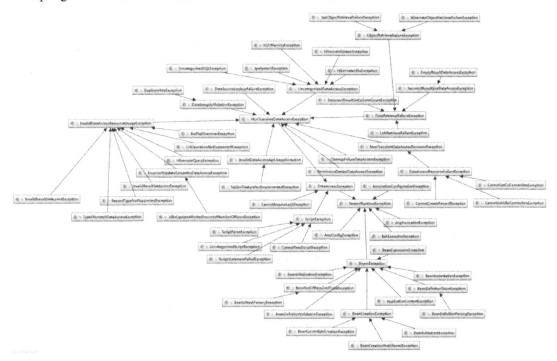

SQLExceptionTranslator 是一个接口，如果你需要在 SQLException 和 org.springframework.dao.DataAccessException 之间做转换，那么必须实现该接口。转换器类的实现可以采用一般通用的做法（比如使用 JDBC 的 SQLState），为了使转换更准确，也可以进行定制（比如使用 Oracle 的 error code）。

SQLErrorCodeSQLExceptionTranslator 是 SQLExceptionTranslator 的默认实现。该实现使用指定数据库厂商的 error code，比采用 SQLState 更精确。转换过程基于一个 JavaBean（类型为 SQLErrorCodes）中的 error code。这个 JavaBean 由 SQLErrorCodesFactory 工厂类创建，其中的内容来自"sql-error-codes.xml"配置文件。该文件中的数据库厂商代码基于 DatabaseMetaData 信息

中的 DatabaseProductName，配合当前数据库的使用。

SQLErrorCodeSQLExceptionTranslator 使用以下匹配规则：

- 首先检查是否存在完成定制转换的子类实现。通常 SQLErrorCodeSQLExceptionTranslator 类可以作为一个具体类使用，不需要进行定制，此时这个规则不适用。
- 接着将 SQLException 的 error code 与错误代码集中的 error code 进行匹配。默认情况下，错误代码集将从 SQLErrorCodesFactory 取得。错误代码集来自 classpath 下的 "sql-error-codes.xml" 配置文件，它们将与数据库 metadata 信息中的 database name 进行映射。
- 最后使用 fallback 翻译器。SQLStateSQLExceptionTranslator 类是默认的 fallback 翻译器。

18.2　config 模块

DefaultBeanDefinitionDocumentReader 使用 NamespaceHandler 接口来处理在"spring.xml"配置文件中自定义的命名空间，如下图所示。

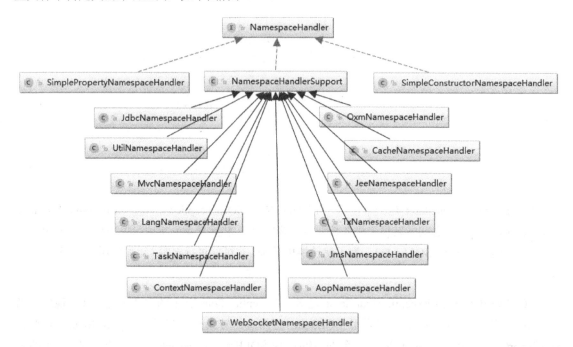

在 JDBC 模块中，我们使用 JdbcNamespaceHandler 来处理 JDBC 配置的命名空间，其代码如下：

```java
public class JdbcNamespaceHandler extends NamespaceHandlerSupport {

    @Override
    public void init() {
        registerBeanDefinitionParser("embedded-database", new
EmbeddedDatabaseBeanDefinitionParser());
        registerBeanDefinitionParser("initialize-database", new
 InitializeDatabaseBeanDefinitionParser());
    }
}
```

EmbeddedDatabaseBeanDefinitionParser()继承了 AbstractBeanDefinitionParser()，解析<embedded-database>元素，并使用 EmbeddedDatabaseFactoryBean 创建一个 BeanDefinition。

这里顺便介绍一下用到的软件包 org.w3c.dom。软件包 org.w3c.dom 为文档对象模型（DOM）提供接口，该模型是 Java API for XML Processing 的组件 API。该 Document Object Model Level 2 Core API 允许程序动态访问和更新文档的内容和结构。涉及的接口如下：

- Attr：Attr 接口表示 Element 对象中的属性。
- CDATASection：用于转义文本块，该文本块包含的字符如果不转义则会被视为标记。
- CharacterData：CharacterData 接口使用属性集合和用于访问 DOM 中字符数据的方法扩展节点。
- Comment：此接口继承自 CharacterData，表示注释的内容，即起始"<!—"和结束"-->"之间的所有字符。
- Document：Document 接口表示整个 HTML 或 XML 文档。
- DocumentFragment：DocumentFragment 是"轻量级"或"最小"Document 对象。
- DocumentType：每个 Document 都有 doctype 属性，该属性的值可以为 null，也可以为 DocumentType 对象。
- DOMConfiguration：DOMConfiguration 接口表示文档的配置，并维护一个可识别的参数表。
- DOMError：DOMError 是一个描述错误的接口。
- DOMErrorHandler：DOMErrorHandler 是在报告处理 XML 数据时发生的错误或在进行某些其他处理（如验证文档）时 DOM 实现可以调用的回调接口。
- DOMImplementation：DOMImplementation 接口为执行独立于文档对象模型的任何特定实例的操作提供了许多方法。
- DOMImplementationList：DOMImplementationList 接口提供对 DOM 实现的有序集合的抽象，没有定义或约束如何实现此集合。

- DOMImplementationSource：此接口允许 DOM 实现程序根据请求的功能和版本提供一个或多个实现。
- DOMLocator：DOMLocator 是一个描述位置（如发生错误的位置）的接口。
- DOMStringList：DOMStringList 接口提供对 DOMString 值的有序集合的抽象，没有定义或约束如何实现此集合。
- Element：Element 接口表示 HTML 或 XML 文档中的一个元素。
- Entity：表示在 XML 文档中解析和未解析的已知实体。
- EntityReference：EntityReference 节点可以用来在树中表示实体引用。
- NamedNodeMap：实现 NamedNodeMap 接口的对象，用于表示可以通过名称访问的节点的集合。
- NameList：NameList 提供对并行的名称和名称空间值对（可以为 null 值）的有序集合的抽象，无须定义或约束如何实现此集合。
- Node：Node 接口是整个文档对象模型的主要数据类型。
- NodeList：NodeList 接口提供对节点的有序集合的抽象，没有定义或约束如何实现此集合。
- Notation：表示在 DTD 中声明的表示法。
- ProcessingInstruction：表示处理指令，该指令可作为在 XML 文档中保留特定处理器信息的方法使用。
- Text：继承自 CharacterData，表示 Element 或 Attr 的文本内容（在 XML 中称为字符数据）。
- TypeInfo：TypeInfo 接口表示从 Element 或 Attr 节点引用的类型，用与文档相关的模式指定。
- UserDataHandler：当使用 Node.setUserData()将一个对象与节点上的键相关联时，当克隆、导入或重命名该对象关联的节点时，应用程序可以提供调用的处理程序。

18.3 core 模块

JdbcTemplate 对象结构如下图所示。

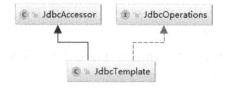

RowMapper 对象结构如下图所示。

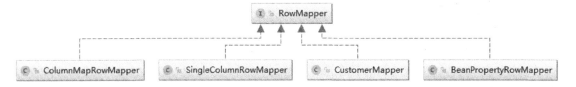

元数据 metaData 模块如下图所示。本节中 Spring 用到了工厂模式，结合代码可以更具体了解。

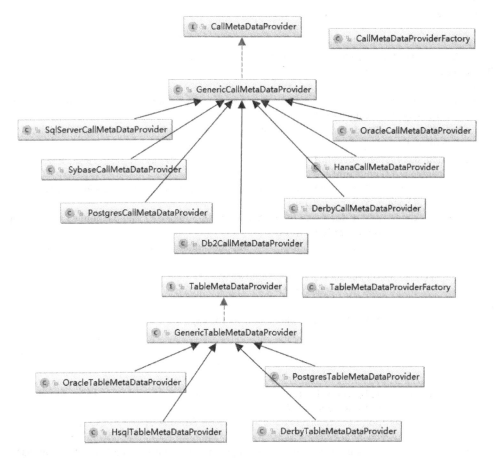

CallMetaDataProviderFactory 创建 CallMetaDataProvider 的工厂类，其代码如下：

```
public static final List<String> supportedDatabaseProductsForProcedures = Arrays.asList(
    "Apache Derby",
```

```java
        "DB2",
        "MySQL",
        "Microsoft SQL Server",
        "Oracle",
        "PostgreSQL",
        "Sybase"
    );

public static final List<String> supportedDatabaseProductsForFunctions = Arrays.asList(
        "MySQL",
        "Microsoft SQL Server",
        "Oracle",
        "PostgreSQL"
    );

static public CallMetaDataProvider createMetaDataProvider(DataSource dataSource, final CallMetaDataContext context) {
    try {
        CallMetaDataProvider result = (CallMetaDataProvider) JdbcUtils.extractDatabaseMetaData(dataSource, databaseMetaData -> {
            String databaseProductName = JdbcUtils.commonDatabaseName(databaseMetaData.getDatabaseProductName());
            boolean accessProcedureColumnMetaData = context.isAccessCallParameterMetaData();
            if (context.isFunction()) {
                if (!supportedDatabaseProductsForFunctions.contains(databaseProductName)) {
                    if (logger.isWarnEnabled()) {
                        logger.warn(databaseProductName + " is not one of the databases fully supported for function calls " +
                                "-- supported are: " + supportedDatabaseProductsForFunctions);
                    }
                    if (accessProcedureColumnMetaData) {
                        logger.warn("Metadata processing disabled - you must specify all parameters explicitly");
                        accessProcedureColumnMetaData = false;
                    }
                }
            }
            else {
                if (!supportedDatabaseProductsForProcedures.contains(databaseProductName)) {
                    if (logger.isWarnEnabled()) {
                        logger.warn(databaseProductName + " is not one of the databases fully supported for procedure calls " +
                                "-- supported are: " + supportedDatabaseProductsForProcedures);
                    }
                    if (accessProcedureColumnMetaData) {
                        logger.warn("Metadata processing disabled - you must specify all parameters explicitly");
```

```java
                accessProcedureColumnMetaData = false;
            }
        }
    }
    CallMetaDataProvider provider;
    if ("Oracle".equals(databaseProductName)) {
        provider = new OracleCallMetaDataProvider(databaseMetaData);
    }
    else if ("DB2".equals(databaseProductName)) {
        provider = new Db2CallMetaDataProvider((databaseMetaData));
    }
    else if ("Apache Derby".equals(databaseProductName)) {
        provider = new DerbyCallMetaDataProvider((databaseMetaData));
    }
    else if ("PostgreSQL".equals(databaseProductName)) {
        provider = new PostgresCallMetaDataProvider((databaseMetaData));
    }
    else if ("Sybase".equals(databaseProductName)) {
        provider = new SybaseCallMetaDataProvider((databaseMetaData));
    }
    else if ("Microsoft SQL Server".equals(databaseProductName)) {
        provider = new SqlServerCallMetaDataProvider((databaseMetaData));
    }
    else if ("HDB".equals(databaseProductName)) {
        provider = new HanaCallMetaDataProvider((databaseMetaData));
    }
    else {
        provider = new GenericCallMetaDataProvider(databaseMetaData);
    }
    if (logger.isDebugEnabled()) {
        logger.debug("Using " + provider.getClass().getName());
    }
    provider.initializeWithMetaData(databaseMetaData);
    if (accessProcedureColumnMetaData) {
        provider.initializeWithProcedureColumnMetaData(databaseMetaData,
            context.getCatalogName(), context.getSchemaName(), context.getProcedureName());
    }
    return provider;
    });
    return result;
}
catch (MetaDataAccessException ex) {
    throw new DataAccessResourceFailureException("Error retrieving database metadata", ex);
}
}
```

TableMetaDataProviderFactory 创建 TableMetaDataProvider 工厂类，其创建过程如下：

```java
static public CallMetaDataProvider createMetaDataProvider(DataSource dataSource, final
CallMetaDataContext context) {
    try {
        CallMetaDataProvider result = (CallMetaDataProvider) JdbcUtils.extractDatabaseMetaData
(dataSource, databaseMetaData -> {
            String databaseProductName = JdbcUtils.commonDatabaseName(databaseMetaData.
getDatabaseProductName());
            boolean accessProcedureColumnMetaData = context.isAccessCallParameterMetaData();
            if (context.isFunction()) {
                if (!supportedDatabaseProductsForFunctions.contains(databaseProductName)) {
                    if (logger.isWarnEnabled()) {
                        logger.warn(databaseProductName + " is not one of the databases fully supported
                            for function calls " +
                            "-- supported are: " + supportedDatabaseProductsForFunctions);
                    }
                    if (accessProcedureColumnMetaData) {
                        logger.warn("Metadata processing disabled - you must specify all parameters
explicitly");
                        accessProcedureColumnMetaData = false;
                    }
                }
            }
            else {
                if (!supportedDatabaseProductsForProcedures.contains(databaseProductName)) {
                    if (logger.isWarnEnabled()) {
                        logger.warn(databaseProductName + " is not one of the databases fully supported
                            for procedure calls " +
                            "-- supported are: " + supportedDatabaseProductsForProcedures);
                    }
                    if (accessProcedureColumnMetaData) {
                        logger.warn("Metadata processing disabled - you must specify all parameters
explicitly");
                        accessProcedureColumnMetaData = false;
                    }
                }
            }
            CallMetaDataProvider provider;
            if ("Oracle".equals(databaseProductName)) {
                provider = new OracleCallMetaDataProvider(databaseMetaData);
            }
            else if ("DB2".equals(databaseProductName)) {
                provider = new Db2CallMetaDataProvider((databaseMetaData));
            }
            else if ("Apache Derby".equals(databaseProductName)) {
```

```java
        provider = new DerbyCallMetaDataProvider((databaseMetaData));
    }
    else if ("PostgreSQL".equals(databaseProductName)) {
        provider = new PostgresCallMetaDataProvider((databaseMetaData));
    }
    else if ("Sybase".equals(databaseProductName)) {
        provider = new SybaseCallMetaDataProvider((databaseMetaData));
    }
    else if ("Microsoft SQL Server".equals(databaseProductName)) {
        provider = new SqlServerCallMetaDataProvider((databaseMetaData));
    }
    else if ("HDB".equals(databaseProductName)) {
        provider = new HanaCallMetaDataProvider((databaseMetaData));
    }
    else {
        provider = new GenericCallMetaDataProvider(databaseMetaData);
    }
    if (logger.isDebugEnabled()) {
        logger.debug("Using " + provider.getClass().getName());
    }
    provider.initializeWithMetaData(databaseMetaData);
    if (accessProcedureColumnMetaData) {
        provider.initializeWithProcedureColumnMetaData(databaseMetaData,
            context.getCatalogName(), context.getSchemaName(), context.getProcedureName());
    }
    return provider;
    });
    return result;
}
catch (MetaDataAccessException ex) {
    throw new DataAccessResourceFailureException("Error retrieving database metadata", ex);
}
}
```

有时候使用 Map 来指定参数值非常好，但是这并不是最简单的方式。Spring 提供了一些其他的 SqlParameterSource 实现类来指定参数值。我们可以看看 BeanPropertySqlParameterSource 类，这是一个非常简便的指定参数的实现类，只要你有一个符合 JavaBean 规范的类就行了，它将使用其中的 getter()方法来获取参数值。

SqlParameter 封装了定义 sql 参数的对象。CallableStateMentCallback、PrePareStateMentCallback、StateMentCallback、ConnectionCallback 回调类分别对应 JdbcTemplate 中的不同处理方法，如下图所示。

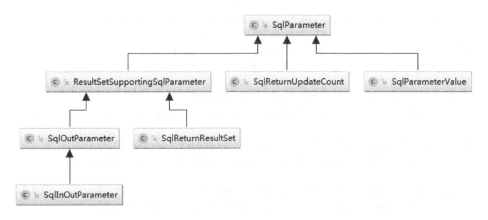

Simple 的实现如下图所示。

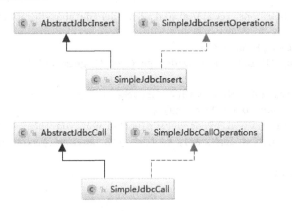

18.4 DataSource

Spring 通过 DataSource 获取数据库的连接。Datasource 是 JDBC 规范的一部分，它通过 ConnectionFactory 获取。一个容器和框架可以在应用代码层隐藏连接池和事务管理。

当使用 Spring 的 JDBC 层时，可以通过 JNDI 来获取 DataSource，也可以通过自己配置的第三方连接池实现来获取。流行的第三方连接池实现有 Apache Jakarta Commons dbcp 和 c3p0。

TransactionAwareDataSourceProxy 作为目标 DataSource 的一个代理，在对目标 DataSource 进行包装的同时，还增加了 Spring 的事务管理能力。在这一点上，这个类的功能非常像 J2EE 服务器所提供的事务化的 JNDI DataSource，如下图所示。

第 18 章 Spring JDBC 源码初探

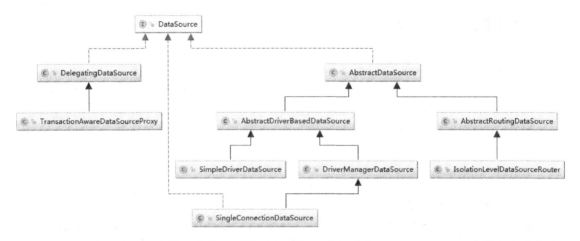

很少用到类，除非现有代码在被调用的时候需要一个标准的 JDBC DataSource 接口实现作为参数。在这种情况下，这个类可以使现有代码参与 Spring 的事务管理。通常最好的做法是使用更高层的抽象来对数据源进行管理，比如 JdbcTemplate 和 DataSourceUtils 等。

注意：DriverManagerDataSource 仅限于测试使用，因为它没有提供连接池的功能，这会导致在多个请求获取连接时性能很差。

18.5 object 模块

object 模块如下图所示。

18.6　JdbcTemplate

JdbcTemplate 类是 core 包的核心类。它替我们完成了资源的创建及释放工作，从而简化了 JDBC 的使用。它还可以帮助我们避免一些常见的错误，比如忘记关闭数据库连接。JdbcTemplate 将完成 JDBC 核心处理流程，比如 SQL 语句的创建、执行，但它会把 SQL 语句的生成及查询结果的提取工作留给我们的应用代码。它可以完成 SQL 查询、更新及调用存储过程，可以对 ResultSet 进行遍历并加以提取。它还可以捕获 JDBC 异常，并将其转换成 org.springframework.dao 包中定义的、通用的、信息更丰富的异常。

使用 JdbcTemplate 进行编码只需要根据明确定义的一组契约来实现回调接口。PreparedStatementCreator 回调接口通过给定的 Connection 创建一个 PreparedStatement，包含 SQL 和任何相关的参数。CallableStatementCreateor 实现同样的处理，只不过它创建的是 CallableStatement。RowCallbackHandler 接口则从数据集的每一行中提取值。

我们可以在 DAO 实现类中通过传递一个 DataSource 引用来完成 JdbcTemplate 的实例化，也可以在 Spring 的 IoC 容器中配置一个 JdbcTemplate 的 Bean 并赋予 DAO 实现类作为一个实例。需要注意的是，DataSource 在 Spring 的 IoC 容器中总是配制成一个 Bean，在第一种情况下 DataSource Bean 将传递给 Service，在第二种情况下 DataSource Bean 将传递给 JdbcTemplate Bean。

18.7　NamedParameterJdbcTemplate

NamedParameterJdbcTemplate 类为 JDBC 操作增加了命名参数的特性支持，而不是使用 "?" 作为参数的占位符。NamedParameterJdbcTemplate 类对 JdbcTemplate 类进行了封装，在底层 JdbcTemplate 完成了多数工作。

第 19 章
基于 Spring JDBC 手写 ORM 框架

19.1 实现思路概述

19.1.1 从 ResultSet 说起

说到 ResultSet，有 Java 开发经验的"小伙伴"自然最熟悉不过了，不过我相信对于大多数人来说也算是"最熟悉的陌生人"。从 ResultSet 取值操作大家都会，比如：

```
private static List<Member> select(String sql) {
    List<Member> result = new ArrayList<>();
    Connection con = null;
    PreparedStatement pstm = null;
    ResultSet rs = null;
    try {
        //1. 加载驱动类
        Class.forName("com.mysql.jdbc.Driver");
        //2. 建立连接
        con = DriverManager.getConnection("jdbc:mysql://127.0.0.1:3306/gp-vip-spring-db-demo",
```

```
    "root","123456");
        //3.创建语句集
        pstm = con.prepareStatement(sql);
        //4.执行语句集
        rs = pstm.executeQuery();
        while (rs.next()){
            Member instance = new Member();
            instance.setId(rs.getLong("id"));
            instance.setName(rs.getString("name"));
            instance.setAge(rs.getInt("age"));
            instance.setAddr(rs.getString("addr"));
            result.add(instance);
        }
        //5.获取结果集
    }catch (Exception e){
        e.printStackTrace();
    }
    //6.关闭结果集、关闭语句集、关闭连接
    finally {
        try {
            rs.close();
            pstm.close();
            con.close();
        }catch (Exception e){
            e.printStackTrace();
        }
    }
    return result;
}
```

以上我们在没有使用框架以前的常规操作。随着业务和开发量的增加，在数据持久层这样的重复代码出现频次非常高。因此，我们就想到将非功能性代码和业务代码进行分离。我们首先想到将 ResultSet 封装数据的代码逻辑分离，增加一个 mapperRow()方法，专门处理对结果的封装，代码如下：

```
private static List<Member> select(String sql) {
    List<Member> result = new ArrayList<>();
    Connection con = null;
    PreparedStatement pstm = null;
    ResultSet rs = null;
    try {
        //1.加载驱动类
        Class.forName("com.mysql.jdbc.Driver");
        //2.建立连接
        con = DriverManager.getConnection("jdbc:mysql://127.0.0.1:3306/gp-vip-spring-db-demo",
"root","123456");
```

```
            //3. 创建语句集
            pstm = con.prepareStatement(sql);
            //4. 执行语句集
            rs = pstm.executeQuery();
            while (rs.next()){
                Member instance = mapperRow(rs,rs.getRow());
                result.add(instance);
            }
            //5. 获取结果集
        }catch (Exception e){
            e.printStackTrace();
        }
        //6. 关闭结果集、关闭语句集、关闭连接
        finally {
            try {
                rs.close();
                pstm.close();
                con.close();
            }catch (Exception e){
                e.printStackTrace();
            }
        }
        return result;
    }

    private static Member mapperRow(ResultSet rs, int i) throws Exception {
        Member instance = new Member();
        instance.setId(rs.getLong("id"));
        instance.setName(rs.getString("name"));
        instance.setAge(rs.getInt("age"));
        instance.setAddr(rs.getString("addr"));
        return instance;
    }
```

但在真实的业务场景中，这样的代码逻辑重复率实在太高，上面的改造只能应用 Member 类，换一个实体类又要重新封装，聪明的程序员肯定不会通过纯体力劳动给每一个实体类写一个 mapperRow() 方法，一定会想到代码复用方案。我们不妨来做这样一个改造。

先创建 Member 类：

```
package com.gupaoedu.vip.orm.demo.entity;

import lombok.Data;

import javax.persistence.Entity;
import javax.persistence.Id;
import javax.persistence.Table;
```

```java
import java.io.Serializable;

@Entity
@Table(name="t_member")
@Data
public class Member implements Serializable {
    @Id private Long id;
    private String name;
    private String addr;
    private Integer age;

    @Override
    public String toString() {
        return "Member{" +
                "id=" + id +
                ", name='" + name + '\'' +
                ", addr='" + addr + '\'' +
                ", age=" + age +
                '}';
    }
}
```

优化 JDBC 操作：

```java
public static void main(String[] args) {
    Member condition = new Member();
    condition.setName("Tom");
    condition.setAge(19);
    List<?> result =  select(condition);
    System.out.println(Arrays.toString(result.toArray()));
}

private static List<?> select(Object condition) {

    List<Object> result = new ArrayList<>();

    Class<?> entityClass = condition.getClass();

    Connection con = null;
    PreparedStatement pstm = null;
    ResultSet rs = null;
    try {
        //1. 加载驱动类
        Class.forName("com.mysql.jdbc.Driver");
        //2. 建立连接
        con = DriverManager.getConnection("jdbc:mysql://127.0.0.1:3306/gp-vip-spring-db-demo?characterEncoding=UTF-8&rewriteBatchedStatements=true","root","123456");
```

```java
//根据类名找属性名
Map<String,String> columnMapper = new HashMap<String,String>();
//根据属性名找字段名
Map<String,String> fieldMapper = new HashMap<String,String>();
Field[] fields = entityClass.getDeclaredFields();
for (Field field : fields) {
    field.setAccessible(true);
    String fieldName = field.getName();
    if(field.isAnnotationPresent(Column.class)){
        Column column = field.getAnnotation(Column.class);
        String columnName = column.name();
        columnMapper.put(columnName,fieldName);
        fieldMapper.put(fieldName,columnName);
    }else {
        //默认就是字段名、属性名一致
        columnMapper.put(fieldName, fieldName);
        fieldMapper.put(fieldName,fieldName);
    }
}

//3. 创建语句集
Table table = entityClass.getAnnotation(Table.class);
String sql = "select * from " + table.name();

StringBuffer where = new StringBuffer(" where 1=1 ");
for (Field field : fields) {
    Object value =field.get(condition);
    if(null != value){
        if(String.class == field.getType()) {
            where.append(" and " + fieldMapper.get(field.getName()) + " = '" + value + "'");
        }else{
            where.append(" and " + fieldMapper.get(field.getName()) + " = " + value + "");
        }
        //其他的在这里就不一一列举，后面我们手写ORM框架时会完善
    }
}
System.out.println(sql + where.toString());
pstm =  con.prepareStatement(sql + where.toString());

//4. 执行语句集
rs = pstm.executeQuery();

//元数据？
//保存了处理真正数值以外的所有附加信息
int columnCounts = rs.getMetaData().getColumnCount();
while (rs.next()){
```

```
            Object instance = entityClass.newInstance();
            for (int i = 1; i <= columnCounts; i++) {
                //实体类属性名,对应数据库表的字段名
                //可以通过反射机制拿到实体类的所有字段

                //从 rs 中取得当前这个游标下的类名
                String columnName = rs.getMetaData().getColumnName(i);
                //有可能是私有的
                Field field = entityClass.getDeclaredField(columnMapper.get(columnName));
                field.setAccessible(true);
                field.set(instance,rs.getObject(columnName));
            }

            result.add(instance);
        }

        //5. 获取结果集
    }catch (Exception e){
        e.printStackTrace();
    }
    //6. 关闭结果集、关闭语句集、关闭连接
    finally {
        try {
            rs.close();
            pstm.close();
            con.close();
        }catch (Exception e){
            e.printStackTrace();
        }
    }

    return result;
}
```

上面巧妙地利用反射机制读取 Class 信息和 Annotation 信息,将数据库表中的列和类中的字段进行关联映射并赋值,以减少重复代码。

19.1.2 为什么需要 ORM 框架

通过前面的讲解,我们已经了解 ORM 框架的基本实现原理。ORM 是指对象关系映射(Object Relation Mapping),映射的不只是对象值,还有对象与对象之间的关系,例如一对多、多对多、一对一这样的表关系。现在市面上 ORM 框架也非常多,有大家所熟知的 Hibernate、Spring JDBC、MyBatis、JPA 等。在这里做一个简单的总结,如下表所示。

名称	特征	描述
Hibernate	全自动（挡）	不需要写一句SQL
MyBatis	半自动（挡）	手自一体，支持简单的映射，复杂关系需要自己写SQL
Spring JDBC	纯手动（挡）	所有的SQL都要自己写，它帮我们设计了一套标准流程

既然市面上有这么多选择，我为什么还要自己写 ORM 框架呢？

这得从我的一次空降担任架构师的经验说起。空降面临最大的难题就是如何取得团队"小伙伴们"的信任。当时，团队总共就 8 人，每个人的水平参差不齐，甚至有些人还没接触过 MySQL，诸如 Redis 等缓存中间件更不用说了。基本只会使用 Hibernate 的 CRUD，而且已经影响到了系统性能。由于工期紧张，没有时间和精力给团队做系统培训，也为了兼顾可控性，于是就产生了自研 ORM 框架的想法。我做了这样的顶层设计，以降低团队"小伙伴们"的存息成本，顶层接口统一参数、统一返回值，具体如下。

（1）规定查询方法的接口模型为：

```
/**
 * 获取列表
 * @param queryRule 查询条件
 * @return
 */
List<T> select(QueryRule queryRule) throws Exception;

/**
 * 获取分页结果
 * @param queryRule 查询条件
 * @param pageNo 页码
 * @param pageSize 每页条数
 * @return
 */
Page<?> select(QueryRule queryRule,int pageNo,int pageSize) throws Exception;

/**
 * 根据 SQL 获取列表
 * @param sql SQL 语句
 * @param args 参数
 * @return
 */
List<Map<String,Object>> selectBySql(String sql, Object... args) throws Exception;

/**
 * 根据 SQL 获取分页
 * @param sql SQL 语句
 * @param pageNo 页码
```

```
 * @param pageSize 每页条数
 * @return
 */
Page<Map<String,Object>> selectBySqlToPage(String sql, Object [] param, int pageNo, int pageSize) throws Exception;
```

（2）规定删除方法的接口模型为：

```
/**
 * 删除一条记录
 * @param entity entity 中的 ID 不能为空，如果 ID 为空，其他条件不能为空，都为空不予执行
 * @return
 */
boolean delete(T entity) throws Exception;

/**
 * 批量删除
 * @param list
 * @return 返回受影响的行数
 * @throws Exception
 */
int deleteAll(List<T> list) throws Exception;
```

（3）规定插入方法的接口模型为：

```
/**
 * 插入一条记录并返回插入后的 ID
 * @param entity 只要 entity 不等于 null，就执行插入
 * @return
 */
PK insertAndReturnId(T entity) throws Exception;

/**
 * 插入一条记录自增 ID
 * @param entity
 * @return
 * @throws Exception
 */
boolean insert(T entity) throws Exception;

/**
 * 批量插入
 * @param list
 * @return 返回受影响的行数
 * @throws Exception
 */
int insertAll(List<T> list) throws Exception;
```

（4）规定修改方法的接口模型为：

```java
/**
 *  修改一条记录
 *  @param entity entity 中的 ID 不能为空，如果 ID 为空，其他条件不能为空，都为空不予执行
 *  @return
 *  @throws Exception
 */
boolean update(T entity) throws Exception;
```

利用这套基础的 API，后面我又基于 Redis、MongoDB、ElasticSearch、Hive、HBase 各封装了一套，以此来降低团队的学习成本，也大大提升了程序的可控性，更方便统一监控。

19.2 搭建基础架构

19.2.1 Page

定义 Page 类的主要目的是为后面的分页查询统一返回结果做顶层支持，其主要功能包括分页逻辑的封装、分页数据。

```java
package javax.core.common;

import java.io.Serializable;
import java.util.ArrayList;
import java.util.List;

/**
 * 分页对象，包含当前页数据及分页信息，如总记录数
 * 能够支持和 JQuery EasyUI 直接对接，能够支持和 BootStrap Table 直接对接
 */
public class Page<T> implements Serializable {

    private static final long serialVersionUID = 1L;
    private static final int DEFAULT_PAGE_SIZE = 20;

    private int pageSize = DEFAULT_PAGE_SIZE; //每页的记录数

    private long start; //当前页第一条数据在 List 中的位置，从 0 开始

    private List<T> rows; //当前页中存放的记录，类型一般为 List

    private long total; //总记录数

    /**
```

```java
 * 构造方法，只构造空页
 */
public Page() {
    this(0, 0, DEFAULT_PAGE_SIZE, new ArrayList<T>());
}

/**
 * 默认构造方法
 *
 * @param start    本页数据在数据库中的起始位置
 * @param totalSize 数据库中总记录条数
 * @param pageSize 本页容量
 * @param rows     本页包含的数据
 */
public Page(long start, long totalSize, int pageSize, List<T> rows) {
    this.pageSize = pageSize;
    this.start = start;
    this.total = totalSize;
    this.rows = rows;
}

/**
 * 取总记录数
 */
public long getTotal() {
    return this.total;
}

public void setTotal(long total) {
    this.total = total;
}

/**
 * 取总页数
 */
public long getTotalPageCount() {
    if (total % pageSize == 0){
        return total / pageSize;
    }else{
        return total / pageSize + 1;
    }
}

/**
 * 取每页数据容量
 */
public int getPageSize() {
```

```java
        return pageSize;
    }

    /**
     * 取当前页中的记录
     */
    public List<T> getRows() {
        return rows;
    }

    public void setRows(List<T> rows) {
        this.rows = rows;
    }

    /**
     * 取该页的当前页码,页码从 1 开始
     */
    public long getPageNo() {
        return start / pageSize + 1;
    }

    /**
     * 该页是否有下一页
     */
    public boolean hasNextPage() {
        return this.getPageNo() < this.getTotalPageCount() - 1;
    }

    /**
     * 该页是否有上一页
     */
    public boolean hasPreviousPage() {
        return this.getPageNo() > 1;
    }

    /**
     * 获取任意一页第一条数据在数据集中的位置,每页条数使用默认值
     *
     * @see #getStartOfPage(int,int)
     */
    protected static int getStartOfPage(int pageNo) {
        return getStartOfPage(pageNo, DEFAULT_PAGE_SIZE);
    }

    /**
     * 获取任意一页第一条数据在数据集中的位置
     *
```

```
 * @param pageNo 从 1 开始的页号
 * @param pageSize 每页记录条数
 * @return 该页第一条数据
 */
public static int getStartOfPage(int pageNo, int pageSize) {
    return (pageNo - 1) * pageSize;
}
}
```

19.2.2 ResultMsg

ResultMsg 类主要是为统一返回结果做的顶层设计，主要包括状态码、结果说明内容和返回数据。

```
package javax.core.common;

import java.io.Serializable;

//底层设计
public class ResultMsg<T> implements Serializable {

    private static final long serialVersionUID = 2635002588308355785L;

    private int status;  //状态码，系统的返回码
    private String msg;    //状态码的解释
    private T data;   //放任意结果

    public ResultMsg() {}

    public ResultMsg(int status) {
        this.status = status;
    }

    public ResultMsg(int status, String msg) {
        this.status = status;
        this.msg = msg;
    }

    public ResultMsg(int status, T data) {
        this.status = status;
        this.data = data;
    }

    public ResultMsg(int status, String msg, T data) {
        this.status = status;
        this.msg = msg;
        this.data = data;
```

```java
}

public int getStatus() {
    return status;
}

public void setStatus(int status) {
    this.status = status;
}

public String getMsg() {
    return msg;
}

public void setMsg(String msg) {
    this.msg = msg;
}

public T getData() {
    return data;
}

public void setData(T data) {
    this.data = data;
}
}
```

19.2.3　BaseDao

作为所有 BaseDao 持久化框架的顶层接口，主要定义增、删、改、查统一的参数列表和返回值。

```java
package javax.core.common.jdbc;

import com.gupaoedu.vip.orm.framework.QueryRule;

import javax.core.common.Page;
import java.util.List;
import java.util.Map;

public interface BaseDao<T,PK> {
    /**
     * 获取列表
     * @param queryRule 查询条件
     * @return
```

```java
 */
List<T> select(QueryRule queryRule) throws Exception;

/**
 * 获取分页结果
 * @param queryRule 查询条件
 * @param pageNo 页码
 * @param pageSize 每页条数
 * @return
 */
Page<?> select(QueryRule queryRule,int pageNo,int pageSize) throws Exception;

/**
 * 根据 SQL 获取列表
 * @param sql SQL 语句
 * @param args 参数
 * @return
 */
List<Map<String,Object>> selectBySql(String sql, Object... args) throws Exception;

/**
 * 根据 SQL 获取分页
 * @param sql SQL 语句
 * @param pageNo 页码
 * @param pageSize 每页条数
 * @return
 */
Page<Map<String,Object>> selectBySqlToPage(String sql, Object [] param, int pageNo, int pageSize) throws Exception;

/**
 * 删除一条记录
 * @param entity entity 中的 ID 不能为空,如果 ID 为空,其他条件不能为空,都为空则不予执行
 * @return
 */
boolean delete(T entity) throws Exception;

/**
 * 批量删除
 * @param list
 * @return 返回受影响的行数
 * @throws Exception
 */
int deleteAll(List<T> list) throws Exception;

/**
 * 插入一条记录并返回插入后的 ID
```

```java
     * @param entity 只要 entity 不等于 null，就执行插入操作
     * @return
     */
    PK insertAndReturnId(T entity) throws Exception;

    /**
     * 插入一条记录自增 ID
     * @param entity
     * @return
     * @throws Exception
     */
    boolean insert(T entity) throws Exception;

    /**
     * 批量插入
     * @param list
     * @return 返回受影响的行数
     * @throws Exception
     */
    int insertAll(List<T> list) throws Exception;

    /**
     * 修改一条记录
     * @param entity entity 中的 ID 不能为空，如果 ID 为空，其他条件不能为空，都为空则不予执行
     * @return
     * @throws Exception
     */
    boolean update(T entity) throws Exception;
}
```

19.2.4 QueryRule

如果用 QueryRule 类来构建查询条件，用户在做条件查询时不需要手写 SQL，实现业务代码与 SQL 解耦。

```java
package com.gupaoedu.vip.orm.framework;

import java.io.Serializable;
import java.util.ArrayList;
import java.util.List;

/**
 * QueryRule，主要功能用于构造查询条件
 */
public final class QueryRule implements Serializable
{
```

```java
private static final long serialVersionUID = 1L;
public static final int ASC_ORDER = 101;
public static final int DESC_ORDER = 102;
public static final int LIKE = 1;
public static final int IN = 2;
public static final int NOTIN = 3;
public static final int BETWEEN = 4;
public static final int EQ = 5;
public static final int NOTEQ = 6;
public static final int GT = 7;
public static final int GE = 8;
public static final int LT = 9;
public static final int LE = 10;
public static final int ISNULL = 11;
public static final int ISNOTNULL = 12;
public static final int ISEMPTY = 13;
public static final int ISNOTEMPTY = 14;
public static final int AND = 201;
public static final int OR = 202;
private List<Rule> ruleList = new ArrayList<Rule>();
private List<QueryRule> queryRuleList = new ArrayList<QueryRule>();
private String propertyName;

private QueryRule() {}

private QueryRule(String propertyName) {
    this.propertyName = propertyName;
}

public static QueryRule getInstance() {
    return new QueryRule();
}

/**
 * 添加升序规则
 * @param propertyName
 * @return
 */
public QueryRule addAscOrder(String propertyName) {
    this.ruleList.add(new Rule(ASC_ORDER, propertyName));
    return this;
}

/**
 * 添加降序规则
 * @param propertyName
 * @return
```

```java
*/
public QueryRule addDescOrder(String propertyName) {
    this.ruleList.add(new Rule(DESC_ORDER, propertyName));
    return this;
}

public QueryRule andIsNull(String propertyName) {
    this.ruleList.add(new Rule(ISNULL, propertyName).setAndOr(AND));
    return this;
}

public QueryRule andIsNotNull(String propertyName) {
    this.ruleList.add(new Rule(ISNOTNULL, propertyName).setAndOr(AND));
    return this;
}

public QueryRule andIsEmpty(String propertyName) {
    this.ruleList.add(new Rule(ISEMPTY, propertyName).setAndOr(AND));
    return this;
}

public QueryRule andIsNotEmpty(String propertyName) {
    this.ruleList.add(new Rule(ISNOTEMPTY, propertyName).setAndOr(AND));
    return this;
}

public QueryRule andLike(String propertyName, Object value) {
    this.ruleList.add(new Rule(LIKE, propertyName, new Object[] { value }).setAndOr(AND));
    return this;
}

public QueryRule andEqual(String propertyName, Object value) {
    this.ruleList.add(new Rule(EQ, propertyName, new Object[] { value }).setAndOr(AND));
    return this;
}

public QueryRule andBetween(String propertyName, Object... values) {
    this.ruleList.add(new Rule(BETWEEN, propertyName, values).setAndOr(AND));
    return this;
}

public QueryRule andIn(String propertyName, List<Object> values) {
    this.ruleList.add(new Rule(IN, propertyName, new Object[] { values }).setAndOr(AND));
    return this;
}

public QueryRule andIn(String propertyName, Object... values) {
```

```java
        this.ruleList.add(new Rule(IN, propertyName, values).setAndOr(AND));
        return this;
    }

    public QueryRule andNotIn(String propertyName, List<Object> values) {
        this.ruleList.add(new Rule(NOTIN, propertyName, new Object[] { values }).setAndOr(AND));
        return this;
    }

    public QueryRule orNotIn(String propertyName, Object... values) {
        this.ruleList.add(new Rule(NOTIN, propertyName, values).setAndOr(OR));
        return this;
    }

    public QueryRule andNotEqual(String propertyName, Object value) {
        this.ruleList.add(new Rule(NOTEQ, propertyName, new Object[] { value }).setAndOr(AND));
        return this;
    }

    public QueryRule andGreaterThan(String propertyName, Object value) {
        this.ruleList.add(new Rule(GT, propertyName, new Object[] { value }).setAndOr(AND));
        return this;
    }

    public QueryRule andGreaterEqual(String propertyName, Object value) {
        this.ruleList.add(new Rule(GE, propertyName, new Object[] { value }).setAndOr(AND));
        return this;
    }

    public QueryRule andLessThan(String propertyName, Object value) {
        this.ruleList.add(new Rule(LT, propertyName, new Object[] { value }).setAndOr(AND));
        return this;
    }

    public QueryRule andLessEqual(String propertyName, Object value) {
        this.ruleList.add(new Rule(LE, propertyName, new Object[] { value }).setAndOr(AND));
        return this;
    }

    public QueryRule orIsNull(String propertyName) {
        this.ruleList.add(new Rule(ISNULL, propertyName).setAndOr(OR));
        return this;
    }

    public QueryRule orIsNotNull(String propertyName) {
```

```java
        this.ruleList.add(new Rule(ISNOTNULL, propertyName).setAndOr(OR));
        return this;
    }

    public QueryRule orIsEmpty(String propertyName) {
        this.ruleList.add(new Rule(ISEMPTY, propertyName).setAndOr(OR));
        return this;
    }

    public QueryRule orIsNotEmpty(String propertyName) {
        this.ruleList.add(new Rule(ISNOTEMPTY, propertyName).setAndOr(OR));
        return this;
    }

    public QueryRule orLike(String propertyName, Object value) {
        this.ruleList.add(new Rule(LIKE, propertyName, new Object[] { value }).setAndOr(OR));
        return this;
    }

    public QueryRule orEqual(String propertyName, Object value) {
        this.ruleList.add(new Rule(EQ, propertyName, new Object[] { value }).setAndOr(OR));
        return this;
    }

    public QueryRule orBetween(String propertyName, Object... values) {
        this.ruleList.add(new Rule(BETWEEN, propertyName, values).setAndOr(OR));
        return this;
    }

    public QueryRule orIn(String propertyName, List<Object> values) {
        this.ruleList.add(new Rule(IN, propertyName, new Object[] { values }).setAndOr(OR));
        return this;
    }

    public QueryRule orIn(String propertyName, Object... values) {
        this.ruleList.add(new Rule(IN, propertyName, values).setAndOr(OR));
        return this;
    }

    public QueryRule orNotEqual(String propertyName, Object value) {
        this.ruleList.add(new Rule(NOTEQ, propertyName, new Object[] { value }).setAndOr(OR));
        return this;
    }

    public QueryRule orGreaterThan(String propertyName, Object value) {
        this.ruleList.add(new Rule(GT, propertyName, new Object[] { value }).setAndOr(OR));
        return this;
```

```java
}

public QueryRule orGreaterEqual(String propertyName, Object value) {
    this.ruleList.add(new Rule(GE, propertyName, new Object[] { value }).setAndOr(OR));
    return this;
}

public QueryRule orLessThan(String propertyName, Object value) {
    this.ruleList.add(new Rule(LT, propertyName, new Object[] { value }).setAndOr(OR));
    return this;
}

public QueryRule orLessEqual(String propertyName, Object value) {
    this.ruleList.add(new Rule(LE, propertyName, new Object[] { value }).setAndOr(OR));
    return this;
}

public List<Rule> getRuleList() {
    return this.ruleList;
}

public List<QueryRule> getQueryRuleList() {
    return this.queryRuleList;
}

public String getPropertyName() {
    return this.propertyName;
}

protected class Rule implements Serializable {
    private static final long serialVersionUID = 1L;
    private int type;  //规则的类型
    private String property_name;
    private Object[] values;
    private int andOr = AND;

    public Rule(int paramInt, String paramString) {
        this.property_name = paramString;
        this.type = paramInt;
    }

    public Rule(int paramInt, String paramString,
            Object[] paramArrayOfObject) {
        this.property_name = paramString;
        this.values = paramArrayOfObject;
        this.type = paramInt;
```

```java
        }

        public Rule setAndOr(int andOr){
            this.andOr = andOr;
            return this;
        }

        public int getAndOr(){
            return this.andOr;
        }

        public Object[] getValues() {
            return this.values;
        }

        public int getType() {
            return this.type;
        }

        public String getPropertyName() {
            return this.property_name;
        }
    }
}
```

19.2.5 Order

Order 类主要用于封装排序规则,代码如下:

```java
package com.gupaoedu.vip.orm.framework;

/**
 * SQL 排序组件
 */
public class Order {
    private boolean ascending; //升序还是降序
    private String propertyName; //哪个字段升序,哪个字段降序

    public String toString() {
        return propertyName + ' ' + (ascending ? "asc" : "desc");
    }

    /**
     * Constructor for Order.
     */
    protected Order(String propertyName, boolean ascending) {
        this.propertyName = propertyName;
```

```java
        this.ascending = ascending;
    }

    /**
     * Ascending order
     *
     * @param propertyName
     * @return Order
     */
    public static Order asc(String propertyName) {
        return new Order(propertyName, true);
    }

    /**
     * Descending order
     *
     * @param propertyName
     * @return Order
     */
    public static Order desc(String propertyName) {
        return new Order(propertyName, false);
    }
}
```

19.3 基于 Spring JDBC 实现关键功能

19.3.1 ClassMappings

ClassMappings 主要定义基础的映射类型，代码如下：

```java
package com.gupaoedu.vip.orm.framework;

import java.lang.reflect.Field;
import java.lang.reflect.Method;
import java.lang.reflect.Modifier;
import java.math.BigDecimal;
import java.sql.Date;
import java.sql.Timestamp;
import java.util.Arrays;
import java.util.HashMap;
import java.util.HashSet;
import java.util.Map;
import java.util.Set;

public class ClassMappings {
```

```java
private ClassMappings(){}

static final Set<Class<?>> SUPPORTED_SQL_OBJECTS = new HashSet<Class<?>>();

    static {
        //只要这里写了，默认支持自动类型转换
        Class<?>[] classes = {
                boolean.class, Boolean.class,
                short.class, Short.class,
                int.class, Integer.class,
                long.class, Long.class,
                float.class, Float.class,
                double.class, Double.class,
                String.class,
                Date.class,
                Timestamp.class,
                BigDecimal.class
        };
        SUPPORTED_SQL_OBJECTS.addAll(Arrays.asList(classes));
    }

    static boolean isSupportedSQLObject(Class<?> clazz) {
        return clazz.isEnum() || SUPPORTED_SQL_OBJECTS.contains(clazz);
    }

    public static Map<String, Method> findPublicGetters(Class<?> clazz) {
        Map<String, Method> map = new HashMap<String, Method>();
        Method[] methods = clazz.getMethods();
        for (Method method : methods) {
            if (Modifier.isStatic(method.getModifiers()))
                continue;
            if (method.getParameterTypes().length != 0)
                continue;
            if (method.getName().equals("getClass"))
                continue;
            Class<?> returnType = method.getReturnType();
            if (void.class.equals(returnType))
                continue;
            if(!isSupportedSQLObject(returnType)){
                continue;
            }
            if ((returnType.equals(boolean.class)
                    || returnType.equals(Boolean.class))
                    && method.getName().startsWith("is")
                    && method.getName().length() > 2) {
                map.put(getGetterName(method), method);
```

```java
            continue;
        }
        if ( ! method.getName().startsWith("get"))
            continue;
        if (method.getName().length() < 4)
            continue;
        map.put(getGetterName(method), method);
    }
    return map;
}

public static Field[] findFields(Class<?> clazz){
    return clazz.getDeclaredFields();
}

public static Map<String, Method> findPublicSetters(Class<?> clazz) {
    Map<String, Method> map = new HashMap<String, Method>();
    Method[] methods = clazz.getMethods();
    for (Method method : methods) {
        if (Modifier.isStatic(method.getModifiers()))
            continue;
        if ( ! void.class.equals(method.getReturnType()))
            continue;
        if (method.getParameterTypes().length != 1)
            continue;
        if ( ! method.getName().startsWith("set"))
            continue;
        if (method.getName().length() < 4)
            continue;
        if(!isSupportedSQLObject(method.getParameterTypes()[0])){
            continue;
        }
        map.put(getSetterName(method), method);
    }
    return map;
}

public static String getGetterName(Method getter) {
    String name = getter.getName();
    if (name.startsWith("is"))
        name = name.substring(2);
    else
        name = name.substring(3);
    return Character.toLowerCase(name.charAt(0)) + name.substring(1);
}

private static String getSetterName(Method setter) {
```

```
        String name = setter.getName().substring(3);
        return Character.toLowerCase(name.charAt(0)) + name.substring(1);
    }
}
```

19.3.2 EntityOperation

EntityOperation 主要实现数据库表结构和对象类结构的映射关系，代码如下：

```java
package com.gupaoedu.vip.orm.framework;

import java.lang.reflect.Field;
import java.lang.reflect.Method;
import java.sql.ResultSet;
import java.sql.ResultSetMetaData;
import java.sql.SQLException;
import java.util.HashMap;
import java.util.Map;
import java.util.TreeMap;
import javax.persistence.Column;
import javax.persistence.Entity;
import javax.persistence.Id;
import javax.persistence.Table;
import javax.persistence.Transient;
import org.apache.log4j.Logger;
import org.springframework.jdbc.core.RowMapper;
import javax.core.common.utils.StringUtils;

/**
 * 实体对象的反射操作
 *
 * @param <T>
 */
public class EntityOperation<T> {
    private Logger log = Logger.getLogger(EntityOperation.class);
    public Class<T> entityClass = null;  // 泛型实体 Class 对象
    public final Map<String, PropertyMapping> mappings;
    public final RowMapper<T> rowMapper;

    public final String tableName;
    public String allColumn = "*";
    public Field pkField;

    public EntityOperation(Class<T> clazz,String pk) throws Exception{
        if(!clazz.isAnnotationPresent(Entity.class)){
            throw new Exception("在" + clazz.getName() + "中没有找到 Entity 注解,不能做 ORM 映射");
```

```java
        }
        this.entityClass = clazz;
        Table table = entityClass.getAnnotation(Table.class);
        if (table != null) {
            this.tableName = table.name();
        } else {
            this.tableName = entityClass.getSimpleName();
        }
        Map<String, Method> getters = ClassMappings.findPublicGetters(entityClass);
        Map<String, Method> setters = ClassMappings.findPublicSetters(entityClass);
        Field[] fields = ClassMappings.findFields(entityClass);
        fillPkFieldAndAllColumn(pk,fields);
        this.mappings = getPropertyMappings(getters, setters, fields);
        this.allColumn = this.mappings.keySet().toString().replace("[", "").replace("]","").replaceAll(" ","");
        this.rowMapper = createRowMapper();
    }

    Map<String, PropertyMapping> getPropertyMappings(Map<String, Method> getters, Map<String, Method> setters, Field[] fields) {
        Map<String, PropertyMapping> mappings = new HashMap<String, PropertyMapping>();
        String name;
        for (Field field : fields) {
            if (field.isAnnotationPresent(Transient.class))
                continue;
            name = field.getName();
            if(name.startsWith("is")){
                name = name.substring(2);
            }
            name = Character.toLowerCase(name.charAt(0)) + name.substring(1);
            Method setter = setters.get(name);
            Method getter = getters.get(name);
            if (setter == null || getter == null){
                continue;
            }
            Column column = field.getAnnotation(Column.class);
            if (column == null) {
                mappings.put(field.getName(), new PropertyMapping(getter, setter, field));
            } else {
                mappings.put(column.name(), new PropertyMapping(getter, setter, field));
            }
        }
        return mappings;
    }

    RowMapper<T> createRowMapper() {
        return new RowMapper<T>() {
```

```java
        public T mapRow(ResultSet rs, int rowNum) throws SQLException {
            try {
                T t = entityClass.newInstance();
                ResultSetMetaData meta = rs.getMetaData();
                int columns = meta.getColumnCount();
                String columnName;
                for (int i = 1; i <= columns; i++) {
                    Object value = rs.getObject(i);
                    columnName = meta.getColumnName(i);
                    fillBeanFieldValue(t,columnName,value);
                }
                return t;
            }catch (Exception e) {
                throw new RuntimeException(e);
            }
        }
    };
}

protected void fillBeanFieldValue(T t, String columnName, Object value) {
    if (value != null) {
        PropertyMapping pm = mappings.get(columnName);
        if (pm != null) {
            try {
              pm.set(t, value);
          } catch (Exception e) {
              e.printStackTrace();
          }
        }
      }
}

private void fillPkFieldAndAllColumn(String pk, Field[] fields) {
    //设定主键
    try {
      if(!StringUtils.isEmpty(pk)){
         pkField = entityClass.getDeclaredField(pk);
         pkField.setAccessible(true);
      }
    } catch (Exception e) {
          log.debug("没找到主键列,主键列名必须与属性名相同");
    }
   for (int i = 0 ; i < fields.length ;i ++) {
      Field f = fields[i];
      if(StringUtils.isEmpty(pk)){
         Id id = f.getAnnotation(Id.class);
         if(id != null){
```

```java
            pkField = f;
            break;
          }
        }
      }
    }
}

public T parse(ResultSet rs) {
    T t = null;
    if (null == rs) {
      return null;
    }
    Object value = null;
    try {
      t = (T) entityClass.newInstance();
      for (String columnName : mappings.keySet()) {
        try {
          value = rs.getObject(columnName);
        } catch (Exception e) {
          e.printStackTrace();
        }
        fillBeanFieldValue(t,columnName,value);
      }
    } catch (Exception ex) {
      ex.printStackTrace();
    }
    return t;
}

public Map<String, Object> parse(T t) {
    Map<String, Object> _map = new TreeMap<String, Object>();
    try {

      for (String columnName : mappings.keySet()) {
        Object value = mappings.get(columnName).getter.invoke(t);
        if (value == null)
          continue;
        _map.put(columnName, value);
      }

    } catch (Exception e) {
      e.printStackTrace();
    }
    return _map;
}

public void println(T t) {
```

```java
        try {
            for (String columnName : mappings.keySet()) {
                Object value = mappings.get(columnName).getter.invoke(t);
                if (value == null)
                    continue;
                System.out.println(columnName + " = " + value);
            }
        } catch (Exception e) {
            e.printStackTrace();
        }
    }
}

class PropertyMapping {

    final boolean insertable;
    final boolean updatable;
    final String columnName;
    final boolean id;
    final Method getter;
    final Method setter;
    final Class enumClass;
    final String fieldName;

    public PropertyMapping(Method getter, Method setter, Field field) {
        this.getter = getter;
        this.setter = setter;
        this.enumClass = getter.getReturnType().isEnum() ? getter.getReturnType() : null;
        Column column = field.getAnnotation(Column.class);
        this.insertable = column == null || column.insertable();
        this.updatable = column == null || column.updatable();
        this.columnName = column == null ? ClassMappings.getGetterName(getter) : ("".equals(column.name()) ? ClassMappings.getGetterName(getter) : column.name());
        this.id = field.isAnnotationPresent(Id.class);
        this.fieldName = field.getName();
    }

    @SuppressWarnings("unchecked")
    Object get(Object target) throws Exception {
        Object r = getter.invoke(target);
        return enumClass == null ? r : Enum.valueOf(enumClass, (String) r);
    }

    @SuppressWarnings("unchecked")
    void set(Object target, Object value) throws Exception {
        if (enumClass != null && value != null) {
            value = Enum.valueOf(enumClass, (String) value);
```

```
            }
            //BeanUtils.setProperty(target, fieldName, value);
            try {
                if(value != null){
                    setter.invoke(target, setter.getParameterTypes()[0].cast(value));
                }
            } catch (Exception e) {
                e.printStackTrace();
                /**
                 * 出错原因如果是 boolean 字段、mysql 字段类型，设置 tinyint(1)
                 */
                System.err.println(fieldName + "--" + value);
            }

        }
}
```

19.3.3　QueryRuleSqlBuilder

QueryRuleSqlBuilder 根据用户构建好的 QueryRule 来自动生成 SQL 语句，代码如下：

```java
package com.gupaoedu.vip.orm.framework;

import java.util.ArrayList;
import java.util.HashMap;
import java.util.List;
import java.util.Map;
import java.util.regex.Matcher;
import java.util.regex.Pattern;
import org.apache.commons.lang.ArrayUtils;
import com.gupaoedu.vip.orm.framework.QueryRule.Rule;
import javax.core.common.utils.StringUtils;


/**
 * 根据 QueryRule 自动构建 SQL 语句
 */
public class QueryRuleSqlBuilder {
    private int CURR_INDEX = 0;  //记录参数所在的位置
    private List<String> properties;  //保存列名列表
    private List<Object> values;  //保存参数值列表
    private List<Order> orders;  //保存排序规则列表

    private String whereSql = "";
    private String orderSql = "";
    private Object [] valueArr = new Object[]{};
    private Map<Object,Object> valueMap = new HashMap<Object,Object>();
```

```java
/**
 * 获得查询条件
 * @return
 */
public String getWhereSql(){
    return this.whereSql;
}

/**
 * 获得排序条件
 * @return
 */
public String getOrderSql(){
    return this.orderSql;
}

/**
 * 获得参数值列表
 * @return
 */
public Object [] getValues(){
    return this.valueArr;
}

/**
 * 获取参数列表
 * @return
 */
public Map<Object,Object> getValueMap(){
    return this.valueMap;
}

/**
 * 创建 SQL 构造器
 * @param queryRule
 */
public QueryRuleSqlBuilder(QueryRule queryRule) {
    CURR_INDEX = 0;
    properties = new ArrayList<String>();
    values = new ArrayList<Object>();
    orders = new ArrayList<Order>();
    for (QueryRule.Rule rule : queryRule.getRuleList()) {
        switch (rule.getType()) {
        case QueryRule.BETWEEN:
            processBetween(rule);
            break;
```

```
            case QueryRule.EQ:
                processEqual(rule);
                break;
            case QueryRule.LIKE:
                processLike(rule);
                break;
            case QueryRule.NOTEQ:
                processNotEqual(rule);
                break;
            case QueryRule.GT:
                processGreaterThen(rule);
                break;
            case QueryRule.GE:
                processGreaterEqual(rule);
                break;
            case QueryRule.LT:
                processLessThen(rule);
                break;
            case QueryRule.LE:
                processLessEqual(rule);
                break;
            case QueryRule.IN:
                processIN(rule);
                break;
            case QueryRule.NOTIN:
                processNotIN(rule);
                break;
            case QueryRule.ISNULL:
                processIsNull(rule);
                break;
            case QueryRule.ISNOTNULL:
                processIsNotNull(rule);
                break;
            case QueryRule.ISEMPTY:
                processIsEmpty(rule);
                break;
            case QueryRule.ISNOTEMPTY:
                processIsNotEmpty(rule);
                break;
            case QueryRule.ASC_ORDER:
                processOrder(rule);
                break;
            case QueryRule.DESC_ORDER:
                processOrder(rule);
                break;
            default:
                throw new IllegalArgumentException("type " + rule.getType() + " not supported.");
```

```
        }
    }
    //拼装where语句
    appendWhereSql();
    //拼装排序语句
    appendOrderSql();
    //拼装参数值
    appendValues();
}

/**
 * 去掉order
 *
 * @param sql
 * @return
 */
protected String removeOrders(String sql) {
    Pattern p = Pattern.compile("order\\s*by[\\w|\\W|\\s|\\S]*", Pattern.CASE_INSENSITIVE);
    Matcher m = p.matcher(sql);
    StringBuffer sb = new StringBuffer();
    while (m.find()) {
        m.appendReplacement(sb, "");
    }
    m.appendTail(sb);
    return sb.toString();
}

/**
 * 去掉select
 *
 * @param sql
 * @return
 */
protected String removeSelect(String sql) {
    if(sql.toLowerCase().matches("from\\s+")){
        int beginPos = sql.toLowerCase().indexOf("from");
        return sql.substring(beginPos);
    }else{
        return sql;
    }
}

/**
 * 处理like
 * @param rule
 */
private void processLike(QueryRule.Rule rule) {
```

```java
        if (ArrayUtils.isEmpty(rule.getValues())) {
            return;
        }
        Object obj = rule.getValues()[0];

        if (obj != null) {
            String value = obj.toString();
            if (!StringUtils.isEmpty(value)) {
                value = value.replace('*', '%');
                obj = value;
            }
        }
        add(rule.getAndOr(),rule.getPropertyName(),"like","%"+rule.getValues()[0]+"%");
    }

    /**
     * 处理 between
     * @param rule
     */
    private void processBetween(QueryRule.Rule rule) {
        if ((ArrayUtils.isEmpty(rule.getValues()))
                || (rule.getValues().length < 2)) {
            return;
        }
        add(rule.getAndOr(),rule.getPropertyName(),"","between",rule.getValues()[0],"and");
        add(0,"","","",rule.getValues()[1],"");
    }

    /**
     * 处理 =
     * @param rule
     */
    private void processEqual(QueryRule.Rule rule) {
        if (ArrayUtils.isEmpty(rule.getValues())) {
            return;
        }
        add(rule.getAndOr(),rule.getPropertyName(),"=",rule.getValues()[0]);
    }

    /**
     * 处理 <>
     * @param rule
     */
    private void processNotEqual(QueryRule.Rule rule) {
        if (ArrayUtils.isEmpty(rule.getValues())) {
            return;
        }
```

```java
    add(rule.getAndOr(),rule.getPropertyName(),"<>",rule.getValues()[0]);
}

/**
 * 处理 >
 * @param rule
 */
private void processGreaterThen(
    QueryRule.Rule rule) {
    if (ArrayUtils.isEmpty(rule.getValues())) {
        return;
    }
    add(rule.getAndOr(),rule.getPropertyName(),">",rule.getValues()[0]);
}

/**
 * 处理>=
 * @param rule
 */
private void processGreaterEqual(
    QueryRule.Rule rule) {
    if (ArrayUtils.isEmpty(rule.getValues())) {
        return;
    }
    add(rule.getAndOr(),rule.getPropertyName(),">=",rule.getValues()[0]);
}

/**
 * 处理<
 * @param rule
 */
private void processLessThen(QueryRule.Rule rule) {
    if (ArrayUtils.isEmpty(rule.getValues())) {
        return;
    }
    add(rule.getAndOr(),rule.getPropertyName(),"<",rule.getValues()[0]);
}

/**
 * 处理<=
 * @param rule
 */
private void processLessEqual(
    QueryRule.Rule rule) {
    if (ArrayUtils.isEmpty(rule.getValues())) {
        return;
    }
}
```

```java
        add(rule.getAndOr(),rule.getPropertyName(),"<=",rule.getValues()[0]);
    }

    /**
     * 处理 is null
     * @param rule
     */
    private void processIsNull(QueryRule.Rule rule) {
        add(rule.getAndOr(),rule.getPropertyName(),"is null",null);
    }

    /**
     * 处理 is not null
     * @param rule
     */
    private void processIsNotNull(QueryRule.Rule rule) {
        add(rule.getAndOr(),rule.getPropertyName(),"is not null",null);
    }

    /**
     * 处理 <>''
     * @param rule
     */
    private void processIsNotEmpty(QueryRule.Rule rule) {
        add(rule.getAndOr(),rule.getPropertyName(),"<>","''");
    }

    /**
     * 处理 =''
     * @param rule
     */
    private void processIsEmpty(QueryRule.Rule rule) {
        add(rule.getAndOr(),rule.getPropertyName(),"=","''");
    }

    /**
     * 处理 in 和 not in
     * @param rule
     * @param name
     */
    private void inAndNotIn(QueryRule.Rule rule,String name){
        if (ArrayUtils.isEmpty(rule.getValues())) {
            return;
        }
        if ((rule.getValues().length == 1) && (rule.getValues()[0] != null)
                && (rule.getValues()[0] instanceof List)) {
```

```java
         List<Object> list = (List) rule.getValues()[0];

      if ((list != null) && (list.size() > 0)){
         for (int i = 0; i < list.size(); i++) {
            if(i == 0 && i == list.size() - 1){
               add(rule.getAndOr(),rule.getPropertyName(),"",name + " (",list.get(i),")");
            }else if(i == 0 && i < list.size() - 1){
               add(rule.getAndOr(),rule.getPropertyName(),"",name + " (",list.get(i),"");
            }
            if(i > 0 && i < list.size() - 1){
               add(0,"",",","",list.get(i),"");
            }
            if(i == list.size() - 1 && i != 0){
               add(0,"",",","",list.get(i),")");
            }
         }
      }
   } else {
      Object[] list =  rule.getValues();
      for (int i = 0; i < list.length; i++) {
         if(i == 0 && i == list.length - 1){
            add(rule.getAndOr(),rule.getPropertyName(),"",name + " (",list[i],")");
         }else if(i == 0 && i < list.length - 1){
            add(rule.getAndOr(),rule.getPropertyName(),"",name + " (",list[i],"");
         }
         if(i > 0 && i < list.length - 1){
            add(0,"",",","",list[i],"");
         }
         if(i == list.length - 1 && i != 0){
            add(0,"",",","",list[i],")");
         }
      }
   }
}
/**
 * 处理 not in
 * @param rule
 */
private void processNotIN(QueryRule.Rule rule){
   inAndNotIn(rule,"not in");
}

/**
 * 处理 in
 * @param rule
 */
```

```java
private void processIN(QueryRule.Rule rule) {
    inAndNotIn(rule,"in");
}

/**
 * 处理 order by
 * @param rule 查询规则
 */
private void processOrder(Rule rule) {
    switch (rule.getType()) {
    case QueryRule.ASC_ORDER:
        //propertyName 非空
        if (!StringUtils.isEmpty(rule.getPropertyName())) {
            orders.add(Order.asc(rule.getPropertyName()));
        }
        break;
    case QueryRule.DESC_ORDER:
        //propertyName 非空
        if (!StringUtils.isEmpty(rule.getPropertyName())) {
            orders.add(Order.desc(rule.getPropertyName()));
        }
        break;
    default:
        break;
    }
}

/**
 * 加入 SQL 查询规则队列
 * @param andOr and 或者 or
 * @param key 列名
 * @param split 列名与值之间的间隔
 * @param value 值
 */
private void add(int andOr,String key,String split ,Object value){
    add(andOr,key,split,"",value,"");
}

/**
 * 加入 SQL 查询规则队列
 * @param andOr and 或则 or
 * @param key 列名
 * @param split 列名与值之间的间隔
 * @param prefix 值前缀
 * @param value 值
 * @param suffix 值后缀
```

```java
    */
    private void add(int andOr,String key,String split,String prefix,Object value,String suffix){
        String andOrStr = (0 == andOr ? "" :(QueryRule.AND == andOr ? " and " : " or "));
        properties.add(CURR_INDEX, andOrStr + key + " " + split + prefix + (null != value ? " ? " : " ") + suffix);
        if(null != value){
            values.add(CURR_INDEX,value);
            CURR_INDEX ++;
        }
    }


    /**
     * 拼装 where 语句
     */
    private void appendWhereSql(){
        StringBuffer whereSql = new StringBuffer();
        for (String p : properties) {
            whereSql.append(p);
        }
        this.whereSql = removeSelect(removeOrders(whereSql.toString()));
    }

    /**
     * 拼装排序语句
     */
    private void appendOrderSql(){
        StringBuffer orderSql = new StringBuffer();
        for (int i = 0 ; i < orders.size(); i ++) {
            if(i > 0 && i < orders.size()){
                orderSql.append(",");
            }
            orderSql.append(orders.get(i).toString());
        }
        this.orderSql = removeSelect(removeOrders(orderSql.toString()));
    }

    /**
     * 拼装参数值
     */
    private void appendValues(){
        Object [] val = new Object[values.size()];
        for (int i = 0; i < values.size(); i ++) {
            val[i] = values.get(i);
            valueMap.put(i, values.get(i));
        }
```

```
      this.valueArr = val;
  }
}
```

19.3.4　BaseDaoSupport

BaseDaoSupport 主要是对 JdbcTemplate 的包装，下面讲一下其重要代码，全部代码请"小伙伴们"自行下载。

先看全局定义：

```
package com.gupaoedu.vip.orm.framework;

...

/**
 * BaseDao 扩展类，主要功能是支持自动拼装 SQL 语句，必须继承方可使用
 * @author Tom
 */
public abstract class BaseDaoSupport<T extends Serializable, PK extends Serializable>
implements BaseDao<T,PK> {
  private Logger log = Logger.getLogger(BaseDaoSupport.class);

  private String tableName = "";

  private JdbcTemplate jdbcTemplateWrite;
  private JdbcTemplate jdbcTemplateReadOnly;

  private DataSource dataSourceReadOnly;
  private DataSource dataSourceWrite;

  private EntityOperation<T> op;

  @SuppressWarnings("unchecked")
  protected BaseDaoSupport(){
    try{
      Class<T> entityClass = GenericsUtils.getSuperClassGenricType(getClass(), 0);
      op = new EntityOperation<T>(entityClass,this.getPKColumn());
      this.setTableName(op.tableName);
    }catch(Exception e){
      e.printStackTrace();
    }
  }

  protected String getTableName() { return tableName; }
  protected DataSource getDataSourceReadOnly() { return dataSourceReadOnly; }
```

```java
    protected DataSource getDataSourceWrite() { return dataSourceWrite; }

    /**
     * 动态切换表名
     */
    protected void setTableName(String tableName) {
        if(StringUtils.isEmpty(tableName)){
            this.tableName = op.tableName;
        }else{
            this.tableName = tableName;
        }
    }

    protected void setDataSourceWrite(DataSource dataSourceWrite) {
        this.dataSourceWrite = dataSourceWrite;
        jdbcTemplateWrite = new JdbcTemplate(dataSourceWrite);
    }

    protected void setDataSourceReadOnly(DataSource dataSourceReadOnly) {
        this.dataSourceReadOnly = dataSourceReadOnly;
        jdbcTemplateReadOnly = new JdbcTemplate(dataSourceReadOnly);
    }

    private JdbcTemplate jdbcTemplateReadOnly() {
        return this.jdbcTemplateReadOnly;
    }

    private JdbcTemplate jdbcTemplateWrite() {
        return this.jdbcTemplateWrite;
    }

    /**
     * 还原默认表名
     */
    protected void restoreTableName(){ this.setTableName(op.tableName); }

    /**
     * 获取主键列名称，建议子类重写
     * @return
     */
    protected abstract String getPKColumn();

    protected abstract void setDataSource(DataSource dataSource);

//此处有省略
}
```

为了照顾程序员的一般使用习惯，查询方法的前缀命名主要有 select、get、load，兼顾 Hibernate 和 MyBatis 的命名风格。

```java
/**
 * 查询函数，使用查询规则
 * 例如以下代码查询条件为匹配的数据
 *
 * @param queryRule 查询规则
 * @return 查询的结果 List
 */
public List<T> select(QueryRule queryRule) throws Exception{
    QueryRuleSqlBuilder bulider = new QueryRuleSqlBuilder(queryRule);
    String ws = removeFirstAnd(bulider.getWhereSql());
    String whereSql = ("".equals(ws) ? ws : (" where " + ws));
    String sql = "select " + op.allColumn + " from " + getTableName() + whereSql;
    Object [] values = bulider.getValues();
    String orderSql = bulider.getOrderSql();
    orderSql = (StringUtils.isEmpty(orderSql) ? "" : (" order by " + orderSql));
    sql += orderSql;
    log.debug(sql);
    return (List<T>) this.jdbcTemplateReadOnly().query(sql, this.op.rowMapper, values);
}

...

/**
 * 根据 SQL 语句执行查询，参数为 Object 数组对象
 * @param sql 查询语句
 * @param args 为 Object 数组
 * @return 符合条件的所有对象
 */
public List<Map<String,Object>> selectBySql(String sql,Object... args) throws Exception{
    return this.jdbcTemplateReadOnly().queryForList(sql,args);
}

...

/**
 * 分页查询函数，使用查询规则<br>
 * 例如以下代码查询条件为匹配的数据
 *
 * @param queryRule 查询规则
 * @param pageNo 页号，从 1 开始
 * @param pageSize 每页的记录条数
 * @return 查询的结果 Page
 */
```

```java
    public Page<T> select(QueryRule queryRule,final int pageNo, final int pageSize) throws Exception{
        QueryRuleSqlBuilder bulider = new QueryRuleSqlBuilder(queryRule);
        Object [] values = bulider.getValues();
        String ws = removeFirstAnd(bulider.getWhereSql());
        String whereSql = ("".equals(ws) ? ws : (" where " + ws));
        String countSql = "select count(1) from " + getTableName() + whereSql;
        long count = (Long) this.jdbcTemplateReadOnly().queryForMap(countSql, values).get("count(1)");
        if (count == 0) {
            return new Page<T>();
        }
        long start = (pageNo - 1) * pageSize;
        //在有数据的情况下,继续查询
        String orderSql = bulider.getOrderSql();
        orderSql = (StringUtils.isEmpty(orderSql) ? " " : (" order by " + orderSql));
        String sql = "select " + op.allColumn +" from " + getTableName() + whereSql + orderSql + " limit " + start + "," + pageSize;
        List<T> list = (List<T>) this.jdbcTemplateReadOnly().query(sql, this.op.rowMapper, values);
        log.debug(sql);
        return new Page<T>(start, count, pageSize, list);
    }
...

    /**
     * 分页查询特殊 SQL 语句
     * @param sql 语句
     * @param param 查询条件
     * @param pageNo 页码
     * @param pageSize 每页内容
     * @return
     */
    public Page<Map<String,Object>> selectBySqlToPage(String sql, Object [] param, final int pageNo, final int pageSize) throws Exception {
        String countSql = "select count(1) from (" + sql + ") a";

        long count = (Long) this.jdbcTemplateReadOnly().queryForMap(countSql,param).get("count(1)");
        if (count == 0) {
            return new Page<Map<String,Object>>();
        }
        long start = (pageNo - 1) * pageSize;
        sql = sql + " limit " + start + "," + pageSize;
        List<Map<String,Object>> list = (List<Map<String,Object>>) this.jdbcTemplateReadOnly().queryForList(sql, param);
        log.debug(sql);
        return new Page<Map<String,Object>>(start, count, pageSize, list);
```

```java
}
/**
 * 获取默认的实例对象
 * @param <T>
 * @param pkValue
 * @param rowMapper
 * @return
 */
private <T> T doLoad(Object pkValue, RowMapper<T> rowMapper){
    Object obj = this.doLoad(getTableName(), getPKColumn(), pkValue, rowMapper);
    if(obj != null){
        return (T)obj;
    }
    return null;
}
```

插入方法，均以 insert 开头：

```java
/**
 * 插入并返回 ID
 * @param entity
 * @return
 */
public PK insertAndReturnId(T entity) throws Exception{
    return (PK)this.doInsertRuturnKey(parse(entity));
}


/**
 * 插入一条记录
 * @param entity
 * @return
 */
public boolean insert(T entity) throws Exception{
    return this.doInsert(parse(entity));
}
/**
 * 批量保存对象.<br>
 *
 * @param list 待保存的对象 List
 * @throws InvocationTargetException
 * @throws IllegalArgumentException
 * @throws IllegalAccessException
 */
public int insertAll(List<T> list) throws Exception {
    int count = 0 ,len = list.size(),step = 50000;
    Map<String, PropertyMapping> pm = op.mappings;
```

```java
        int maxPage = (len % step == 0) ? (len / step) : (len / step + 1);
        for (int i = 1; i <= maxPage; i ++) {
            Page<T> page = pagination(list, i, step);
            String sql = "insert into " + getTableName() + "(" + op.allColumn + ") values ";// ("
+ valstr.toString() + ")";
            StringBuffer valstr = new StringBuffer();
            Object[] values = new Object[pm.size() * page.getRows().size()];
            for (int j = 0; j < page.getRows().size(); j ++) {
                if(j > 0 && j < page.getRows().size()){ valstr.append(","); }
                valstr.append("(");
                int k = 0;
                for (PropertyMapping p : pm.values()) {
                    values[(j * pm.size()) + k] = p.getter.invoke(page.getRows().get(j));
                    if(k > 0 && k < pm.size()){ valstr.append(","); }
                    valstr.append("?");
                    k ++;
                }
                valstr.append(")");
            }
            int result = jdbcTemplateWrite().update(sql + valstr.toString(), values);
            count += result;
        }

        return count;
    }

private Serializable doInsertRuturnKey(Map<String,Object> params){
        final List<Object> values = new ArrayList<Object>();
        final String sql = makeSimpleInsertSql(getTableName(),params,values);
        KeyHolder keyHolder = new GeneratedKeyHolder();
        final JdbcTemplate jdbcTemplate = new JdbcTemplate(getDataSourceWrite());
        try {

            jdbcTemplate.update(new PreparedStatementCreator() {
                public PreparedStatement createPreparedStatement(

                    Connection con) throws SQLException {
                    PreparedStatement ps = con.prepareStatement(sql,Statement.RETURN_GENERATED_KEYS);

                    for (int i = 0; i < values.size(); i++) {
                      ps.setObject(i+1, values.get(i)==null?null:values.get(i));

                    }
                    return ps;
                }
```

```java
      }, keyHolder);
    } catch (DataAccessException e) {
      log.error("error",e);
    }



    if (keyHolder == null) { return ""; }



    Map<String, Object> keys = keyHolder.getKeys();
    if (keys == null || keys.size() == 0 || keys.values().size() == 0) {
      return "";
    }
    Object key = keys.values().toArray()[0];
    if (key == null || !(key instanceof Serializable)) {
      return "";
    }
    if (key instanceof Number) {
      //Long k = (Long) key;
      Class clazz = key.getClass();
//      return clazz.cast(key);
      return (clazz == int.class || clazz == Integer.class) ? ((Number) key).intValue() : ((Number)key).longValue();


    } else if (key instanceof String) {
      return (String) key;
    } else {
      return (Serializable) key;
    }


  }
/**
 * 插入
 * @param params
 * @return
 */
  private boolean doInsert(Map<String, Object> params) {
    String sql = this.makeSimpleInsertSql(this.getTableName(), params);
    int ret = this.jdbcTemplateWrite().update(sql, params.values().toArray());
    return ret > 0;
  }
```

删除方法，均以 delete 开头：

```java
/**
 * 删除对象.<br>
 *
 * @param entity 待删除的实体对象
 */
public boolean delete(T entity) throws Exception {
    return this.doDelete(op.pkField.get(entity)) > 0;
}

/**
 * 删除对象.<br>
 *
 * @param list 待删除的实体对象列表
 * @throws InvocationTargetException
 * @throws IllegalArgumentException
 * @throws IllegalAccessException
 */
public int deleteAll(List<T> list) throws Exception {
    String pkName = op.pkField.getName();
    int count = 0 ,len = list.size(),step = 1000;
    Map<String, PropertyMapping> pm = op.mappings;
    int maxPage = (len % step == 0) ? (len / step) : (len / step + 1);
    for (int i = 1; i <= maxPage; i ++) {
        StringBuffer valstr = new StringBuffer();
        Page<T> page = pagination(list, i, step);
        Object[] values = new Object[page.getRows().size()];

        for (int j = 0; j < page.getRows().size(); j ++) {
            if(j > 0 && j < page.getRows().size()){ valstr.append(","); }
            values[j] = pm.get(pkName).getter.invoke(page.getRows().get(j));
            valstr.append("?");
        }

        String sql = "delete from " + getTableName() + " where " + pkName + " in (" + valstr.toString() + ")";
        int result = jdbcTemplateWrite().update(sql, values);
        count += result;
    }
    return count;
}

/**
 * 根据 id 删除对象。如果有记录则删之，没有记录也不报异常<br>
 * 例如：删除主键唯一的记录
 *
```

```java
 * @param id 序列化id
 */
protected void deleteByPK(PK id) throws Exception {
    this.doDelete(id);
}

/**
 * 删除实例对象,返回删除记录数
 * @param tableName
 * @param pkName
 * @param pkValue
 * @return
 */
private int doDelete(String tableName, String pkName, Object pkValue) {
    StringBuffer sb = new StringBuffer();
    sb.append("delete from ").append(tableName).append(" where ").append(pkName).append(" = ?");
    int ret = this.jdbcTemplateWrite().update(sb.toString(), pkValue);
    return ret;
}
```

修改方法,均以 update 开头:

```java
/**
 * 更新对象.<br>
 *
 * @param entity 待更新对象
 * @throws IllegalAccessException
 * @throws IllegalArgumentException
 */
public boolean update(T entity) throws Exception {
    return this.doUpdate(op.pkField.get(entity), parse(entity)) > 0;
}

/**
 * 更新实例对象,返回删除记录数
 * @param pkValue
 * @param params
 * @return
 */
private int doUpdate(Object pkValue, Map<String, Object> params){
    String sql = this.makeDefaultSimpleUpdateSql(pkValue, params);
    params.put(this.getPKColumn(), pkValue);
    int ret = this.jdbcTemplateWrite().update(sql, params.values().toArray());
    return ret;
}
```

19.4　动态数据源切换的底层原理

这里简单介绍一下 AbstractRoutingDataSource 的基本原理。实现数据源切换的功能就是自定义一个类扩展 AbstractRoutingDataSource 抽象类，其实相当于数据源的路由中介，可以实现在项目运行时根据相应 key 值切换到对应的 DataSource 上。先看看 AbstractRoutingDataSource 类的源码：

```java
public abstract class AbstractRoutingDataSource extends AbstractDataSource implements
InitializingBean {
/*只列出部分代码*/
    @Nullable
    private Map<Object, Object> targetDataSources;
    @Nullable
    private Object defaultTargetDataSource;
    private boolean lenientFallback = true;
    private DataSourceLookup dataSourceLookup = new JndiDataSourceLookup();
    @Nullable
    private Map<Object, DataSource> resolvedDataSources;
    @Nullable
    private DataSource resolvedDefaultDataSource;

    ...

    public Connection getConnection() throws SQLException {
        return this.determineTargetDataSource().getConnection();
    }

    public Connection getConnection(String username, String password) throws SQLException {
        return this.determineTargetDataSource().getConnection(username, password);
    }

    ...

    protected DataSource determineTargetDataSource() {
        Assert.notNull(this.resolvedDataSources, "DataSource router not initialized");
        Object lookupKey = this.determineCurrentLookupKey();
        DataSource dataSource = (DataSource)this.resolvedDataSources.get(lookupKey);
        if(dataSource == null && (this.lenientFallback || lookupKey == null)) {
            dataSource = this.resolvedDefaultDataSource;
        }

        if(dataSource == null) {
            throw new IllegalStateException("Cannot determine target DataSource for lookup key [" + lookupKey + "]");
        } else {
```

```
            return dataSource;
        }
    }

    @Nullable
    protected abstract Object determineCurrentLookupKey();
}
```

可以看出，AbstractRoutingDataSource类继承了AbstractDataSource类，并实现了InitializingBean。AbstractRoutingDataSource类的getConnection()方法调用了determineTargetDataSource()的该方法。这里重点看 determineTargetDataSource()方法的代码，它使用了 determineCurrentLookupKey()方法，它是 AbstractRoutingDataSource 类的抽象方法，也是实现数据源切换扩展的方法。该方法的返回值就是项目中所要用的 DataSource 的 key 值，得到该 key 值后就可以在 resolvedDataSource 中取出对应的 DataSource，如果找不到 key 对应的 DataSource 就使用默认的数据源。

自定义类扩展AbstractRoutingDataSource类时要重写determineCurrentLookupKey()方法来实现数据源切换。

19.4.1 DynamicDataSource

DynamicDataSource 类封装自定义数据源，继承原生 Spring 的 AbstractRoutingDataSource 类的数据源动态路由器。

```
package javax.core.common.jdbc.datasource;

import org.springframework.jdbc.datasource.lookup.AbstractRoutingDataSource;
/**
 * 动态数据源
 */
public class DynamicDataSource extends AbstractRoutingDataSource {
  private DynamicDataSourceEntry dataSourceEntry;
    @Override
  protected Object determineCurrentLookupKey() {
      return this.dataSourceEntry.get();
  }
  public void setDataSourceEntry(DynamicDataSourceEntry dataSourceEntry) {
      this.dataSourceEntry = dataSourceEntry;
  }
  public DynamicDataSourceEntry getDataSourceEntry(){
      return this.dataSourceEntry;
  }
}
```

19.4.2　DynamicDataSourceEntry

DynamicDataSourceEntry 类实现对数据源的操作功能，代码如下：

```java
package javax.core.common.jdbc.datasource;

import org.aspectj.lang.JoinPoint;

/**
 * 动态切换数据源
 */
public class DynamicDataSourceEntry {

    //默认数据源
    public final static String DEFAULT_SOURCE = null;

    private final static ThreadLocal<String> local = new ThreadLocal<String>();

    /**
     * 清空数据源
     */
    public void clear() {
        local.remove();
    }

    /**
     * 获取当前正在使用的数据源的名字
     *
     * @return String
     */
    public String get() {
        return local.get();
    }

    /**
     * 还原指定切面的数据源
     *
     * @param joinPoint
     */
    public void restore(JoinPoint join) {
        local.set(DEFAULT_SOURCE);
    }

    /**
     * 还原当前切面的数据源
```

```java
     */
    public void restore() {
        local.set(DEFAULT_SOURCE);
    }

    /**
     * 设置已知名字的数据源
     *
     * @param dataSource
     */
    public void set(String source) {
        local.set(source);
    }

    /**
     * 根据年份动态设置数据源
     * @param year
     */
    public void set(int year) {
        local.set("DB_" + year);
    }
}
```

19.5 运行效果演示

19.5.1 创建 Member 实体类

创建 Member 实体类代码如下:

```java
package com.gupaoedu.vip.orm.demo.entity;

import lombok.Data;

import javax.persistence.Entity;
import javax.persistence.Id;
import javax.persistence.Table;
import java.io.Serializable;

@Entity
@Table(name="t_member")
@Data
public class Member implements Serializable {
    @Id private Long id;
    private String name;
```

```java
    private String addr;
    private Integer age;

    @Override
    public String toString() {
        return "Member{" +
                "id=" + id +
                ", name='" + name + '\'' +
                ", addr='" + addr + '\'' +
                ", age=" + age +
                '}';
    }
}
```

19.5.2 创建 Order 实体类

创建 Order 实体类代码如下：

```java
package com.gupaoedu.vip.orm.demo.entity;

import lombok.Data;

import javax.persistence.Column;
import javax.persistence.Entity;
import javax.persistence.Table;
import java.io.Serializable;

@Entity
@Table(name="t_order")
@Data
public class Order implements Serializable {
    private Long id;
    @Column(name="mid")
    private Long memberId;
    private String detail;
    private Long createTime;
    private String createTimeFmt;

    @Override
    public String toString() {
        return "Order{" +
                "id=" + id +
                ", memberId=" + memberId +
                ", detail='" + detail + '\'' +
                ", createTime=" + createTime +
```

```
            ", createTimeFmt='" + createTimeFmt + '\'' +
            '}';
    }
}
```

19.5.3　创建 MemberDao

创建 MemberDao 代码如下：

```java
package com.gupaoedu.vip.orm.demo.dao;

import com.gupaoedu.vip.orm.demo.entity.Member;
import com.gupaoedu.vip.orm.framework.BaseDaoSupport;
import com.gupaoedu.vip.orm.framework.QueryRule;
import org.springframework.stereotype.Repository;

import javax.annotation.Resource;
import javax.sql.DataSource;
import java.util.List;

@Repository
public class MemberDao extends BaseDaoSupport<Member,Long> {

    @Override
    protected String getPKColumn() {
        return "id";
    }

    @Resource(name="dataSource")
    public void setDataSource(DataSource dataSource){
        super.setDataSourceReadOnly(dataSource);
        super.setDataSourceWrite(dataSource);
    }


    public List<Member> selectAll() throws Exception{
        QueryRule queryRule = QueryRule.getInstance();
        queryRule.andLike("name","Tom%");
        return super.select(queryRule);
    }
}
```

19.5.4　创建 OrderDao

创建 OrderDao 代码如下：

```java
package com.gupaoedu.vip.orm.demo.dao;

import com.gupaoedu.vip.orm.demo.entity.Order;
import com.gupaoedu.vip.orm.framework.BaseDaoSupport;
import org.springframework.stereotype.Repository;

import javax.annotation.Resource;
import javax.core.common.jdbc.datasource.DynamicDataSource;
import javax.sql.DataSource;
import java.text.SimpleDateFormat;
import java.util.Date;


@Repository
public class OrderDao extends BaseDaoSupport<Order, Long> {

    private SimpleDateFormat yearFormat = new SimpleDateFormat("yyyy");
    private SimpleDateFormat fullDataFormat = new SimpleDateFormat("yyyy-MM-dd HH:mm:ss");
    private DynamicDataSource dataSource;
    @Override
    protected String getPKColumn() {return "id";}

    @Resource(name="dynamicDataSource")
    public void setDataSource(DataSource dataSource) {
        this.dataSource = (DynamicDataSource)dataSource;
        this.setDataSourceReadOnly(dataSource);
        this.setDataSourceWrite(dataSource);
    }

    /**
     * @throws Exception
     *
     */
    public boolean insertOne(Order order) throws Exception{
        //约定优于配置
        Date date = null;
        if(order.getCreateTime() == null){
            date = new Date();
            order.setCreateTime(date.getTime());
        }else {
            date = new Date(order.getCreateTime());
        }
        Integer dbRouter = Integer.valueOf(yearFormat.format(date));
        System.out.println("自动分配到【DB_" + dbRouter + "】数据源");
        this.dataSource.getDataSourceEntry().set(dbRouter);

        order.setCreateTimeFmt(fullDataFormat.format(date));
```

```
        Long orderId = super.insertAndReturnId(order);
        order.setId(orderId);
        return orderId > 0;
    }

}
```

19.5.5 修改 db.properties 文件

修改 db.properties 文件代码如下：

```
#sysbase database mysql config

#mysql.jdbc.driverClassName=com.mysql.jdbc.Driver
#mysql.jdbc.url=jdbc:mysql://127.0.0.1:3306/gp-vip-spring-db-demo?characterEncoding=UTF-8&rewriteBatchedStatements=true
#mysql.jdbc.username=root
#mysql.jdbc.password=123456


db2018.mysql.jdbc.driverClassName=com.mysql.jdbc.Driver
db2018.mysql.jdbc.url=jdbc:mysql://127.0.0.1:3306/gp-vip-spring-db-2018?characterEncoding=UTF-8&rewriteBatchedStatements=true
db2018.mysql.jdbc.username=root
db2018.mysql.jdbc.password=123456

db2019.mysql.jdbc.driverClassName=com.mysql.jdbc.Driver
db2019.mysql.jdbc.url=jdbc:mysql://127.0.0.1:3306/gp-vip-spring-db-2019?characterEncoding=UTF-8&rewriteBatchedStatements=true
db2019.mysql.jdbc.username=root
db2019.mysql.jdbc.password=123456

#alibaba druid config
dbPool.initialSize=1
dbPool.minIdle=1
dbPool.maxActive=200
dbPool.maxWait=60000
dbPool.timeBetweenEvictionRunsMillis=60000
dbPool.minEvictableIdleTimeMillis=300000
dbPool.validationQuery=SELECT 'x'
dbPool.testWhileIdle=true
dbPool.testOnBorrow=false
dbPool.testOnReturn=false
dbPool.poolPreparedStatements=false
dbPool.maxPoolPreparedStatementPerConnectionSize=20
```

```
dbPool.filters=stat,log4j,wall
```

19.5.6　修改 application-db.xml 文件

修改 application-db.xml 文件代码如下：

```xml
<bean id="datasourcePool" abstract="true" class="com.alibaba.druid.pool.DruidDataSource" init-method="init" destroy-method="close">
    <property name="initialSize" value="${dbPool.initialSize}" />
    <property name="minIdle" value="${dbPool.minIdle}" />
    <property name="maxActive" value="${dbPool.maxActive}" />
    <property name="maxWait" value="${dbPool.maxWait}" />
    <property name="timeBetweenEvictionRunsMillis" value="${dbPool.timeBetweenEvictionRunsMillis}" />
    <property name="minEvictableIdleTimeMillis" value="${dbPool.minEvictableIdleTimeMillis}" />
    <property name="validationQuery" value="${dbPool.validationQuery}" />
    <property name="testWhileIdle" value="${dbPool.testWhileIdle}" />
    <property name="testOnBorrow" value="${dbPool.testOnBorrow}" />
    <property name="testOnReturn" value="${dbPool.testOnReturn}" />
    <property name="poolPreparedStatements" value="${dbPool.poolPreparedStatements}" />
    <property name="maxPoolPreparedStatementPerConnectionSize" value="${dbPool.maxPoolPreparedStatementPerConnectionSize}" />
    <property name="filters" value="${dbPool.filters}" />
</bean>

<bean id="dataSource" parent="datasourcePool">
    <property name="driverClassName" value="${db2019.mysql.jdbc.driverClassName}" />
    <property name="url" value="${db2019.mysql.jdbc.url}" />
    <property name="username" value="${db2019.mysql.jdbc.username}" />
    <property name="password" value="${db2019.mysql.jdbc.password}" />
</bean>

<bean id="dataSource2018" parent="datasourcePool">
    <property name="driverClassName" value="${db2018.mysql.jdbc.driverClassName}" />
    <property name="url" value="${db2018.mysql.jdbc.url}" />
    <property name="username" value="${db2018.mysql.jdbc.username}" />
    <property name="password" value="${db2018.mysql.jdbc.password}" />
</bean>

<bean id="dynamicDataSourceEntry" class="javax.core.common.jdbc.datasource.DynamicDataSourceEntry" />

<bean id="dynamicDataSource" class="javax.core.common.jdbc.datasource.DynamicDataSource" >
    <property name="dataSourceEntry" ref="dynamicDataSourceEntry"></property>
    <property name="targetDataSources">
        <map>
            <entry key="DB_2019" value-ref="dataSource"></entry>
            <entry key="DB_2018" value-ref="dataSource2018"></entry>
```

```xml
        </map>
    </property>
    <property name="defaultTargetDataSource" ref="dataSource" />
</bean>
```

19.5.7 编写测试用例

编写测试用例代码如下：

```java
package com.gupaoedu.vip.orm.test;

import com.gupaoedu.vip.orm.demo.dao.MemberDao;
import com.gupaoedu.vip.orm.demo.dao.OrderDao;
import com.gupaoedu.vip.orm.demo.entity.Member;
import com.gupaoedu.vip.orm.demo.entity.Order;
import org.junit.Ignore;
import org.junit.Test;
import org.junit.runner.RunWith;
import org.springframework.beans.factory.annotation.Autowired;
import org.springframework.test.context.ContextConfiguration;
import org.springframework.test.context.junit4.SpringJUnit4ClassRunner;

import java.text.SimpleDateFormat;
import java.util.Arrays;
import java.util.Date;
import java.util.List;

@ContextConfiguration(locations = {"classpath:application-context.xml"})
@RunWith(SpringJUnit4ClassRunner.class)
public class OrmTest {

    private SimpleDateFormat sdf = new SimpleDateFormat("yyyyMMddHHmmdd");

    @Autowired private MemberDao memberDao;

    @Autowired private OrderDao orderDao;

    //ORM（对象关系映射，Object Relation Mapping）
    //Hibernate、Spring JDBC、MyBatis、JPA 中的一对多、多对多、一对一关系

    //为什么有 MyBatis 还要自己手写 ORM 框架呢？
    //1. 用 MyBatis 可控性无法保证
    //2. 我又不敢用 Hibernate，高级玩家玩的
    //3. 没有时间自己从 0 到 1 写一个 ORM 框架
    //4. 站在巨人的肩膀上再升级，做二次开发

    //约定优于配置
```

```
//先制定顶层接口,参数返回值全部统一
//List<?> Page<?>  select(QueryRule queryRule)
//Int delete(T entity) entity 中的 ID 不能为空,如果 ID 为空,其他条件不能为空,都为空不予执行
//ReturnId  insert(T entity) 只要 entity 不等于 null
//Int update(T entity) entity 中的 ID 不能为空,如果 ID 为空,其他条件不能为空,都为空不予执行
//在之后的应用中,我们可以基于这些接口继续扩展,例如:
//基于 JDBC 封装一套
//基于 Redis 封装一套
//基于 MongoDB 封装一套
//基于 ElasticSearch 封装一套
//基于 Hive 封装一套
//基于 HBase 封装一套

@Test
public void testSelectAllForMember(){
    try {
        List<Member> result = memberDao.selectAll();
        System.out.println(Arrays.toString(result.toArray()));
    } catch (Exception e) {
        e.printStackTrace();
    }
}

@Test
@Ignore
public void testInsertMember(){
    try {
        for (int age = 25; age < 35; age++) {
            Member member = new Member();
            member.setAge(age);
            member.setName("Tom");
            member.setAddr("Hunan Changsha");
            memberDao.insert(member);
        }
    }catch (Exception e){
        e.printStackTrace();
    }

}

    @Test
// @Ignore
    public void testInsertOrder(){
        try {
            Order order = new Order();
            order.setMemberId(1L);
```

```
            order.setDetail("历史订单");
            Date date = sdf.parse("20180201123456");
            order.setCreateTime(date.getTime());
            orderDao.insertOne(order);
        }catch (Exception e){
            e.printStackTrace();
        }
    }
}
```

  上面的代码完整地演示了自研 ORM 框架的原理，以及数据源动态切换的基本原理，并且了解了 Spring JdbcTemplate 的 API 应用。希望通过本章的学习，"小伙伴们"在日常工作中能够有更好的解决问题的思路，提高工作效率。

第 6 篇
Spring 经验分享

第 20 章　Spring 5 新特性总结

第 21 章　关于 Spring 的经典高频面试题

第 20 章
Spring 5 新特性总结

Spring 5 于 2017 年 9 月发布了通用版本,它是自 2013 年 12 月以来第一个主要的 Spring 版本。它提供了一些人们期待已久的改进,还采用了一种全新的编程范例,以反应式原则为基础。

这个版本是很长时间以来最令人激动的版本。Spring 5 兼容 Java™8 和 JDK 9,它集成了反应式流,以方便后续提供一种颠覆性方法来实现端点和 Web 应用程序开发。

当然,反应式编程不仅是此版本的主题,还是令许多程序员激动不已的重大特性。人们对能够针对负载波动进行无缝扩展的容灾和响应式服务的需求在不断增加,Spring 5 很好地满足了这一需求。

下面介绍 Java SE 8 和 Java EE 7 API 升级的基本内容、Spring 5 的新反应式编程模型、对 HTTP/2 的支持,以及 Spring 通过 Kotlin 对函数式编程的全面支持。还会简要介绍测试和性能增强,最后介绍对 Spring 核心和容器的一般性修订。

20.1 升级到 Java SE 8 和 Java EE 7

以前的 Spring 一直在支持一些弃用的 Java 版本,而 Spring 5 已从"旧包袱"中解放出来。为了充分利用 Java 8 的特性,它的代码库已进行了改进,而且要求将 Java 8 作为最低的 JDK 版本。

Spring 5 在类路径（和模块路径）上完全兼容 Java 9，而且它通过了 JDK 9 测试套件的测试。对 Java 9 爱好者而言，这是一个好消息。

在 API 级别上，Spring 5 兼容 Java EE 8 技术，满足对 Servlet 4.0、Bean Validation 2.0 和全新的 JSON Binding API 的需求。对 Java EE API 的最低要求为 V7，该版本引入了针对 Servlet、JPA 和 Bean Validation API 的次要版本。

20.2 反应式编程模型

Spring 5 最令人兴奋的新特性是它的反应式编程模型。Spring 5 基于一种反应式基础而构建，而且是完全异步和非阻塞的。只需少量的线程，新的事件循环执行模型就可以垂直扩展。

Spring 5 采用反应式流来提供在反应式组件中传播负压的机制。负压是一个确保来自多个生产者的数据不会让使用者不堪重负的概念。

Spring WebFlux 是 Spring 5 的反应式核心，它为开发人员提供了两种为 Spring Web 编程而设计的编程模型：基于注解的模型和 Functional Web Framework（WebFlux.fn）。

基于注解的模型是 Spring Web MVC 的现代替代方案，该模型基于反应式基础而构建，而 Functional Web Framework 是基于@Controller 注解的编程模型的替代方案。这些模型都通过同一种反应式规则来运行，后者调整非阻塞 HTTP 来适应反应式流 API。

20.3 使用注解进行编程

Web MVC 程序员应该对 Spring 5 的基于注解的编程模型非常熟悉，Spring 5 调整了 Web MVC 的@Controller 编程模型，采用了相同的注解。

在下面的代码中 BookController 类提供了两个方法，分别响应针对某个图书列表的 HTTP 请求，以及针对具有给定 id 的图书的 HTTP 请求。请注意 Mono 和 Flux 等对象。这些对象是实现反应式流规范中的 Publisher 接口的反应式类型，它们的职责是处理数据流。Mono 对象处理一个仅含 1 个元素的流，而 Flux 表示一个包含 N 个元素的流。

```
@RestController
public class BookController {            //反应式控制器

    @GetMapping("/book")
    Flux<Book> list() {
```

```
        return this.repository.findAll();
    }

    @GetMapping("/book/{id}")
    Mono<Book> findById(@PathVariable String id) {
        return this.repository.findOne(id);
    }
}
```

以上是针对 Spring Web 编程的注解,下面我们使用函数式 Web 框架来解决同一个问题。

20.4 函数式编程

Spring 5 的函数式方法将请求委托给处理函数,这些函数接收一个服务器请求实例并返回一种反应式类型。来看一段代码,创建 BookHandler 类,其中 listBooks() 和 getBook() 方法相当于 Controller 中的功能。

```
public class BookHandler {

    public Mono<ServerResponse> listBooks(ServerRequest request) {
        return ServerResponse.ok()
                .contentType(APPLICATION_JSON)
                .body(repository.allPeople(), Book.class);
    }

    public Mono<ServerResponse> getBook(ServerRequest request) {
        return repository.getBook(request.pathVariable("id"))
                .then(book -> ServerResponse.ok()
                        .contentType(APPLICATION_JSON)
                        .body(fromObject(book)))
                .otherwiseIfEmpty(ServerResponse.notFound().build());
    }
}
```

通过路由函数来匹配 HTTP 请求参数与媒体类型,将客户端请求路由到处理函数。下面的代码展示了图书资源端点 URI 将调用委托给合适的处理函数:

```
BookHandler handler = new BookHandler();

RouterFunction<ServerResponse> personRoute =
        route(
                GET("/books/{id}")
                        .and(accept(APPLICATION_JSON)), handler::getBook)
                .andRoute(
```

```
                GET("/books")
                        .and(accept(APPLICATION_JSON)), handler::listBooks);
```

这些示例背后的数据存储也支持完整的反应式体验，该体验是通过 Spring Data 对反应式 Couchbase、Reactive MongoDB 和 Cassandra 的支持来实现的。

20.5　使用 REST 端点执行反应式编程

新的编程模型脱离了传统的 Spring Web MVC 模型，引入了一些很不错的新特性。

举例来说，WebFlux 模块为 RestTemplate 提供了一种完全非阻塞、反应式的替代方案，名为 WebClient。下面创建一个 WebClient，并调用 books 端点来请求一本给定 id 为 1234 的图书。

```
//通过 WebClient 调用 REST 端点
Mono<Book> book = WebClient.create("http://localhost:8080")
    .get()
    .url("/books/{id}", 1234)
    .accept(APPLICATION_JSON)
    .exchange(request)
    .then(response -> response.bodyToMono(Book.class));
```

20.6　支持 HTTP/2

HTTP/2 提高了传输性能，减少了延迟，并提高了应用程序的吞吐量，从而提供了丰富的 Web 体验。

Spring 5 提供专门的 HTTP/2 特性支持，还支持人们期望出现在 JDK 9 中的新 HTTP 客户端。尽管 HTTP/2 的服务器推送功能已通过 Jetty Servlet 引擎的 ServerPushFilter 类向 Spring 开发人员公开很长一段时间了，但如果发现 Spring 5 中开箱即用地提供了 HTTP/2 性能增强，Web 优化者们一定会为此欢呼雀跃。

Spring 5.1 提供 Servlet 4.0，HTTP/2 新特性将由 Tomcat 9.0、Jetty 9.3 和 Undertow 1.4 原生提供。

20.7　Kotlin 和 Spring WebFlux

Kotlin 是一种来自 JetBrains 的面向对象语言，支持函数式编程。它的主要优势之一是与 Java

有非常高的互操作性。通过引入对 Kotlin 的专门支持，Spring 5 全面吸纳了这一优势。它的函数式编程风格与 Spring WebFlux 模块完美匹配，它的新路由 DSL 利用了函数式 Web 框架及干净且符合语言习惯的代码。可以像下面代码中这样简单地表达端点路由：

```
//Kotlin 用于定义端点的路由 DSL
@Bean
fun apiRouter() = router {
    (accept(APPLICATION_JSON) and "/api").nest {
        "/book".nest {
            GET("/", bookHandler::findAll)
            GET("/{id}", bookHandler::findOne)
        }
        "/video".nest {
            GET("/", videoHandler::findAll)
            GET("/{genre}", videoHandler::findByGenre)
        }
    }
}
```

使用 Kotlin 1.1.4 以上版本时，还添加了对 Kotlin 的不可变类的支持（通过带默认值的可选参数），以及对完全支持 null 的 API 的支持。

20.8 使用 Lambda 表达式注册 Bean

作为传统 XML 和 JavaConfig 的替代方案，现在可以使用 Lambda 表达式注册 Spring Bean，使 Bean 可以实际注册为提供者。下面代码中使用 Lambda 表达式注册了一个 Book Bean：

```
GenericApplicationContext context = new GenericApplicationContext();
context.registerBean(Book.class, () -> new
        Book(context.getBean(Author.class))
);
```

20.9 Spring Web MVC 支持最新的 API

全新的 WebFlux 模块提供了许多新的、令人兴奋的功能，但 Spring 5 也迎合了愿意继续使用 Spring MVC 的开发人员的需求。Spring 5 中更新了"模型-视图-控制器"框架，以兼容 WebFlux 和最新版的 Jackson 2.9 和 Protobuf 3.0，甚至包括对新的 Java EE 8 JSON-Binding API 的支持。

除了 HTTP/2 特性的基础服务器实现，Spring Web MVC 还通过 MVC 控制器方法的一个参数来支持 Servlet 4.0 的 PushBuilder。最后，Web MVC 全面支持 Reactor 3.1 的 Flux 和 Mono 对象，

以及 RxJava 1.3 和 RxJava 2.1，它们被视为来自 MVC 控制器方法的返回值。这项支持的最终目的是支持 Spring Data 中新的反应式 WebClient 和反应式存储库。

20.10　使用 JUnit 5 执行条件和并发测试

1. JUnit 和 Spring 5

Spring 5 全面接纳了函数式范例，并支持 JUnit 5 及其新的函数式测试风格。还提供了对 JUnit 4 的向后兼容性，以确保不会破坏旧代码。

Spring 5 的测试套件通过多种方式得到了增强，但最明显的是它对 JUnit 5 的支持。现在可以在单元测试中利用 Java 8 中提供的函数式编程特性。以下代码演示了这一支持：

```
@Test
void givenStreamOfInts_SumShouldBeMoreThanFive() {
    assertTrue(Stream.of(20, 40, 50)
            .stream()
            .mapToInt(i -> i)
            .sum() > 110, () -> "Total should be more than 100");
}
```

2. 迁移到 JUnit 5

如果你对升级到 JUnit 5 持观望态度，Steve Perry 的分两部分的深入剖析教程将说服你进行尝试。

Spring 5 继承了 JUnit 5 在 Spring TestContext Framework 内实现多个扩展 API 的灵活性。举例，开发人员可以使用 JUnit 5 的条件测试执行注解@EnabledIf 和@DisabledIf 来自动计算一个 SpEL（Spring Expression Language）表达式，并适当地启用或禁用测试。借助这些注解，Spring 5 支持以前很难实现的复杂的条件测试方案。Spring TextContext Framework 现在能够并发执行测试。

3. 使用 Spring WebFlux 执行集成测试

Spring Test 现在包含一个 WebTestClient，后者支持对 Spring WebFlux 服务器端点执行集成测试。WebTestClient 使用模拟请求和响应来避免耗尽服务器资源，并能直接绑定到 WebFlux 服务器的基础架构。

WebTestClient 可绑定到真实的服务器，或者使用控制器或函数。在下面的代码中，WebTestClient 被绑定到 localhost：

```
WebTestClient testClient = WebTestClient
```

```
        .bindToServer()
        .baseUrl("http://localhost:8080")
        .build();
```

下面的代码将 WebTestClient 绑定到 RouterFunction：

```
RouterFunction bookRouter = RouterFunctions.route(
        RequestPredicates.GET("/books"),
        request -> ServerResponse.ok().build()
);
WebTestClient
        .bindToRouterFunction(bookRouter)
        .build().get().uri("/books")
        .exchange()
        .expectStatus().isOk()
        .expectBody().isEmpty();
```

20.11 包清理和弃用

Spring 5 终止了对一些过时 API 的支持。遭此厄运的有 Hibernate 3 和 Hibernate 4，为了支持 Hibernate 5，它们遭到了弃用。另外，对 Portlet、Velocity、JasperReports、XMLBeans、JDO 和 Guava 的支持也已终止。

包级别上的清理工作仍在继续。Spring 5 不再支持 beans.factory.access、jdbc.support.nativejdbc、mock.staticmock（来自 spring-aspects 模块）或 web.view.tiles2M。Tiles 3 现在是 Spring 的最低要求。

20.12 Spring 核心和容器的一般更新

Spring 5 改进了扫描和识别组件的方法，使大型项目的性能得到提升。目前，扫描是在编译时执行的，而且向 META-INF/spring.components 文件中的索引文件添加了组件坐标。该索引是通过一个为项目定义的特定于平台的应用程序构建任务来生成的。

标有来自 javax 包的注解的组件会添加到索引中，任何带@Index 注解的类或接口都会添加到索引中。Spring 的传统类路径扫描方式没有被删除，而是保留下来作为一种后备选择。有许多针对大型代码库的明显性能优势，托管许多 Spring 项目的服务器也会缩短启动时间。

Spring 5 还添加了对@Nullable 的支持，后者可用于指示可选的注入点。使用者现在必须准备接受 null 值。此外，还可以使用此注解来标记可以为 null 的参数、字段和返回值。@Nullable 主要用于 IntelliJ IDEA 等 IDE，但也可用于 Eclipse 和 FindBugs，它使得在编译时处理 null 值变得更方便，无须在运行时发送 NullPointerExceptions。

Spring Logging 还提升了性能，自带开箱即用的 Commons Logging 桥接器。现在已通过资源抽象支持防御性编程，为 getFile 访问提供了 isFile 指示器。

20.13 我如何看 Spring 5

Spring 5 的首要特性是新的反应式编程模型，这代表着对提供可无缝扩展、基于 Spring 的响应式服务的重大保障。随着人们对 Spring 5 的采用，反应式编程有望成为使用 Java 语言的 Web 和企业应用程序开发的未来。

未来的 Spring 将继续体现这一承诺，因为 Spring Security、Spring Data 和 Spring Integration 有望采用反应式编程的特征和优势。

总之，Spring 5 代表着一次大受 Spring 开发人员欢迎的华丽转变，同时也为其他框架指出了一条发展之路。Spring 5 的升级也为 Spring Boot、Spring Cloud 提供了非常丰富的经验，Spring 不只是一个框架，已然成了一个编程生态。

第 21 章
关于 Spring 的经典高频面试题

21.1 什么是 Spring 框架，Spring 框架有哪些主要模块

Spring 框架是一个为 Java 应用程序开发提供综合、广泛的基础性支持的 Java 平台。Spring 帮助开发者解决了开发中基础性的问题，使得开发人员可以专注于应用程序的开发。Spring 框架本身也是按照设计模式精心打造的，这使得我们可以在开发环境中安心地集成 Spring 框架，不必担心 Spring 是如何在后台工作的。主要模块内容介绍可以参考之前章节。

21.2 使用 Spring 框架能带来哪些好处

下面列举了一些使用 Spring 框架带来的主要好处。

（1）Dependency Injection（DI）使得构造器和 JavaBean properties 文件中的依赖关系一目了然。

（2）与 EJB 容器相比较，IoC 容器更加趋向于轻量级。这样一来使用 IoC 容器在有限的内存和 CPU 资源的情况下进行应用程序的开发和发布就变得十分有利。

（3）Spring 并没有闭门造车，Spring 利用了已有的技术，比如 ORM 框架、logging 框架、J2EE、Quartz 和 JDK Timer，以及其他视图技术。

（4）Spring 框架是按照模块的形式来组织的。由包和类的编号就可以看出其所属的模块，开发者只需选用需要的模块即可。

（5）要测试一个用 Spring 开发的应用程序十分简单，因为测试相关的环境代码都已经囊括在框架中了。更加简单的是，利用 JavaBean 形式的 POJO 类，可以很方便地利用依赖注入来写入测试数据。

（6）Spring 的 Web 框架也是一个精心设计的 Web MVC 框架，为开发者在 Web 框架的选择上提供了一个除主流框架（比如 Struts）和过度设计的、不流行 Web 框架以外的选择。

（7）Spring 提供了一个便捷的事务管理接口，适用于小型的本地事务处理（比如在单 DB 的环境下）和复杂的共同事务处理（比如利用 JTA 的复杂 DB 环境）。

21.3 什么是控制反转（IoC），什么是依赖注入

（1）控制反转是应用于软件工程领域的，在运行时被装配器对象用来绑定耦合对象的一种编程技巧，对象之间的耦合关系在编译时通常是未知的。在传统的编程方式中，业务逻辑的流程是由应用程序中早已被设定好关联关系的对象来决定的。在使用控制反转的情况下，业务逻辑的流程是由对象关系图来决定的，该对象关系图由装配器负责实例化，这种实现方式还可以将对象之间的关联关系的定义抽象化。绑定的过程是通过"依赖注入"实现的。

（2）控制反转是一种以给予应用程序中目标组件更多控制为目的设计范式，并在实际工作中起到了有效的作用。

（3）依赖注入是在编译阶段尚未知所需的功能是来自哪个的类的情况下，将其他对象所依赖的功能对象实例化的模式。这就需要一种机制来激活相应的组件以提供特定的功能，所以依赖注入是控制反转的基础。否则如果在组件不受框架控制的情况下，框架又怎么知道要创建哪个组件呢？

21.4 在 Java 中依赖注入有哪些方式

（1）构造器注入。

（2）Setter 方法注入。

（3）接口注入。

21.5　BeanFactory 和 ApplicationContext 有什么区别

BeanFactory 可以理解为含有 Bean 集合的工厂类。BeanFactory 包含了 Bean 的定义，以便在接收到客户端请求时将对应的 Bean 实例化。

BeanFactory 还能在实例化对象时生成协作类之间的关系。此举将 Bean 自身从 Bean 客户端的配置中解放出来。BeanFactory 还包含 Bean 生命周期的控制，调用客户端的初始化方法（Initialization Method）和销毁方法（Destruction Method）。

从表面上看，ApplicationContext 如同 BeanFactory 一样具有 Bean 定义、Bean 关联关系的设置及根据请求分发 Bean 的功能。但 ApplicationContext 在此基础上还提供了其他功能。

（1）提供了支持国际化的文本消息。

（2）统一的资源文件读取方式。

（3）已在监听器中注册的 Bean 的事件。

以下是三种较常见的 ApplicationContext 实现方式。

（1）ClassPathXmlApplicationContext：从 ClassPath 的 XML 配置文件中读取上下文，并生成上下文定义。应用程序上下文从程序环境变量中取得。

```
ApplicationContext context = new ClassPathXmlApplicationContext("application.xml");
```

（2）FileSystemXmlApplicationContext：由文件系统中的 XML 配置文件读取上下文。

```
ApplicationContext context = new FileSystemXmlApplicationContext( "application.xml" );
```

（3）XmlWebApplicationContext：由 Web 应用的 XML 文件读取上下文。

21.6　Spring 提供几种配置方式来设置元数据

Spring 提供以下三种配置方式来设置元数据：

（1）基于 XML 的配置。

（2）基于注解的配置。

（3）基于 Java 的配置。

21.7 如何使用 XML 配置方式配置 Spring

在 Spring 框架中，依赖和服务需要专门的配置文件实现，一般用 XML 格式的配置文件。这些配置文件的格式采用公共的模板，由一系列的 Bean 定义和专门的应用配置选项组成。

Spring XML 配置的主要目的是使所有的 Spring 组件都可以用 XML 文件的形式来进行配置。这意味着不会出现其他的 Spring 配置类型（比如声明配置方式或基于 Java Class 的配置方式）。

Spring 的 XML 配置方式是使用被 Spring 命名空间所支持的一系列的 XML 标签来实现的。Spring 主要的命名空间有 context、beans、jdbc、tx、aop、mvc 和 aso。例如：

```xml
<beans>
    <!-- JSON Support -->
    <bean name="viewResolver"
        class="org.springframework.web.servlet.view.BeanNameViewResolver"/>
    <bean name="jsonTemplate"
        class="org.springframework.web.servlet.view.json.MappingJackson2JsonView"/>
    <bean id="restTemplate" class="org.springframework.web.client.RestTemplate"/>
</beans>
```

下面这个 web.xml 仅配置了 DispatcherServlet，最简单的配置便能满足应用程序配置运行时组件的需求。

```xml
<web-app>
    <display-name>Archetype Created Web Application</display-name>
    <servlet>
        <servlet-name>spring</servlet-name>
        <servlet-class>
            org.springframework.web.servlet.DispatcherServlet
        </servlet-class>
        <load-on-startup>1</load-on-startup>
    </servlet>
    <servlet-mapping>
        <servlet-name>spring</servlet-name>
        <url-pattern>/</url-pattern>
    </servlet-mapping>
</web-app>
```

21.8 Spring 提供哪些配置形式

Spring 对 Java 配置的支持是由@Configuration 注解和@Bean 注解来实现的。由@Bean 注解的方法将会实例化、配置和初始化一个新对象，这个对象将由 Spring 的 IoC 容器来管理。@Bean 声明所起到的作用与元素类似。被@Configuration 所注解的类则表示这个类的主要目的是作为 Bean 定义的资源。被@Configuration 声明的类可以通过在同一个类内部调用@bean 方法来设置嵌入 Bean 的依赖关系。

最简单的@Configuration 声明类请参考下面的代码：

```java
@Configuration
public class AppConfig{
    @Bean
    public MyService myService() {
        return new MyServiceImpl();
    }
}
```

与上面的@Beans 配置文件相同的 XML 配置文件如下：

```xml
<beans>
    <bean id="myService" class="com.gupaoedu.services.MyServiceImpl"/>
</beans>
```

上述配置方式的实例化方式如下：

```java
public static void main(String[] args) {
    ApplicationContext ctx = new AnnotationConfigApplicationContext(AppConfig.class);
    MyService myService = ctx.getBean(MyService.class);
    myService.doStuff();
}
```

要使用组件扫描，仅需用@Configuration 进行注解即可：

```java
@Configuration
@ComponentScan(basePackages = "com.gupaoedu")
public class AppConfig  {
}
```

在上面的例子中，com.gupaoedu 包首先会被扫描到，然后在容器内查找被@Component 声明的类，找到后将这些类按照 Spring Bean 定义进行注册。

如果你要在 Web 应用开发中选用上述配置方式，需要用 AnnotationConfigWebApplicationContext 类来读取配置文件，可以用来配置 Spring 的 Servlet 监听器 ContrextLoaderListener 或者 Spring MVC 的 DispatcherServlet。

例如：

```xml
<web-app>
  <context-param>
    <param-name>contextClass</param-name>
    <param-value>
      org.springframework.web.context.support.AnnotationConfigWebApplicationContext
    </param-value>
  </context-param>
  <context-param>
    <param-name>contextConfigLocation</param-name>
    <param-value>com.gupaoedu.AppConfig</param-value>
  </context-param>
  <listener>
    <listener-class>org.springframework.web.context.ContextLoaderListener</listener-class>
  </listener>
  <servlet>
    <servlet-name>dispatcher</servlet-name>
    <servlet-class>org.springframework.web.servlet.DispatcherServlet</servlet-class>
    <init-param>
      <param-name>contextClass</param-name>
      <param-value>
        org.springframework.web.context.support.AnnotationConfigWebApplicationContext
      </param-value>
    </init-param>
    <init-param>
      <param-name>contextConfigLocation</param-name>
      <param-value>com.gupaoedu.web.MVCConfig</param-value>
    </init-param>
  </servlet>
  <servlet-mapping>
    <servlet-name>dispatcher</servlet-name>
    <url-pattern>/web/*</url-pattern>
  </servlet-mapping>
</web-app>
```

21.9　怎样用注解的方式配置 Spring

Spring 在 2.5 版本以后开始支持用注解的方式配置依赖注入。可以用注解的方式来替代 XML 方式的 Bean 描述，可以将 Bean 描述转移到组件类的内部，只需要在相关类上、方法上或者字段声明上使用注解即可。注解注入将会被容器在 XML 注入之前处理，所以后者会覆盖前者对于同一个属性的处理结果。

注解装配在 Spring 中是默认关闭的，需要在 Spring 文件中进行配置才能使用基于注解的装配模式。如果你想要在应用程序中使用注解的方式，请参考如下配置：

```
<beans>
  <context:annotation-config/>
</beans>
```

配置完成以后，就可以用注解的方式在 Spring 中向属性、方法和构造方法中自动装配变量。

下面是几种比较重要的注解类型。

（1）@Required：该注解应用于设值方法。

（2）@Autowired：该注解应用于设值方法、非设值方法、构造方法和变量。

（3）@Qualifier：该注解和 @Autowired 注解搭配使用，用于消除特定 Bean 自动装配的歧义。

（4）JSR-250 Annotations：Spring 支持基于 JSR-250 注解的注解，即 @Resource、@PostConstruct 和 @PreDestroy。

21.10　请解释 Spring Bean 的生命周期

Spring Bean 的生命周期简单易懂。在一个 Bean 实例被初始化时，需要执行一系列初始化操作以使其达到可用的状态。同样，当一个 Bean 不再被调用时需要进行相关的析构操作，并从 Bean 容器中移除。

Spring Bean Factory 负责管理在 Spring 容器中被创建的 Bean 的生命周期。Bean 的生命周期由两组回调方法组成。

（1）初始化之后调用的回调方法。

（2）销毁之前调用的回调方法。

Spring 提供了以下 4 种方式来管理 Bean 的生命周期事件：

（1）InitializingBean 和 DisposableBean 回调接口。

（2）针对特殊行为的其他 Aware 接口。

（3）Bean 配置文件中的 customInit() 方法和 customDestroy() 方法。

（4）@PostConstruct 和 @PreDestroy 注解方式。

使用 customInit()和 customDestroy()方法管理 Bean 生命周期的代码样例如下：

```
<beans>
  <bean id="demoBean" class="com.gupaoedu.task.DemoBean"
      init-Method="customInit" destroy-Method="customDestroy">
  </bean>
</beans>
```

21.11　Spring Bean 作用域的区别是什么

　　Spring 容器中的 Bean 可以分为 5 个作用域。所有作用域的名称都是自说明的，但是为了避免混淆，还是让我们来解释一下。

　　（1）singleton：这种 Bean 作用域是默认的，这种作用域确保不管接收到多少个请求，每个容器中只有一个 Bean 实例，单例模式由 Bean Factory 自身来维护。

　　（2）prototype：prototype 作用域与 singleton 作用域相反，为每一个 Bean 请求提供一个实例。

　　（3）request：在请求 Bean 作用域内为每一个来自客户端的网络请求创建一个实例，在请求完成以后，Bean 会失效并被垃圾回收器回收。

　　（4）Session：与 request 作用域类似，确保每个 Session 中有一个 Bean 实例，在 Session 过期后，Bean 会随之失效。

　　（5）global-session：global-session 和 Portlet 应用相关。当应用部署在 Portlet 容器中时，它包含很多 Portlet。如果想让所有的 Portlet 共用全局存储变量，那么这个全局存储变量需要存储在 global-session 中。全局作用域与 Servlet 中的 Session 作用域效果相同。

21.12　什么是 Spring Inner Bean

　　在 Spring 中，无论何时，当 Bean 仅被调用了一个属性时，一个明智的做法是将这个 Bean 声明为内部 Bean。内部 Bean 可以用 setter 注入"属性"和用构造方法注入"构造参数"的方式来实现。

　　比如，在应用程序中一个 Customer 类引用了一个 Person 类，我们要创建一个 Person 类的实例，然后在 Customer 内部使用。

```
public class Customer{
```

```
    private Person person;
}
public class Person{
    private String name;
    private String address;
    private int age;
}
```

内部 Bean 的声明方式如下:

```xml
<bean id="CustomerBean" class="com.gupaoedu.common.Customer">
    <property name="person">
        <bean class="com.gupaoedu.common.Person">
            <property name="name" value="lokesh" />
            <property name="address" value="India" />
            <property name="age" value="34" />
        </bean>
    </property>
</bean>
```

21.13 Spring 中的单例 Bean 是线程安全的吗

Spring 并没有对单例 Bean 进行任何多线程的封装处理。关于单例 Bean 的线程安全和并发问题需要开发者自行解决。但实际上,大部分 Spring Bean 并没有可变的状态(比如 Serview 类和 DAO 类),所以在某种程度上,Spring 的单例 Bean 是线程安全的。如果你的 Bean 有多种状态(比如 View Model 对象),就需要自行保证线程安全。

最容易的解决办法就是将多态 Bean 的作用域由"singleton"变更为"prototype"。

21.14 请举例说明如何在 Spring 中注入一个 Java 集合

Spring 提供了以下 4 种集合类的配置元素:

(1) <list>标签用来装配可重复的 list 值。

(2) <set>标签用来装配没有重复的 set 值。

(3) <map>标签用来注入键和值,可以为任何类型的键值对。

(4) <props>标签支持注入键和值都是字符串类型的键值对。

下面看一个具体的例子：

```xml
<beans>
  <bean id="javaCollection" class="com.gupaoedu.JavaCollection">
    <property name="customList">
      <list>
        <value>INDIA</value>
        <value>Pakistan</value>
        <value>USA</value>
        <value>UK</value>
      </list>
    </property>
    <property name="customSet">
      <set>
        <value>INDIA</value>
        <value>Pakistan</value>
        <value>USA</value>
        <value>UK</value>
      </set>
    </property>
    <property name="customMap">
      <map>
        <entry key="1" value="INDIA"/>
        <entry key="2" value="Pakistan"/>
        <entry key="3" value="USA"/>
        <entry key="4" value="UK"/>
      </map>
    </property>
    <property name="customProperies">
      <props>
        <prop key="admin">admin@gupaoedu.com</prop>
        <prop key="support">support@gupaoedu.com</prop>
      </props>
    </property>
  </bean>
</beans>
```

21.15 如何向 Spring Bean 中注入 java.util.Properties

第一种方法是使用如下代码所示的标签：

```xml
<bean id="adminUser" class="com.gupaoedu.common.Customer">
  <property name="emails">
    <props>
      <prop key="admin">admin@gupaoedu.com</prop>
```

```xml
        <prop key="support">support@gupaoedu.com</prop>
    </props>
  </property>
</bean>
```

也可用"util:"命名空间从 Properties 文件中创建一个 Properties Bean，然后利用 setter 方法注入 Bean 的引用。

21.16 请解释 Spring Bean 的自动装配

在 Spring 框架中，在配置文件中设定 Bean 的依赖关系是一个很好的机制，Spring 容器还可以自动装配合作关系 Bean 之间的关联关系。这意味着 Spring 可以通过向 BeanFactory 中注入的方式自动搞定 Bean 之间的依赖关系。自动装配可以设置在每个 Bean 上，也可以设置在特定的 Bean 上。

下面的 XML 配置文件表明了如何根据名称将一个 Bean 设置为自动装配模式：

```xml
<bean id="employeeDAO" class="com.gupaoedu.EmployeeDAOImpl" autowire="byName" />
```

除了 Bean 配置文件中提供的自动装配模式，还可以使用@Autowired 注解来自动装配指定的Bean。在使用@Autowired 注解之前需要按照如下的配置方式在 Spring 配置文件中进行配置：

```xml
<context:annotation-config />
```

也可以通过在配置文件中配置 AutowiredAnnotationBeanPostProcessor 达到相同的效果：

```xml
<bean class="org.springframework.beans.factory.annotation.AutowiredAnnotationBeanPostProcessor"/>
```

配置好以后就可以使用@Autowired 来标注了：

```java
@Autowired
public EmployeeDAOImpl ( EmployeeManager manager ) {
    this.manager = manager;
}
```

21.17 自动装配有哪些局限性

自动装配有如下局限性。

- 重写：你仍然需要使用< property >设置指明依赖，这意味着总要重写自动装配。
- 原生数据类型：你不能自动装配简单的属性，如原生类型、字符串和类。

- 模糊特性：自动装配总是没有自定义装配精确，因此如果可能尽量使用自定义装配。

21.18　请解释各种自动装配模式的区别

在 Spring 中共有 5 种自动装配模式，让我们逐一分析。

（1）no：这是 Spring 的默认设置，在该设置下自动装配是关闭的，开发者需要自行在 Bean 定义中用标签明确地设置依赖关系。

（2）byName：该模式可以根据 Bean 名称设置依赖关系。当向一个 Bean 中自动装配一个属性时，容器将根据 Bean 的名称自动在配置文件中查询一个匹配的 Bean。如果找到就装配这个属性，如果没找到就报错。

（3）byType：该模式可以根据 Bean 类型设置依赖关系。当向一个 Bean 中自动装配一个属性时，容器将根据 Bean 的类型自动在配置文件中查询一个匹配的 Bean。如果找到就装配这个属性，如果没找到就报错。

（4）constructor：和 byType 模式类似，但是仅适用于有与构造器相同参数类型的 Bean，如果在容器中没有找到与构造器参数类型一致的 Bean，那么将会抛出异常。

（5）autodetect：该模式自动探测使用 constructor 自动装配或者 byType 自动装配。首先会尝试找合适的带参数的构造器，如果找到就是用构造器自动装配，如果在 Bean 内部没有找到相应的构造器或者构造器是无参构造器，容器就会自动选择 byType 模式。

21.19　请举例解释@Required 注解

在产品级别的应用中，IoC 容器可能声明了数十万个 Bean，Bean 与 Bean 之间有着复杂的依赖关系。设值注解方法的短板之一就是验证所有的属性是否被注解是一项十分困难的操作。可以通过设置"dependency-check"来解决这个问题。

在应用程序的生命周期中，你可能不大愿意花时间验证所有 Bean 的属性是否按照上下文文件正确配置，或者你宁可验证某个 Bean 的特定属性是否被正确设置。即使用"dependency-check"属性也不能很好地解决这个问题，在这种情况下需要使用@Required 注解。

可用如下的方式来标明 Bean 的设值方法：

```
public class EmployeeFactoryBean extends AbstractFactoryBean<Object> {
```

```
private String designation;
public String getDesignation() {
  return designation;
}
@Required
public void setDesignation(String designation) {
  this.designation = designation;
}
```

RequiredAnnotationBeanPostProcessor 是 Spring 中的后置处理器，用来验证被@Required 注解的 Bean 属性是否被正确设置了。在使用 RequiredAnnotationBeanPostProcesso 验证 Bean 属性之前，要在 IoC 容器中对其进行注册：

```
<bean
class="org.springframework.beans.factory.annotation.RequiredAnnotationBeanPostProcessor"
/>
```

但是如果没有属性被用@Required 注解过，后置处理器会抛出一个 BeanInitializationException 异常。

21.20　请举例说明@Qualifier 注解

@Qualifier 注解意味着可以在被标注 Bean 的字段上自动装配。@Qualifier 注解可以用来取消 Spring 不能取消的 Bean 应用。

21.21　构造方法注入和设值注入有什么区别

请注意以下明显的区别：

（1）设值注入支持大部分依赖注入，如果我们仅需要注入 int、string 和 long 型的变量，不要用设值方法注入。对于基本类型，如果没有注入，可以为基本类型设置默认值。构造方法注入不支持大部分依赖注入，因为在调用构造方法时必须传入正确的构造参数，否则会报错。

（2）设值注入不会重写构造方法的值。如果我们对同一个变量同时使用了构造方法注入和设值注入，那么构造方法将不能覆盖设值注入的值。很明显，因为构造方法只在对象被创建时被调用。

（3）在使用设值注入时还不能保证某种依赖是否已经被注入，也就是说，这时对象的依赖关系有可能是不完整的。而在另一种情况下，构造器注入则不允许生成依赖关系不完整的对象。

（4）在设值注入时如果对象 A 和对象 B 互相依赖，在创建对象 A 时 Spring 会抛出 ObjectCurrentlyInCreationException 异常，因为在对象 B 被创建之前对象 A 是不能被创建的，反之亦然。Spring 用设值注入解决了循环依赖问题，因为对象的设值方法是在对象被创建之前被调用的。

21.22 Spring 中有哪些不同类型的事件

Spring 的 ApplicationContext 提供了支持事件和代码中监听器的功能。

我们可以创建 Bean 来监听在 ApplicationContext 中发布的事件。对于 ApplicationEvent 类和在 ApplicationContext 接口中处理的事件，如果一个 Bean 实现了 ApplicationListener 接口，当一个 ApplicationEvent 被发布以后，Bean 会自动被通知。

```java
public class AllApplicationEventListener implements ApplicationListener<ApplicationEvent> {
    @Override
    public void onApplicationEvent(ApplicationEvent applicationEvent) {
        //process event
    }
}
```

Spring 提供了以下 5 种标准的事件。

（1）上下文更新事件（ContextRefreshedEvent）：该事件会在 ApplicationContext 被初始化或者更新时发布。也可以在调用 ConfigurableApplicationContext 接口中的 refresh()方法时被触发。

（2）上下文开始事件（ContextStartedEvent）：当容器调用 ConfigurableApplicationContext 的 Start()方法开始或重新开始容器时触发该事件。

（3）上下文停止事件（ContextStoppedEvent）：当容器调用 ConfigurableApplicationContext 的 Stop()方法停止容器时触发该事件。

（4）上下文关闭事件（ContextClosedEvent）：当 ApplicationContext 被关闭时触发该事件。容器被关闭时，其管理的所有单例 Bean 都被销毁。

（5）请求处理事件（RequestHandledEvent）：在 Web 应用中，当一个 HTTP 请求（Request）结束时触发该事件。

除了上面介绍的事件，还可以通过扩展 ApplicationEvent 类来自定义事件：

```java
public class CustomApplicationEvent extends ApplicationEvent {
  public CustomApplicationEvent ( Object source, final String msg ){
    super(source);
    System.out.println("Created a Custom event");
  }
}
```

为了监听这个事件，还需要创建一个监听器：

```java
public class CustomEventListener implements ApplicationListener < CustomApplicationEvent >{
  @Override
  public void onApplicationEvent(CustomApplicationEvent applicationEvent) {
  }
}
```

之后通过 ApplicationContext 接口的 publishEvent()方法来发布自定义事件：

```java
CustomApplicationEvent customEvent = new CustomApplicationEvent(applicationContext, "Test message" );
applicationContext.publishEvent(customEvent);
```

21.23　FileSystemResource 和 ClassPathResource 有什么区别

在 FileSystemResource 中需要给出 spring-config.xml 文件在项目中的相对路径或者绝对路径。在 ClassPathResource 中 Spring 会在 ClassPath 中自动搜寻配置文件，所以要把 ClassPathResource 文件放在 ClassPath 下。

如果将 spring-config.xml 保存在了 src 目录下，只需给出配置文件的名称即可，因为 src 是默认的路径。

简而言之，ClassPathResource 在环境变量中读取配置文件，FileSystemResource 在配置文件中读取配置文件。

21.24　Spring 中用到了哪些设计模式

Spring 中使用了大量的设计模式，下面列举了一些比较有代表性的设计模式。

（1）代理模式：在 AOP 和 remoting 中被用得比较多。

（2）单例模式：在 Spring 配置文件中定义的 Bean 默认为单例模式。

（3）模板模式：用来解决代码重复问题，比如 RestTemplate、JmsTemplate、JpaTemplate。

（4）委派模式：Spring 提供了 DispatcherServlet 来对请求进行分发。

（5）工厂模式：BeanFactory 用来创建对象的实例，贯穿于 BeanFactory 和 ApplicationContext 接口。

（6）代理模式：代理模式 AOP 思想的底层实现技术，Spring 中采用 JDK Proxy 和 CGLib 类库。

21.25　在 Spring 中如何更有效地使用 JDBC

使用 Spring JDBC 可以使得资源管理及错误处理的代价减小。开发人员只需通过 statements 和 queries 语句从数据库中存取数据。Spring 通过模板类能更有效地使用 JDBC，也就是所谓的 JdbcTemplate。

21.26　请解释 Spring 中的 IoC 容器

Spring 中的 org.springframework.beans 包和 org.springframework.context 包构成了 Spring IoC 容器的基础。

BeanFactory 接口提供了一个先进的配置机制，使得任何类型的对象的配置都成为可能。ApplicationContex 接口对 BeanFactory（是一个子接口）进行了扩展，在 BeanFactory 的基础上添加了其他功能，比如与 Spring 的 AOP 更容易集成，也提供了处理 Message Resource 的机制（用于国际化），以及事件传播及应用层的特别配置，比如针对 Web 应用的 WebApplicationContext。

21.27　在 Spring 中可以注入 null 或空字符串吗

完全可以。

反侵权盗版声明

电子工业出版社依法对本作品享有专有出版权。任何未经权利人书面许可，复制、销售或通过信息网络传播本作品的行为；歪曲、篡改、剽窃本作品的行为，均违反《中华人民共和国著作权法》，其行为人应承担相应的民事责任和行政责任，构成犯罪的，将被依法追究刑事责任。

为了维护市场秩序，保护权利人的合法权益，我社将依法查处和打击侵权盗版的单位和个人。欢迎社会各界人士积极举报侵权盗版行为，本社将奖励举报有功人员，并保证举报人的信息不被泄露。

举报电话： (010)88254396；(010)88258888
传　　真： (010)88254397
E – mail： dbqq@phei.com.cn
通信地址： 北京市万寿路 173 信箱
　　　　　电子工业出版社总编办公室
邮　　编： 100036